Objective Forestry

For All Competitive Examination

3rd Edition

ABOUT THE AUTHORS

Dr. K.T. Parthiban is a Professor borne at the faculty of Forestry a involved in forestry teaching, research and extension activities for t past 24 years. He has conceived the value chain models for Industri Agroforestry and promoted in association with wood based industrie. He is also pioneer in development of varieties in trees. He has handled numerous projects funded by DST, DBT, ICAR, NOVOD Board and several wood based industries. He has published 12 books and over 100 research papers. He has conceived and implemented India's First Consortium of industrial Agroforestry and Agroforestry Business Incubator and resolving the issues in Agroforestry sector. Dr.K.T.Parthiban travelled widely in European, South American and Asian countries and contributed towards Forestry Research and Development.

Dr. S. Vennila is borne at the faculty of forestry and currently working as a Assistant Professor at FC&RI, Mettupalayam. She has written 05 books, 10 book chapters and more than 30 research papers in the sphere of forestry which were published in the peer reviewed national and international journals.

Dr. S. Umesh Kanna Subramani is an Associate Professor of Forestry specialized in Tree Improvement and involved in Education, Research and Extension for the past 14 years. He has guided 15 research scholars and responsible for the release of 05 HYSR varieties amenable for Industrial Agroforestry. He has authored more than 20 books and published 85 research papers in the peer reviewed international and national journals.

Dr. P. Durairasu I.F.S is currently the PCCF, heading the State Forest Department and involved in conservation and management of the entire State forest resources. He has served as a Dean at Forest College and Research Institute, TNAU, Mettupalayam during the period between 2010 and 2014 and contributed significantly towards Forestry Education, Research and Extension activities. He is instrumental in promotion of Tree Insurance Scheme and establishment of Consortium of Industrial Agroforestry.

Objective Forestry

For All Competitive Examination

3rd Edition

K.T. Parthiban
S. Vennila
S. Umesh Kanna
P. Durairasu

Forest College and Research Institute
Tamil Nadu Agricultural University
Mettupalayam - 641 301

Jodhpur
5 A, New Pali Road, P.O. Box 91
Jodhpur - 342 001 Rajasthan (India)
E-mail: info@scientificpub.com
Website: www.scientificpubonline.com

New Delhi
4806/24, Ansari Road, Daryaganj,
New Delhi - 110 002 (India)

Print : 2026

First Edition : 2015
Second Edition : 2016
Third Edition : 2020
Third Edition (Reprint) : 2023, 2025

ISBN: 978-93-89184-35-8
eISBN: 978-93-89184-36-5

Printed in India

Preface

3rd Edition

The Professional Forestry Education has started its inception in the year 1985 in five State Agricultural Universities under the ages of Indian Council of Agricultural Research. Since then, many State Agricultural Universities have started imparting forestry education as a four years professional degree programme. The National Forest Policy (1988) have directed the State Agricultural Universities to strengthen the forestry education and also indicated that the country would need massive human resources to cater to the needs of management and conservation of pristine resources of the country. As on today, over 15 state agricultural universities are offering forestry education in the country. This besides, central universities and some of the conventional universities are also imparting forestry as a subject of interest. This demanded development of competitive books to cater to the academic needs of graduating scholars besides satisfying the demands of other competitive exams.

All competitive exams in the country particularly Central (UPSC) and State Service Exams, ICAR-JRF, SRF and Scientist Exams etc. are predominantly conducted through objective and subjective mode of examinations. Forestry is a broader science and incorporates wide range of subjects which include fundamental, management, applied and advanced courses. The available competitive books on objective forestry have not included all facets of forestry and are also not updated with the recent developments.

The authors have already presented the 1st edition of the book in the year 2015 and 2nd edition in the year 2016 which received wider attraction across the country. These two editions have benefited wide range of scholars which resulted in successful graduation programme and witnessed successful aspirants of state and central forest service examinations. Almost all of my students during the last four years have benefited through this book particularly the book have made the demands of TNPSC exams. Several of my students and few teachers have extended suggestions to include subjects like Introduction to Forestry, Global Forest Resources, Forest survey and Engineering etc in the proposed revised edition. Accordingly, the authors have comprehensively developed objective type questions for these chapters and incorporated to benefit the graduating students and others who are aspiring for competitive examinations. The authors are also thankful to Scientific Publishers for successfully publishing the first two editions and ready for 3rd edition with updated question bank.

We thank all the readers and again request the readers to extend all support to the 3rd edition and offer suggestions for improving the contents of the book in the future editions.

Authors

Contents

1

INTRODUCTORY FORESTRY

KEY POINTS

- Forest is an area set aside for the production of timber and / or other forest produce, or maintained under woody vegetation for certain indirect benefits which it provides, e.g., climatic or protective
- Forest is an area of land proclaimed to be a forest under a forest law
- Forest is a plant community predominantly of trees and other woody vegetation, usually with a closed canopy
- Forestry is the theory and practice of all that constitutes creation, conservation, scientific management of forests and utilization of their resources
- Pole is a young tree from the time when the lower branches begin to fall off to the time when the rate of height growth begins to slow down and crown expansions become marked.
- Seedling is a plant grown from seed till it attains a height of about one meter.
- Sapling is a young tree from the time when it reaches about one meter (3 feet) in height till the lower branches begin to fall.
- Fuel wood species - Acacia spp., *Casuarina* spp., Prosopis spp.,, Eucalyptus spp., *Leucaena leucocephala*, etc.
- Major timber species - Teak, Sal, Rosewood, Pterocarpus, Terminalias, Cedar
- Major fodder trees - Acacia, Albizia, Hardwickia, Pterocarpus, Robinia etc.
- India has a potential of 175 million tonnes of carbon to be sequestrated annually through land use and forestry operations.
- Stump is a specially prepared seedling in which shoots are cut above the collar region and roots are pruned leaving primary roots at defined length.
- Spacing is defined as the distance between the plants put out in a plantation or standing crop.
- Tending is defined as an operation carried out for the benefit of a forest crop at any stage of its life between the seedling and the mature stages.
- Weeding is defined as a tending operation done in the seedlings stage in nursery or in a forest crop
- Girdling is defined as cutting through bark and other living layers of woods in a continuous incision all around the pole of a tree.

- Improvement felling refers to the removal of less valuable trees in a crop in the interest of valuable individuals. It is applicable to mixed as well as irregular crops.
- Pruning refers to the removal of live or dead branches or multiple leaders from standing trees for the improvement of the trees or its timber.
- Bud pruning is a special kind of pruning which is defined as the rubbing off the lateral buds to prevent the development of branches as a measure to obtain knot-free timber.
- The total industrial wood demand for the year 2020 - 153 million m^3
- Fuel wood alone amounts to approximately 90 per cent total wood production in India.
- The total estimated wood consumption in India - 69 million cum per year.
- A rotation period is a planned number of years or the time period between the establishment of a stand of trees and to the harvestable stage of the same stand or the final felling.
- Short rotation is the term refers to species that are fast growing and are harvested in a very short period of time i.e., within 5 - 6 years.
- Medium rotation species are those that are harvested between 7 and 15 years where the species acquires the harvestable property at that age.
- Long rotation crops have the rotation age of more than 15 years, where the species is harvested only after 20 years of age.
- Bamboo production in India - 5.38 million m^3
- Total Timber production - 70.9 million m^3
- Annual production of timber from TOF - 44.34 million m^3
- Total wood production - 456.15 million m^3
- Total Annual wood consumption - 402.00 million m^3
- Total Annual Timber consumption - 69.00 million m^3

MULTIPLE CHOICE QUESTIONS

1. The word FOREST is derived from the Latin word
   (a) Foris (b) Fores
   (c) Forex (d) Forix
2. The first Conservator of Indian Forests.
   (a) Trevor (b) Troupe
   (c) Watson (d) Hartig
3. Reserved forest are constituted under the provision of ________
   (a) National Forest Policy, 1988
   (b) Forest Policy, 1952
   (c) Indian Forest Act, 1927
   (d) Forest Conservation Act, 1980
4. The first Inspector General of Indian Forests ________
   (a) Dietrich Brandis
   (b) Watson
   (c) Chathu Menon
   (d) O'Conolly
5. In India, Plantation forestry was started in the year ________
   (a) 1942 (b) 1842
   (c) 1947 (d) 1847
6. First Teak plantation in India is established at ________
   (a) Nilambur (b) West Bengal
   (c) Siwalik (d) Gorakhpur

7. A general appearance of forest community is called ________
(a) Physiognomy
(b) Physiology
(c) Phenology
(d) Phenotypes

8. ______ is the first forest policy of independent India
(a) Indian Forest policy 1950
(b) Indian Forest policy 1951
(c) Indian Forest Policy 1952
(d) Indian Forest Policy 1953

9. Tallest tree species of the world is seen at ________
(a) Tropical seasonal forests
(b) Temperate forests
(c) Mangrove Forests
(d) Subtropical Forest

10. Indian forests are classified into forest types in the year ________
(a) 1947 (b) 1935
(c) 1952 (d) 1988

11. First Indian Forest Act was enacted during ____
(a) 1882 (b) 1865
(c) 1894 (d) 1852

12. The head of a circle in Forest Administration
(a) Ranger (b) Forester
(c) Conservator (d) DFO

13. The lowest Administrative unit in Forest Administration
(a) Section (b) Beat
(c) Range (d) Division

14. Forest is source of ______
(a) Wood (b) Food
(c) Timber (d) All the above

15. Function of Forest: Pick the odd man out
(a) Productive function
(b) Ameliorative function
(c) Deforestation function
(d) Protective function

16. The yield of teak plantation in one hectare is 200 tonnes. The approximate carbon sequested by the plantation is ______
(a) 25 tonnes (b) 50 tonnes
(c) 75 tonnes (d) 100 tonnes

17. Protective role of forest: Pick the odd man out
(a) Soil improvement
(b) Carbon Sink
(c) Ecological Security
(d) NTFP Production

18. Indian Forests are classified into forest types based on ________classification
(a) Mayr
(b) Koppen
(c) Seth and Khan
(d) Champion and Seth

19. ________ is one of the mega biodiversity country is in the world.
(a) USA (b) India
(c) England (d) Korea

20. A unit of vegetation which possesses (broad) characteristics in physiognomy and structure sufficiently pronounced to permit its differentiation from other such units
(a) Forest type
(b) Forest
(c) Forestry
(d) Agroforestry

21. The systems of classification of vegetation: Pick the odd man out
(a) Vegetation
(b) Climate
(c) Ecosystem
(d) Energy

22. The plant grown from seed till it attains a height of about one metre
(a) Seedling (b) Sapling
(c) Pole (d) Tree

23. Choose the correct order: Stages of Growth and Development
(a) Seedling, Sapling, Tree, Pole
(b) Seedling, Sapling, Pole, Tree
(c) Sapling, Seedling, Tree, Pole
(d) Tree, Pole, Sapling, Seedling

24. Formation of new organs in a plant is referred as
(a) Growth (b) Development
(c) Shoot (d) Root

25. The diameter increment is attributable to
(a) Vertical Growth
(b) Horizontal Growth
(c) Meristematic Growth
(d) None

26. Factors of locality: Pick the odd man out
(a) Climatic factor
(b) Edaphic factor
(c) Biotic factor
(d) Vegetation factor

27. Biotic factor is related to
(a) Temperature (b) Plants
(c) Soil (d) Rainfall

28. Edaphic factor is related to
(a) Temperature (b) Wind
(c) Soil (d) Rainfall

29. The influence of living organisms on the vegetation is due to
(a) Climatic factors
(b) Edaphic factors
(c) Biotic factors
(d) Topographic factors

30. _______ is the movement and exchange of organic and inorganic matter back into the production of matter.
(a) Nitrogen cycling
(b) Bio-geochemical cycle
(c) Nutrient cycling
(d) Water cycle

31. Nutrient cycling: Pick odd man out
(a) Nitrogen cycling
(b) Phosphorus cycle
(c) Carbon cycling
(d) Water cycle

32. Micronutrient : Pick odd man out
(a) Ca (b) Zn
(c) Cu (d) Mn

33. N, P and K are called as
(a) Macro nutrients
(b) Micro nutrients
(c) Secondary nutrients
(d) Herbicides

34. One of the suggested measure for reducing emission level is utilizing ____
(a) Coal (b) Petrol
(c) Diesel (d) Biofuel

35. The major impacts of global warming
(a) Melting of polar ice
(b) Rising of sea level
(c) Erratic rainfall pattern
(d) All of these

36. Climate change is mainly caused due to
(a) Natural Forest area
(b) Plantations
(c) Greenhouse gases
(d) Biodiversity

37. The most ideal mitigation measures for climate change
(a) More utilization of fossil fuels
(b) Afforestation
(c) Industrial Development
(d) Urbanization

38. JFM first implemented at ________
(a) Jabalpur (b) Ayyalur
(c) Jhansi (d) Arabari

39. National Forest Policy (N.F.P), 1952 envisages that ________ per cent of the forest in the hilly region.
(a) 33 (b) 60
(c) 20 (d) 30

40. Tree suitable for Plywood production
(a) *Pungam*
(b) *Santalum album*
(c) *Casuarina*
(d) *Melia dubia*

41. Global warming is attributable primarily due to
(a) CO (b) SO_2
(c) CO_2 (d) No_2

42. ____________ is forestry for the people, by the people and of the people.
(a) Urban Forestry
(b) Production Forestry
(c) Social Forestry
(d) Protection Forestry

43. World Wetlands Day
(a) 2nd February
(b) 20th March
(c) 23rd May
(d) 5th June

44. World Sparrow Day
(a) 2nd February
(b) 20th March
(c) 23rd May
(d) 5th June

45. World Turtle Day
(a) 2nd February
(b) 20th March
(c) 23rd May
(d) 5th June

46. An example of a multipurpose tree species.
(a) Teak (b) Neem
(c) Sandal (d) Redsanders

47. World Wood day
(a) 20th March
(b) 21st March
(c) 20th April
(d) 21st April

48. ITTO Stands for
(a) International Timber Trade Organization
(b) International Tropical Timber Organization
(c) Indian Timber Trade Organization
(d) Indian Tropical Timber Organization

49. The Headquarters of FAO
(a) Italy (b) England
(c) France (d) Finland

50. Bamboo exempted from trees list by amendment in IFAct, 1927 in the year
(a) 2000 (b) 2006
(c) 2010 (d) 2017

51. Identify the Potential fodder trees
(a) Kadam (b) Teak
(c) Casuarina (d) Eucalyptus

52. Leucaena is preferred species for
(a) Dendroenergy
(b) Timber
(c) Plywood
(d) Match splint

53. FSI Stands for
(a) Forest Survey of India
(b) Forest Survey Institute
(c) Forest Society of India
(d) Social Forest Institute

54. Species suitable for plywood
(a) *Melia dubia*
(b) *Casuarina equisetifolia*
(c) *Albizia lebbeck*
(d) *Prosopis juliflora*

55. Short Rotation Species (Pick the odd man out)
(a) Teak (b) Casuarina
(c) Sandal (d) Redsander

56. Jatropha is a
(a) Plywood species
(b) Pulpwood species
(c) Biofuel species
(d) Matchwood species

57. The wood of Albizia is suitable for
(a) Wood Antique
(b) Pulpwood
(c) Plywood
(d) Packing cases

58. World Elephant Day is celebrated on
(a) 12th August (b) 16th September
(c) 18th May (d) 13th June

59. __________ is the nitrogen fixing non legume tree
(a) *Alnus nepalensis*
(b) *Larix decidua*
(c) *Fagus sylvatica*
(d) *Ulmus americana*

60. The tree with the rotation age of 6 to 10 years is termed as
(a) Short rotation
(b) Medium rotation
(c) Long rotation
(d) Very long rotation

61. __________ is the productivity of Indian Forests
(a) 0.5 m^3/ha/yr
(b) 1 m^3/ha/yr
(c) 5 m^3/ha/yr (d) 10 m^3/ha/yr

62. FRI (Forest Research Institute) located at
(a) Uttarakhand (b) Dehradun
(c) Faridabad (d) Haryana

63. "Vanamahotsav" the planting festival is Initiated by
(a) P.K. Nair
(b) K.M. Munshi
(c) Captain Watson
(d) M.D. Chaturvedi

64. Topographic factor (Pick the odd man out)
(a) Aspect (b) Altitude
(c) Slope (d) Temperature

65. The Forest cover in India as per 2017 classification _____
(a) 24.16% (b) 21.54%
(c) 23.18% (d) 23.78%

66. The increase in temperature of the atmosphere in earth surface
(a) Desertification
(b) Global Warming
(c) Climate moderation
(d) Climate change

67. Social forestry is attributable to
(a) Dietrich Brandis
(b) Westoby
(c) Chathu Menon
(d) Watson

68. Forest Conservation Act came into force
(a) 1974 (b) 1980
(c) 2006 (d) 2001

69. _____ is the state having largest forest cover in India
(a) Madhya Pradesh
(b) Uttar Pradesh
(c) Maharashtra
(d) Karnataka

70. _________ is the short form of The Scheduled Tribes and Other Traditional Forest Dwellers Act, 2006
(a) Forest Conservation Act
(b) Forest Protection Act
(c) Forest Right Act
(d) Indian Tribal Act

71. The plant whose stem is always green and tender and height is usually not more than one meter
(a) Herb (b) Shrub
(c) Tree (d) Pole

72. A woody perennial plant differing from a perennial herb in its persistent and woody stem and less definitely from a tree in its low stature and its habit of branching from the base is
(a) Herb (b) Shrub
(c) Tree (d) Pole

73. A large woody perennial plant having a single well defined stem (bole or trunk) and a more or less definite crown is

(a) Herb (b) Shrub
(c) Tree (d) Pole

74. A young tree from the time when it reaches about one meter (3 feet) in height till the lower branches begin to fall

(a) Sapling (b) Seedling
(c) Stem (d) Pole

75. The principal axis of plant from which buds and shoots are developed is

(a) Sapling (b) Seedling
(c) Stem (d) Pole

76. A young tree from the time when the lower branches begin to fall off to the time when the rate of height growth begins to slow down and crown expansion become marked.

(a) Sapling (b) Seedling
(c) Stem (d) Pole

77. The plant grown from seed till it attains a height of about one meter is

(a) Sapling (b) Seedling
(c) Stem (d) Pole

78. The upper branchy part of a tree above the pole is

(a) Bole (b) Clean Bole
(c) Crown (d) Pole

79. The part of the bole that is free of branches

(a) Bole (b) Clean Bole
(c) Crown (d) Pole

80. Captain Watson appointed as first Conservator of forest in the year

(a) 1759 (b) 1806
(c) 1857 (d) 1906

81. Indian Forest Act passed in the year

(a) 1785 (b) 1878
(c) 1865 (d) 1934

82. Scientific forest management started in India during the year

(a) 1864 (b) 1875
(c) 1907 (d) 1909

83. First Indian Forest policy is enunciated in the year

(a) 1894 (b) 1927
(c) 1945 (d) 1948

84. Indian Forest Act enacted during

(a) 1894 (b) 1927
(c) 1945 (d) 1948

85. The estimated wood demand for the year 2020 is (as per FAO)

(a) 100 mill m^3 (b) 152 mill m^3
(c) 110 mill m^3 (d) 169 mill m^3

86. The Tree Outside Forest area in India as per the FSI 2017

(a) 72546 sq km (b) 57643 sq km
(c) 10675 sq km (d) 93815 sq km

87. The tree is termed as fast grower if the productivity is

(a) > 10 m^3/ha/yr (b) < 10 m^3/ha/yr
(c) 5-10 m^3/ha/yr (d) 1-5 m^3/ha/yr

88. The wood demand for short rotation species during 2020 (as per FAO)

(a) 87.70 mil. m^3 (b) 65.10 mil. m^3
(c) 52.52 mil. m^3 (d) 93.65 mil. m^3

89. The current per capita consumption of paper in India

(a) 5.5 kg (b) 7.5 kg
(c) 15.5 kg (d) 11 kg

90. The National Biofuel Policy suggested the blending of biofuel during 2017

(a) 5% (b) 10%
(c) 20% (d) 30%

91. CIAF stands for

(a) Consortium of Industrial Agroforestry
(b) Cogeneration of Industrial Agroforestry
(c) Consortium of Industries and Allied sectors
(d) Consortium of Industrial Areas

92. The percapita electricity consumption in India
(a) >1000 kWh (b) >2000 kWh
(c) >3000 kWh (d) > 4000 kWh

93. Annual requirement of wood for plywood production in Tamil Nadu
(a) 8 lakh tonnes (b) 10 lakh tonnes
(c) 12 lakh tonnes (d) 14 lakh tonnes

94. The agroforestry area in India.
(a) 11.15 million ha
(b) 13.75 million ha
(c) 10.46 million ha
(d) 12.54 million ha

95. The National Agroforestry Forest Policy was enunciated during the year ______
(a) 2014 (b) 2013
(c) 2012 (d) 2011

96. The plantlet produced through clonal propagation is termed as____________.
(a) Seedling (b) Sapling
(c) Ramet (d) Ortet

97. The state having highest percentage of forest cover in India with respect to the geographical area as per the ISFR 2017 assessment
(a) Lakshadweep
(b) Mizoram
(c) Arunachal Pradesh
(d) Andaman & Nicobar Island

98. ________ state is the plywood hub in South India
(a) Kerala (b) Tamil Nadu
(c) Karnataka (d) Andhra Pradesh

99. The Forest area in India is ____________
(a) 708273 sq km (b) 658231 sq km
(c) 963547 sq km (d) 456548 sq km

100. Annual Timber consumption in India is ________
(a) 120 million cum
(b) 69 million cum
(c) 56 million cum
(d) 152 million cum

101. CRIDA is located at ________
(a) Chennai (b) Jhansi
(c) Dehradun (d) Hyderabad

ANSWERSHEET
1. Introductory Forestry

1. **(a)**	2. **(c)**	3. **(c)**	4. **(a)**	5. **(b)**	6. **(a)**	7. **(a)**	8. **(c)**	9. **(b)**	10. **(b)**
11. **(b)**	12. **(c)**	13. **(b)**	14. **(d)**	15. **(c)**	16. **(d)**	17. **(d)**	18. **(d)**	19. **(b)**	20. **(a)**
21. **(d)**	22. **(a)**	23. **(b)**	24. **(b)**	25. **(b)**	26. **(d)**	27. **(b)**	28. **(c)**	29. **(c)**	30. **(c)**
31. **(d)**	32. **(a)**	33. **(a)**	34. **(d)**	35. **(d)**	36. **(c)**	37. **(b)**	38. **(d)**	39. **(b)**	40. **(d)**
41. **(c)**	42. **(c)**	43. **(a)**	44. **(b)**	45. **(c)**	46. **(b)**	47. **(a)**	48. **(b)**	49. **(a)**	50. **(d)**
51. **(a)**	52. **(a)**	53. **(a)**	54. **(a)**	55. **(b)**	56. **(c)**	57. **(a)**	58. **(a)**	59. **(a)**	60. **(b)**
61. **(a)**	62. **(b)**	63. **(b)**	64. **(d)**	65. **(b)**	66. **(b)**	67. **(b)**	68. **(b)**	69. **(a)**	70. **(c)**
71. **(a)**	72. **(b)**	73. **(c)**	74. **(a)**	75. **(c)**	76. **(d)**	77. **(b)**	78. **(c)**	79. **(b)**	80. **(b)**
81. **(c)**	82. **(a)**	83. **(a)**	84. **(b)**	85. **(b)**	86. **(d)**	87. **(a)**	88. **(a)**	89. **(d)**	90. **(c)**
91. **(a)**	92. **(a)**	93. **(c)**	94. **(a)**	95. **(a)**	96. **(c)**	97. **(a)**	98. **(a)**	99. **(a)**	100. **(b)**
101. **(d)**									

2

GLOBAL FOREST RESOURCES

KEY NOTES

- Global net forest loss - 129 million ha (between 1990 and 2015)
- Global annual forest loss - 7.6 million ha
- Total global forest area - 3999 million ha
- Total global wooded land - 1204 million ha
- Global Natural Forest – 3695 million ha
- Global Planted Forest - 291 million ha
- Global Forest growing Stock - 531 billion m^3 or 129 m^3 per ha
- Forest area of Africa – 624 million ha
- Forest area of Asia – 593 million ha
- Forest area of Europe – 1015 million ha
- Forest area of North & Central America – 751 million ha
- Forest area of Ocenania – 174 million ha
- Forest area of South America – 842 million ha
- Forest Monitoring and assessment report was prepared by FAO
- Highest Forest area -
  - 1st Russia (814.931 million ha)
  - 2nd Brazil (493.538 million ha)
  - 3rd Canada (347.069 million ha)
  - 4th USA (310.09 million ha)
  - 5th China (208.321 million ha)
- India ranked 10th position in forest area (70.682 million ha)
- The world forest are classified into 9 major types
- Amazon forest contribute 1/3 of the tropical forest in the world
- Savannahs - Tropical grass lands in which trees are distributed sporadically throughout the tropical region.
- Savannahs are distributed predominantly in Africa, Australia and south East Asia.
- Desert – An area which receives less than 10 inches of rainfall coupled with extreme dryness
- Deserts are predominantly distributed in Africa, Australia, North America and Asia.

- Chaparrals are found in California (North America), around Mediterranean Sea and along the southern coast of Australia.
- 80% of forestland is owned by edijos or comunidades forestales in Mexico.
- Mexican forests contain most pine and oak species in the world.
- Tundra means rolling treeless plains.
- Mexico's total timber production is only about 2% of the US's and 5% of Canada's production volumes.
- Central America is part of the Mesoamerican biodiversity hotspot, boasting 7% of the world's biodiversity.
- The cloud forest of Sierra de las Minas, Guatemala, is the largest forest in Central America.
- Secondary forests of Africa is dominated by parasol trees (*Musanga ceropiodes)*
- Queensland has the largest area of Australia's forest (51.0 million hectares—41% of Australia's forest)
- More than 80 per cent of Australia's native forest area is dominated by Eucalypt and Acacia forests
- Russia has 1/4 of the world's reserves of wood
- The total forest area of Russia - 885 million hectares, representing 45% of the total area of the country.
- China is the world's third largest country (9.6 million square kilometers).
- The area of natural forest in China is 121.84 million ha
- Shunan Bamboo Forest (Shunan Zhuhai) is the oldest and biggest bamboo forest in China (120 square kilometers)
- China is homed with around 400 types of bamboo
- Forest area of Japan - 62.3 million acres covering 2/3 of the country's mountainous land area.
- FSI - preparation of State Forest Report once in every two years
- FSI provides assessment of latest forest cover in the country.
- FSI develops thematic map on 1:50000 scale
- ICFRE was established in the year 1986
- Headquarters of ICFRE - Dehradun
- ICFRE has nine regional research institutes and five centres in the country.
- Botanical Survey of India (BSI) was established in 1890
- The objectives of BSI was exploring the plant resources of the country and identifying plant species with economic virtue.
- Central Agroforestry Research Institute (NRCAF) was established in 1988 at Jhansi, Uttar Pradesh
- The functioning of zoos in India is regulated by an autonomous statutory body called Central Zoo Authority
- The CZA consists of a Chairman, ten members and a Member Secretary.
- Central Research Institute for Dryland Agriculture was established in 1985.
- The Forest Research Institute (FRI) is an institute of the Indian Council of Forestry Research and Education.

- FRI is located at Dehradun in Uttarakhand
- FRI is one of the oldest forestry institutions.
- FRI was founded in 1878 as the British Imperial Forest School by Dietrich Brandis.
- In 1906, FRI was reestablished as the Imperial Forest Research Institute, under the British Imperial Forestry Service.

MULTIPLE CHOICE QUESTIONS

1. Scientific Name of larch
(a) *Larix sibirica*
(b) *Quercus robur*
(c) *Populus deltoides*
(d) *Betula nigra*

2. Congolese forests
(a) China (b) Central Africa
(c) Australia (d) Russia

3. Largest Forest cover in the world
(a) China (b) Central Africa
(c) Australia (d) Russia

4. Central America is part of the Mesoamerican biodiversity hotspot, boasting ____ of the world's biodiversity.
(a) 7% (b) 17%
(c) 27% (d) 37%

5. ______ has the largest area of Australia's forest
(a) South Australia
(b) New South Wales
(c) Queensland
(d) Western Australia

6. Bamboo Forest was predominantly found in
(a) China (b) Australia
(c) Russia (d) Japan

7. Eucalyptus Forest is robust in
(a) China (b) Australia
(c) Russia (d) Japan

8. Identify the Coniferous Tree
(a) *Betula platyphylla*
(b) *Pinus roxburghii*
(c) *Quercus glauca*
(d) *Tectona grandis*

9. Identify the Birch
(a) *Betula platyphylla*
(b) *Pinus roxburghii*
(c) *Quercus glauca*
(d) *Tectona grandis*

10. World Agroforestry Centre is located at ____
(a) Peru (b) Uganda
(c) Kenya (d) Italy

11. Shunan Bamboo Forest (Shunan Zhuhai) is the oldest and biggest bamboo forest in
(a) China (b) Japan
(c) India (d) Russia

12. Forest Area of Japan
(a) 52.3 million acres
(b) 72.3 million acres
(c) 62.3 million acres
(d) 42.3 million acres

13. The main vegetation features of Central Africa is
(a) Rain Forest
(b) Central African Mangrove
(c) Swamp and Flooded Forests
(d) Cross river forest

14. Secondary forests of Africa is dominated by _____ tree
(a) Khaya spp. (b) *Prunus africana*
(c) *Musanga ceropiodes*
(d) *Araucaria angustifolia*

15. Australia's native forests are grouped into ___
(a) Twelve types (b) Sixteen types
(c) Eight types (d) Six types

16. Australia's native forest area is dominated by
(a) Acacia forest only
(b) Eucalypt forest only
(c) Eucalypt and Acacia forest
(d) Casuarina forest only

17. Chir pine
(a) *Pinus wallichiana*
(b) *Pinus sylvestris*
(c) *Pinus roxburghii*
(d) *Pinus nigra*

18. The total geographical area of India
(a) 121.36 million ha
(b) 258.85 million ha
(c) 328.73 million ha
(d) 468.22 million ha

19. Identify the oak species
(a) *Pinus roxburghii*
(b) *Acer platanoides*
(c) *Alnus incana*
(d) *Quercus robur*

20. Identify the Larch species
(a) *Lagerstroemia speciosa*
(b) *Leucaena leucocephala*
(c) *Larix occidentalis*
(d) *Lannea coromendalica*

21. Identify the birch species
(a) *Betula papyrifera*
(b) *Butea monosperma*
(c) *Boswellia serrata*
(d) *Bombax ceiba*

22. Blue pine
(a) *Pinus wallichiana*
(b) *Pinus nigra*
(c) *Pinus sylvestris*
(d) *Pinus roxburghii*

23. Identify the Spruce
(a) *Picea smithiana*
(b) *Prosopis juliflora*
(c) *Pinus roxburghii*
(d) *Populus tremuloides*

24. Identify the fir
(a) *Alnus nepalensis*
(b) *Abies pindrow*
(c) *Alstonia scholaris*
(d) *Artocarpus heterophyllus*

25. National Wildlife week
(a) 1st week of August
(b) 1st week of October
(c) 2nd week of March
(d) 2nd week of June

26. International year of Forest
(a) 2011 (b) 2013
(c) 2012 (d) 2014

27. IUFRO is located in
(a) Denmark (b) Berlin
(c) Finland (d) Austria

28. The species grown in desert and xeric shrublands
(a) *Eucalyptus globulus*
(b) *Rhizophora mangle*
(c) *Cedrus deodara*
(d) *Prosopis juiliflora*

29. The scientific name of Douglas fir
(a) *Abies pindrow*
(b) *Abies concolor*
(c) *Pseudotsuga nigra*
(d) *Pseudotsuga menziesii*

30. China has homed with ____ number of bamboo species
(a) 200 species (b) 300 species
(c) 400 species (d) 500 species

31. *Avicennia racemosa* is a
(a) Red mangrove
(b) Black mangrove
(c) White mangrove
(d) Green mangrove

32. The net forest loss during the period between 1990 and 2015 has been estimated at __________ in the world.
(a) 129 million ha (b) 92 million ha
(c) 152 million ha (d) 189 million ha

33. The total forest area in the world based on 2015 assessment
(a) 3999million ha (b) 1204 million ha
(c) 284 million ha (d) 3695 million ha

34. The Natural forest area in the world based on 2015 assessment
(a) 3999million ha (b) 1204 million ha
(c) 284 million ha (d) 3695 million ha

35. The Planted forest area in the world based on 2015 assessment
(a) 658 million ha (b) 291 million ha
(c) 284 million ha (d) 695 million ha

36. The growing stock of forest in the world
(a) 623 billion m^3 (b) 531 billion m^3
(c) 389 billion m^3 (d) 547 billion m^3

37. The forests area in Russia
(a) 814.93 million ha
(b) 493.54 million ha
(c) 347.07 million ha
(d) 310.09 million ha

38. The forests area in China
(a) 214.44 million ha
(b) 208.32 million ha
(c) 310.09 million ha
(d) 319.79 million ha

39. _______ forests are occupied in the area between tropic of cancer and tropic of Capricorn.
(a) Tropical Forest
(b) Temperate Forest
(c) Alpine forest
(d) Tundra Forest

40. ________ are the tropical grass lands in which trees are distributed sporadically throughout the tropical region.
(a) Chaparral (b) Savannahs
(c) Desert (d) Tundra

41. _______ grass lands are known as Prairie
(a) Chaparral grass land
(b) Temperate grass land
(c) Polar grass land
(d) Tundra grass land

42. The United States and Canada together have about ________of the world's total forest cover.
(a) 8.5% (b) 12.5%
(c) 15.5% (d) 20.5%

43. 80% of forestland is common property land owned by edijos or comunidades forestales in _________ country
(a) Mexico (b) US
(c) Canada (d) Africa

44. The forest area of Australia
(a) 100 million hectares
(b) 125 million hectares
(c) 150 million hectares
(d) 175 million hectares

45. Australia has _______ largest forest cover in the world.
(a) 7th (b) 9th
(c) 11th (d) 13th

46. The forest area of China
(a) 174.91 million hectares
(b) 156.32 million hectares
(c) 198.33 million hectares
(d) 132.23 million hectares

47. The forest area of Africa
(a) 624 million ha
(b) 751 million ha
(c) 593 million ha
(d) 1015 million ha

48. The forest area of Asia
(a) 624 million ha
(b) 751 million ha
(c) 593 million ha
(d) 1015 million ha

49. The forest area of Europe
(a) 624 million ha
(b) 751 million ha
(c) 593 million ha
(d) 1015 million ha

50. The forest area of North and Central America
(a) 624 million ha
(b) 751 million ha
(c) 593 million ha
(d) 1015 million ha

51. The forest area of Ocenania countries
(a) 174 million ha
(b) 624 million ha
(c) 593 million ha
(d) 1015 million ha

52. The forest area of South America
(a) 707 million ha
(b) 174 million ha
(c) 842 million ha
(d) 593 million ha

53. The world forest cover was assessed by
(a) FRI (b) FSI
(c) FAO (d) ICRAF

54. The world forest cover assessment was conducted once in every ______ years
(a) 3 Years (b) 5 Years
(c) 7 Years (d) 10 Years

55. The total forest cover of India
(a) 70803 sq.km
(b) 72365 sq.km
(c) 65983 sq.km
(d) 85246 sq.km

56. The total forest and tree cover of India as per the 2017 assessment of FSI
(a) 21.63% (b) 24.54%
(c) 22.85% (d) 23.56%

57. Three Leaf pine
(a) *Pinus wallichiana*
(b) *Pinus nigra*
(c) *Pinus sylvestris*
(d) *Pinus roxburghii*

58. Five leaf pine
(a) *Pinus wallichiana*
(b) *Pinus nigra*
(c) *Pinus sylvestris*
(d) *Pinus roxburghii*

59. Center for International Forestry Research is located at ______
(a) Italy (b) Indonesia
(c) China (d) Kenya

60. International Tropical Timber Organization is located at ______
(a) Japan (b) China
(c) Germany (d) Africa

61. IUCN was established in
(a) 1942 (b) 1956
(c) 1948 (d) 1964

62. The World Wide Fund for Nature (WWF) is founded during
(a) 1961 (b) 1963
(c) 1965 (d) 1967

63. International Network for Bamboo and Rattan (INBAR) is located at ______
(a) Japan (b) China
(c) Germany (d) Africa

64. International Network for Bamboo and Rattan (INBAR) is established in ______
(a) 1995 (b) 1997
(c) 2000 (d) 2003

65. FSC Stands for
(a) Forest Stand Council
(b) Forest Stand Certification
(a) Forest Stewardship Council
(b) Forest Stewardship Certification

66. Forest Stewardship Council (FSC) is established in the year
(a) 1993 (b) 1997
(c) 2008 (d) 2013

67. Intergovernmental Panel on Climate Change (IPCC) was first established in ____

(a) 1978 (b) 1988
(c) 1998 (d) 2008

68. The old name of IUCN was ____

(a) International Union for the Protection of Nature
(b) World Conservation Union
(c) Both a and b
(d) IUCN has no old name

69. Identify the Acidophilous oak

(a) *Quercus robur*
(b) *Sorbus aucuparia*
(c) *Picea abies*
(d) *Betula pendula*

70. The species found in Mexican forests is

(a) Teak and Sandal
(b) Poplars and Pines
(c) Pine and Oak species
(d) Pines only

71. The alternate name for Californian chaparral is ____

(a) Mediterranean Shrub
(b) Hot desert
(c) Cold desert
(d) Tropical savanna

72. Identify the Douglas fir

(a) Pseudotsuga (b) Abies
(c) Pines (d) Oaks

73. The scientific name of Red mangrove

(a) *Rhizophora mangle*
(b) *Avicennia germinans*
(c) *Laguncularia racemosa*
(d) *Conocarpus erectus*

74. The scientific name of White mangrove

(a) *Rhizophora mangle*
(b) *Avicennia germinans*
(c) *Laguncularia racemosa*
(d) *Conocarpus erectus*

75. Identify the Scotch pine

(a) *Pinus wallichiana*
(b) *Pinus nigra*
(c) *Pinus sylvestris*
(d) *Pinus roxburghii*

76. Identify the Aspen

(a) *Populus tremuloides*
(b) *Ulmus species*
(c) *Alnus incana*
(d) *Pinus nigra*

77. Identify the elm species

(a) *Populus tremuloides*
(b) *Ulmus species*
(c) *Alnus incana*
(d) *Picea smithiana*

78. Identify the alder species

(a) *Populus tremuloides*
(b) *Ulmus species*
(c) *Alnus incana*
(d) *Picea smithiana*

79. Identify the Blue pine

(a) *Pinus wallichiana*
(b) *Pinus nigra*
(c) *Pinus sylvestris*
(d) *Pinus roxburghii*

80. Identify the Spruce

(a) *Populus tremuloides*
(b) *Ulmus species*
(c) *Alnus incana*
(d) *Picea smithiana*

81. Identify the Deodar

(a) *Pinus patula*
(b) *Cedrus deodara*
(c) *Alnus incana*
(d) *Acer palmatum*

82. Identify the maple

(a) *Acer palmatum*
(b) *Cedrus deodara*
(c) *Pinus nigra*
(d) *Fagus crenata*

83. Identify the beech
(a) *Acer palmatum*
(b) *Cedrus deodara*
(c) *Pinus nigra*
(d) *Fagus crenata*

84. International Year of Ecotourism (IYE)
(a) 2000 (b) 2002
(c) 2004 (d) 2006

85. International Year of Biodiversity
(a) 2000 (b) 2005
(c) 2010 (d) 2013

86. International Year of Forests
(a) 2010 (b) 2011
(c) 2012 (d) 2013

87. International Year of Soils
(a) 2011 (b) 2013
(c) 2015 (d) 2017

88. International Energy Day
(a) 3rd May (b) 3rd June
(c) 3rd July (d) 3rd August

89. World Biodiversity Day
(a) 18th May (b) 22nd May
(c) 18th March (d) 22nd March

90. World Pollution Day
(a) 8th April (b) 9th May
(c) 10th June (d) 11th July

91. International Mangrove Day
(a) 1st August (b) 11th September
(c) 26th July (d) 29th October

92. International Tiger Day
(a) 14th December (b) 18th May
(c) 27th October (d) 29th July

93. World Elephant Day
(a) 12th August (b) 11th September
(c) 26th May (d) 26th February

94. World Ozone Day
(a) 3rd January (b) 30th November
(c) 21st February (d) 16th September

95. World Energy Conservation Day
(a) 6th October (b) 14th November
(c) 13th April (d) 16th September

96. World Soil Day
(a) 6th March (b) 6th May
(c) 5th December (d) 5th September

97. United Nations Decade on Biodiversity
(a) 2011-2020 (b) 2008-2020
(c) 2015-2020 (d) 2013-2020

98. International Day for Natural Disaster Reduction
(a) 13rd October (b) 30rd June
(c) 17rd July (d) 23rd August

99. International Year of Deserts and Desertification
(a) 2004 (b) 2005
(c) 2006 (d) 2007

100. Total mangrove forest cover in India as per the ISFR, 2017
(a) 4921 Sq.km
(b) 5200 Sq.km
(c) 4500 Sq.km
(d) 5100 Sq.km

101. The tree species wisely used for resin tapping in India
(a) Chir pine (b) Blue pine
(c) Khasi pine (d) Deodar

ANSWERSHEET
2. Global Forest Resources

1. (a)	2. (b)	3. (d)	4. (a)	5. (c)	6. (a)	7. (b)	8. (b)	9. (a)	10. (c)
11. (a)	12. (c)	13. (a)	14. (c)	15. (c)	16. (c)	17. (c)	18. (c)	19. (d)	20. (c)
21. (a)	22. (a)	23. (a)	24. (b)	25. (b)	26. (a)	27. (d)	28. (d)	29. (d)	30. (c)
31. (b)	32. (a)	33. (a)	34. (d)	35. (b)	36. (b)	37. (a)	38. (b)	39. (a)	40. (b)
41. (b)	42. (c)	43. (a)	44. (b)	45. (a)	46. (a)	47. (a)	48. (c)	49. (d)	50. (b)
51. (a)	52. (c)	53. (c)	54. (b)	55. (a)	56. (b)	57. (d)	58. (a)	59. (b)	60. (a)
61. (c)	62. (a)	63. (b)	64. (b)	65. (c)	66. (a)	67. (b)	68. (c)	69. (a)	70. (c)
71. (a)	72. (a)	73. (a)	74. (c)	75. (c)	76. (a)	77. (b)	78. (c)	79. (a)	80. (d)
81. (b)	82. (a)	83. (d)	84. (b)	85. (c)	86. (b)	87. (c)	88. (a)	89. (b)	90. (d)
91. (c)	92. (d)	93. (a)	94. (d)	95. (b)	96. (c)	97. (a)	98. (a)	99. (c)	100. (a)
101. (a)									

3

Silviculture

KEY POINTS

- Silviculture is the art and science of cultivating forest crops
- The study of life history and general characteristics of forest trees and crops with particular reference to environmental factors, as the basis for the practice of silviculture is silvics
- Uninfluenced by human activity – Virgin forest
- Based on the objectives, forestry is classified as Protection forestry, Commercial forestry and Social forestry
- Intensive forestry - object of maximum volume and quality of products per unit are through the application of the best silviculture techniques
- Multiple use forestry - simultaneous use of a forest area for two or more purposes
- Social forestry - outside the conventional forest area
- Farm forestry - raising rows of trees on bunds or boundaries of field and individual trees on private agriculture land as well as creation of wind breaks
- Extension forestry - is the practice of forestry in areas devoid of tree growth
- Mixed forestry - practice of forestry for raising fodder grass with scattered fodder trees, fruit trees and fuelwood trees on suitable wastelands, panchayat land and village commons
- Linear strip plantations - fast growing species on linear strips of land on the sides of public roads, canals and railway lines
- Recreational forestry - the object of raising flowering trees and shrubs mainly to serve as recreation forests for the urban and rural population
- Aesthetic forestry - the practice of forestry with the object of developing or maintaining a forest of high scenic value
- Herb - stem is always green and tender and height is usually less than one metre
- Shrub is usually not more than 6 metres in height
- Tree is usually more than 6 metres in height
- Crown is the upper branchy part of a tree above the bole

Crown shape and size

- Cylindrical - *Phoenix, Cocos* and *Borassus*

- Conical - chir, deodar and conifers
- Spherical - *Mangifera indica, Azadirachta indica, Tamarindus indica, Madhuca indica*
- Broad and flat topped - *Albizzia stipulata*
- Crown development is the expansion of crown measured as crown length and crown width
- Leaves undergo a striking change in colour before falling from the tree; such colours are called autumn tints
- **Deciduous** – *Acacia catechu, Adina cordifolia, Ailanthus excelsa, Bombax ceiba, Garuga pinnata, Holoptelia integrifolia, Lannea coromandelica, Melia azaderach, Schleichera oleosa, Terminalia tomentosa.*
- **Evergreen** – *Abies pindrow, Cedrus deodara, Cupressus torulosa, Hopea parviflora, Mallotus philippinensis, Mangifera indica, Michelia champaca, Picea smithiana, Pinus wallichiana, Pterospermum acerifolium.*
- The decrease in diameter of the stem of a tree or of a log from the base upwards, is known as taper
- Buttresses formation tree species - *Acrocarpus fraxinifolius, Bombax ceiba, Pterocarpus dalbergiodes, Terminalia myriocarpa*
- Fluting which is defined as 'irregular involutions and swellings on the bole just above the basal swell e.g., teak
- Prop-roots – *Ficus bengalensis*
- Stilt roots – *Rhizophora*
- Pneumatophore – Knee shaped or spikelike projection Eg. *Heretiera, Bruguiera*
- Lignotubers is underground swelling e.g. *Eucalyptus*
- *Cassia tora* is a non nodules producing legume species
- Monoecious – *Pinus, Abies pindrow*
- Dioecious – *Salix, Mallotus, Taxus, Diospyros, Cedrus deodara, Pistacia*
- Polygamous – *Sterculia, Garuga, Rhus*
- Mycorrhizae is a mutualistic association between soil fungi and plant roots
- Mycorrhizae are divided into three distinct groups depending on whether or not there is fungal penetration of the root cells: endomycorrhizas, ectomycorrhizas and ectendo mycorrhizas
- Ectomycorrhizae – form mantle over the rootlets
- Endomycorrhizae – Penetrate the cells of the cortex
- Mycorrhiza is found in *Pinus, Picea, Abies, Cedrus, Cupressus, Taxus, Populus, Aurocaria, Salix, Podocarpus* and *Eucalyptus*, etc

Pollination

- By wind (Anemophily) – *Conifers,Betula, Alnus, Juglans*
- By birds (Zoophily) – *Bombax, Butea, Erythrina, Anthocephalus, Oroxylum* (the last two are pollinated by bats).
- By insects (Entomophily) – Sal, Teak, Sissoo

Flowering of Bamboo species

- Gregarious flowering – flowering of almost all the individuals over considerable area followed by death of clumps
- Sporadic flowering – few culms in a clump or a few clumps in a locality
- Annual flowering bamboo species – *Arundinaria wightiana, Bambusa lineata, Ochlandra stridula*

Pollination after ripening period

- Teak - 3 months
- Blue pine – 24-26 months
- Stand is an aggregation of trees occupying a specific area sufficiently uniform in composition (species), age arrangement and condition to be distinguishable from the forest on adjoining areas

Classification of Forests

- Based on method of regeneration forest are classified into two viz., High forest and coppice forest
- Based on age : Even-aged or regular forest, Uneven-aged or irregular forest
- Based on composition : Pure or mixed forest
- Based on objects of management: Production forest, protection forest, farm forest, fuel forest, recreation forest
- Based on ownership and legal status: State forest, communal forest and panchayat forest
- Basis of growing stock : Normal and abnormal forest
- Crop height – Average height of regular crop
- Top height – Average height of dominant trees (250 biggest diameter) in stand
- Crop diameter – Diameter corresponding to the mean basal area of a uniform, usally pur crop
- Crop density - the relative completeness of the stocking expressed as a decimal coefficient, taking normal number of trees, basal area or volume as unity
- 42% of the incoming radiation is reflected, and this is known as albedo
- Without light, plants become pale yellow and have long thin internodes, a condition known as etiolation
- Red and blue colours are effective in photosynthesis
- Less than 2% of the light energy is used in photosynthesis
- Latitude is the distance of a place, north or south of equator, measured as an angle, and whose apex is at the center of the earth
- Radiation frost is defined as the frost occurring on nights with a clear sky, produced by loss of heat by radiation
- Pool frost is defined as the accumulation to a considerable depth of heavy cold air flowing down into natural depressions from adjoining areas.
- Advective frost is a frost produced by cold air brought from elsewhere

- Example of leaf mosaic is Acer species

Zone	Mean Annual Temperature
• Tropical Forest	> 24°C
• Sub tropical	17°C - 18°C
• Temperate	7°C - 17°C
• Alpine Forest	< 7°C

- Frost heaving or frost lifting is lifting of seedlings or small plants above their normal soil level
- Frost free season – the period between the last injurious frost in spring and the first injurious frost in the autumn
- Frost hardy – *Acacia catechu, Anogeissus pendula, Dalbergia sissoo, Diospyros melanoxylon, Hardwickia binata, Madhuca indica, Pinus roxburghii, Schleichera oleosa, Toona ciliata, Ziziphus jujuba.*
- Moderately frost-hardy – *Adina cordiforlia, Anogeissus latifolia, Bombax ceiba, Dalbergia latifolia, Gmelina arborea, Morus alba, Pongamia glabra.*
- Frost tender – *Acacia arabica, Azadirachta indica, Boswellia serrata, Tectona grandis, Terminalia arjuna, Terminalia tomentosa*
- Hardening off is the natural process by which plants become adapted to drought, cold or heat
- Modified or adjusted climatic factors are called bioclimatic
- The direction towards which a slope faces called aspect
- The relation of a site to weather conditions, especially sun and wind called Exposure
- Endodynamorphic - Soils whose properties are influenced mainly by the parent material
- Ectodynamorphic - Soils whose properties are influenced mainly by factors other than parent material
- Sedentary soil is a soil formed from parent material in situ
- Secondary soil a soil derived from pre existing soils by translocation and redeposition
- Kankar pan is commonly found in alluvial soils in U.P and black cotton soils of Deccan
- Silica pan usually formed in laterites and lateritic soils in tropical climates
- Coarse textured soils viz., Sand, Sandy loam
- Medium textured soils viz., loams and silt loams
- Fine textured soils viz., clays and clayey loams
- Coarse textured soils is called light soils
- Fine textured soils is called heavy soils
- Xeric – applied to a habitat that is dry
- Mesic – applied to a habitat that is neither dry nor wet
- Hydric – applied to a habitat that is wet
- Xerophyte – a plant that can subsist in drý situation.
- Mesophyte – a plant whose normal habitat is neither very wet nor very dry.
- Hygrophyte – a plant which grows in moist sites
- Hydrophyte – a plant which grows wholly or partly immersed in water

- Total stem parasite – *Cuscuta reflexa* and *Cassytha filiformis*
- Partial stem parasite – Viscum, Loranthus
- Total root parasite – Broom rape
- Partial root parasite – Sandal
- Epiphyte is a plant growing on, but not nourished by another plant Eg. Ficus, Orchids
- Example of Large woody climbers – Liane or liana
- Example of abnoxious weeds is *Lantana camara*
- Thoreau first used the term forest succession
- Clements theory of plant succession is widely accepted
- Migration is the mass movement of plants from one place to another
- Ecesis is the whole process whereby a plant establishes itself in a new area from germination or its equivalent to reproductions whether sexual or asexual
- Primary succession is the succession that takes place in areas which did not bear vegetation before
- Climax is the culmination stage in plant succession for a given environment
- Retrogression or regression is the reversion to some earlier stage of succession consequent on the introduction of an adverse factor
- Monoclimax theory is given by Clements
- Vegetation gradient or climax pattern is given by Whittaker
- Mosic theory – Aubreville
- Example of climatic climax – Southern Tropical wet evergreen forest
- Example of edaphic climax – Wet Bamboo breaks
- Pre climax- Very moist teak forest
- Post climax – Sal forest, presence of evergreen and semi evergreen species
- Biotic or sub climax – Grassy riverain subclimax
- Regeneration is the renewal of a forest crop by natural or artificial means
- Methods of Regeneration of forest crops are 1. Natural regeneration 2. Artificial regeneration and 3. Natural regeneration supplemented by artificial regeneration
- Natural regeneration is the renewal of a forest crop by self-sown seed or by coppice or root suckers
- Artificial regeneration is the renewal of a forest crop by sowing, planting or other artificial methods
- Regeneration obtained from seed forms a crop, it is called a seedling crop
- Regeneration obtained by coppice forms a crop, it is called coppice crop
- Seedling coppice is the coppice shoots arising from the base of seedlings that have been cut or burnt back
- **Stool coppice** is the coppice arising from the stool or a living stump wilding is a natural seedling used in forest planting
- The seed produced by the trees is dispersed by the agency of wind, water, gravity, birds and animals
  - By wind – Conifers, *Acer, Betula, Rhododendron, Populus, Alnus, Salix, Dipterocarps, Terminalias, Dalbergia, Acacia catechu, Adina, Bombax, Holoptelia,*

Apocynacea and *Asclepiadaceae, Casuarina, Cedrela, Chloroxylon, Pterocarpus marsupium,* etc.

- By water – *Trewia,* mangrove species, *Dalbergia,* teak, etc.
- By gravity – Oaks, Juglans regia, Aesculus, etc.
- By birds – Prunus, Mulberry, Broussonetia, Trema, Diospyros melanoxylon, etc.
- By animals – *Acacia Arabica, Prosopis juliflora, Ziziphus, Anthocephalus,* etc.

- Germinative capacity is the percentage, by number, of seeds in a given sample that actually germinate, irrespective of time
- High Germinative capacity (90-100 days) species - *Albizzia lebbek, Anacardium occidentale, Bauhinia variegate, Cassia siamea, Dalbergia sissoo*
- Germinative energy is the percentage, by number, of seeds in a given sample that have germinated upto the time when the rate of germination reaches its peak
- Plant percent which is the percentage of the number of the seeds in a sample that develop into seedlings at the end of the first growing season
- Interval between the seedling felling and final felling on a particular area is called regeneration interval
- Regeneration survey is a survey for the assessment of established and unestablished regeneration generally by sample enumeration

Regeneration category	Symbol	Value
Established	E	5.00
Woody shoots	W	4.00
Woody shoots browsed	w^+	4.00
Whippy unestablished shoots	u^+	2.00
Whippy unestablished shoots	U	1.00
Subwhippy shoots	s^+	0.50
Subwhippy shoots	S	0.25
Recruit	R	0.00

Total value of regeneration categories in the rectangle	Description of the condition of regeneration	Colour in the map
81 to 100	Excellent	Green
61 to 80	Good	Blue
41 to 60	Moderate	Yellow
21 to 40	Fair	Red
0 to 20	Deficient	Blank (no colour)

- Coppice strongly – *Acacia catechu, Azadirachta indica, Butea monosperma, Casia fistula, Dalbergia* spp., *Diospyros tomentosa, Emblica officinalis, Eucalyptus* Spp., *Melia azedarach, Morus alba, Prosopis juliflora, Salix* spp., *Shorea robusta, Syzygium cumini, Tectona grandis,* etc.
- Coppice fairly –*Chloroxylon swietinia, Hardwickia binata, Pterocarpus marsupium, Quercus spp. Terminalia belerica, Terminalia tomentosa,* etc.

- Coppice badly – *Adina cordifolia, Bombax ceiba, Casuarina equisetifolia, Madhuca latifolia, Populus ciliate,* etc.
- Do not coppice – *Abies pindrow, Cedrus deodara, Picea smithiana, Pinus roxburghii, Pinus wallichinana,* etc.
- Pollard is a tree whose stem has been cut off in order to obtain a flush of shoots, usually above the height to which the browsing animals can reach Eg. Salix, *Hardwickia binata* , *Grewia oppositifolia*
- Lopping means cutting of branches of a tree Eg. *Butea monosperma, Acacia Arabica, Acacia catechu*
- Examples of methods of supplementing natural regeneration is Teak – Stump (root and shoot cutting) planting and Deodar – Naked root planting.
- Nursery is as an area where plants are raised for eventual planting out
- Reforestation - restocking of a felled or otherwise cleared woodland by artificial means
- Afforestation is the establishment of a forest by artificial means on an area from which forest vegetation has always or long been absent
- Indicator plant is any plant which by its presence, increase or decrease, indicates the quality of the site
- Fast growing species the height increment is 60 cm per annum in the earlier stages of its life and which gives a minimum yield of $10m^a$ per hectare per annum
- Exotic species which is raised outside its natural range of distribution
- Raised mixed crop, mixture may be divided into two viz., Permanent, Temporary mixture
- Seed orchard is a plantation of genetically superior trees isolated to reduce pollination from genetically inferior ones, and intensively managed to produce frequent, abundant, and easily harvested seed.
- Seedling seed orchard is a seed orchard which has been raised from the seedlings obtained from seeds of plus trees
- Clonal seed orchards are preferred for most species as against seedling seed orchards because the former start producing seed earlier
- Seed orchards produce genetically improved and superior seed
- Beating up or casualty replacement is restocking blanks in an artificially regenerated area with fresh sowings or plantings
- Nurse crop is a crop of trees or shrubs grown to foster the growth of another and more important tree crop in its early stages
- Cover crop is a subsidiary crop of low plants introduced in a plantation to afford soil cover between or below the main crop
- Plantation journal is a register in which all details of works carried out in a plantation and other points of interest concerning it, are recorded from time to time
- Thinning is a felling made in an immature stand for the purpose of improving the growth and form of the trees
- Ordinary thinning is also called as 'low thinning' or 'German thinning' or 'thinning from the below
- Crown thinning is known as French thinning or thinning from the above or high thinning
- Free thinning is otherwise called as 'single stem silviculture', or 'elite thinning"

- Advance thinning known as Craib advance thinning
- Thinning in an irregular crop is usually called selection thinning
- Thinning cycle is a planned interval in years which elapsed between successive thinnings in the same crop
- Thinning cycles are not regular
- Pruning is the elimination of branches in order to obtain trees with clean bole
- The elimination of branches by physical and biotic agencies of the environment is called natural pruning
- Removal of branches from the seated portions of the tree by mechanical means is referred as artificial pruning
- Silvicultural System defined as a method of silvicultural procedure worked out in accordance with accepted sets of silvicultural principles, by which crops constituting forests are tended, harvested and replaced by new crops of distinctive forms
- Silvicultural Systems are classified into two main categories or groups viz., High Forest Systems and Coppice Systems.
- Clear-felling systems - 'the mature crop is removed in one operation'
- Shelter wood systems - the mature crop is removed in a series of operations, viz., seeding felling, secondary fellings and the last is the final felling
- Final felling is defined as the removal of the last seed or shelter trees after regeneration has been effected under a Shelterwood System
- Taungya - method of raising forest plantations in combination with field crops
- The Taungya may be divided into three viz., departmental, leased or village Taungya
- The regeneration fellings are carried out with the object of making uniform openings in the overwood, the system is called Uniform System
- The regeneration fellings are done in scattered groups which serve as foci to induce and/or to extend groups of existing regeneration, gradually to merge with each other later on, the system is called *Group System*
- Periodic Block is the part or parts of forest set aside to be regenerated or otherwise treated during a specified period.
- Area of Periodic Block = F.S. x P/ R
- The Shelterwood-strip is a silvicultural system in which regeneration fellings are done in the form of strips successively from one side of the compartment, progressing against the direction of wind
- Wagner's Blendersaumschlag, Strip and Group System is a modification of Shelterwood-strip System
- The Indian Irregular Shelterwood System had been described by Glover as the Punjab Shelterwood System
- The number of shoots is reduced only to one per stool; the operation being known as the singling out operation
- Shelter wood Coppice System is a modification of the Simple Coppice System
- The Coppice Selection System is a silvicultural system in which fellings are carried out on the principles of Selection System but regeneration is obtained by coppice
- Conversion is a change from one silvicultural system or one set of species to another

MULTIPLE CHOICE QUESTIONS

1. The art of growing and reproducing a forest is called:
(a) Silviculture (b) Mensuration
(c) Ecology (d) None of the above

2. The scientific basis for the practice of silviculture is
(a) Silvics (b) Forest Management
(c) Forest Biology (d) None

3. The word FOREST is derived from the Latin word
(a) Foris (b) Fores
(c) Forex (d) Forix

4. Number of major forest types available in India is
(a) 16 (b) 17
(c) 18 (d) 19

5. Food, Fuel, Fodder, Fibre obtained from the forests is its
(a) Protective role
(b) Productive role
(c) Ameliorative role
(d) Developmental role

6. The forest products can be broadly classified into
(a) Major Timber Products
(b) Non Timber Forest Products
(c) Both a & b
(d) Secondary Timber Products

7. Site index is a term used by foresters to explain the relationship of ______ in forest stands.
(a) Height to age
(b) Diameter to age
(c) Diameter to height
(d) Volume to age

8. Within the forest, trees that have tall, well formed crowns or tops that receive light from above and the sides are called:
(a) Suppressed (b) Co-dominant
(c) Dominant (d) Intermediate

9. Which of the following is not an element of an ecosystem?
(a) Mineral matter
(b) Producers & consumers
(c) Decomposers
(d) All of the above are elements of an ecosystem

10. The ability of a tree to withstand shade is called:
(a) Intolerance (b) Tolerance
(c) Hardiness (d) All of the above

11. The most limiting element of all the environmental factors on the growth of trees is:
(a) Carbon dioxide (b) Moisture
(c) Day length (d) All of the above

12. ---------- are the perennial plants, which have distinct leafless period for some part of the year.
(a) Evergreen species
(b) Deciduous species
(c) Exotic species (d) None

13. A forest which has been reproduced through sprouting is called a ____ forest.
(a) Epicormic (b) Stocked
(c) Coppice (d) All of the above

14. A stand of trees produced after a clear cut is most likely to be:
(a) Uneven aged (b) Even aged
(c) All aged (d) All of the above

15. Based on crown classification which of the following is defined as a smaller tree with a crown below the upper level of the forest canopy and receiving only scattered sunlight from above?
(a) Intermediate (b) Co-dominant
(c) Dominant (d) Overtopped

16. Stocking level is best described as:
    (a) Closeness of tree crown
    (b) Numbers of trees/acre
    (c) Percent utilization of the site potential
    (d) B & C

17. ________ indicates the productive capacity of a specific area of forest land for a particular tree species.
    (a) Stand density
    (b) Site quality or index
    (c) Tree hardiness
    (d) All of the above

18. The periodic removal of individual trees in order to produce better yields is referred as:
    (a) Clearcutting
    (b) Seed-tree cutting
    (c) Selective cutting
    (d) Shelterwood cutting

19. Tree seeds may be dispersed by which of the following methods?
    (a) Wind (b) Birds and animals
    (c) Water (d) All of the above

20. Identification of trees according to species is called ______________.
    (a) Forestry (b) Silvics
    (c) Ecology (d) Dendrology

21. The following method of timber harvesting often appears to be the most aesthetically drastic:
    (a) Clear cutting
    (b) Seed tree cutting
    (c) Selection cutting
    (d) Shelterwood cutting

22. Which of the following is not a function of tree roots?
    (a) Serve as an anchor for the tree
    (b) Absorb water from the soil
    (c) Conduct photosynthesis
    (d) All of the above

23. Which of the following are ways in which coniferous trees reproduce?
    (a) Root suckers (b) Stump sprouts
    (c) Direct seeding (d) All of the above

24. The most important characteristic(s) to use in identifying a tree is(are):
    (a) Leaves (b) Bark
    (c) Twigs (d) All of the above

25. The principal objective of intermediate cuttings include:
    (a) Improvement of the existing stand
    (b) Manipulation or regulation of tree and stand growth
    (c) Creation of stand conditions unfavorable to insects and diseases
    (d) All of the above

26. Leaves do all of the following during the day except _____.
    (a) Manufacture plant food
    (b) Give beauty
    (c) Give off oxygen
    (d) Give off carbon dioxide

27. Which is NOT a common term when referring to coniferous trees?
    (a) Softwoods (b) Evergreens
    (c) Hardwoods (d) Pines

28. Deciduous trees are NOT referred to as ______
    (a) Broadleaf (b) Softwood
    (c) Hardwood (d) All of the above

29. Which group of conifers has its needles in a cluster of from two to five?
    (a) Pines (b) Spruces
    (c) Cedars (d) Firs

30. Which tree leaves are least likely to turn red in fall?
    (a) Sugar maple (b) Silver maple
    (c) Red maple (d) Red oak

31. New terminal growth on pines is called
    (a) Pinettes (b) Needlettes
    (c) Spikes (d) Candles

32. The proper depth to plant seedlings is _______

(a) The same depth as they were in the nursery
(b) Slightly deeper than they were in the nursery
(c) Slightly shallower than they were in the nursery
(d) 12 inches

33. Basal area of a woodlot represents:

(a) Area occupied by the tree in square feet
(b) Crown density
(c) Stand density
(d) Taper of trunk

34. When referring to trees as tolerant or intolerant, a tolerant species is one that will withstand

(a) Drought (b) Cold
(c) Shade (d) Flooding

35. A tree sprout arising from the root is called a _____.

(a) Slash (b) Bole
(c) Sucker (d) Sapling

36. Another term for the trunk of a tree is ____

(a) Bole (b) Club
(c) Skid (d) Pike

37. The leaves and branches of a tree are referred to as _____ of the tree.

(a) Dominant (b) Bole
(c) Crown (d) Diameter

38. Over crowding reduces _____ of the tree.

(a) Stand density (b) Height
(c) Weeding (d) Diameter

39. The term ______ is defined as the reproduction and management of the forest to obtain continuous high yields of forest products.

(a) Dendrology
(b) Ecology
(c) Silviculture
(d) Diameter breast height

40. Which would be the cheapest way to buy seedlings?

(a) Bare root
(b) Root Trainer
(c) Container grown
(d) From a garden center

41. The best time of the year to plant seedlings is ______.

(a) Early spring (b) Early summer
(c) Early fall (d) Early winter

42. The most common cause for seedling death is _______.

(a) Poor seedling stock
(b) Roots have dried out before planting
(c) Diseases
(d) Insects

43. Temporary planting of several trees in long narrow rows is called ______.

(a) Hedging (b) Trenching
(c) Heeling-in (d) Stocking

44. Another name for the hand planting bar is the ______.

(a) Wedge prism (b) Plunger
(c) Dibble (d) Increment borer

45. The closeness of the tree tops is called ____

(a) Closed canopy
(b) Open canopy
(c) Crown density
(d) Carrying capacity

46. The scientific management of forests for the continued output of products is called _________.

(a) Dendrology
(b) Forestry
(c) Ecology
(d) Forest mensuration

47. The art of producing and caring for a forest is called ____.

(a) Dendrology (b) Silviculture
(c) Arboriculture (d) Ecology

48. A method of harvesting that removes all of the trees in the stand is called _______.
(a) Clearing (b) Clear cutting
(c) Total cutting (d) Total removal

49. A cutting which removes trees of poor quality, undesirable species or those in poor condition is called.
(a) Improvement cutting
(b) Upgrading
(c) Thinning
(d) Defect clearing

50. Cutting of burnt or damaged trees before they become completely worthless is called ______.
(a) Improvement cutting
(b) Release cutting
(c) Salvage cutting
(d) Scalping

51. Cutting of large trees which throw shade on the young trees is called.
(a) Thinning
(b) Selection cutting
(c) Shade tree cutting
(d) Release cutting

52. Removal of mature trees which are the highest quality and most valuable is known as ______.
(a) Clear cutting
(b) Selection cutting & economic high grading
(c) Type cutting
(d) Mature cutting

53. Removing trees damaged by fire, insects or disease is called _____.
(a) Salvage cut (b) Clean cutting
(c) Clear cut (d) Sanitation cut

54. Removal of mature timber in a series of cuttings so that the remaining trees might reseed the area is called
(a) Selection cutting
(b) Natal cutting
(c) Shelterwood cutting
(d) Density cutting

55. Removal of the lower branches of a tree is called _____.
(a) Thinning
(b) Cutting back
(c) Pinching
(d) Pruning

56. Which is not considered in stand density?
(a) Basal area
(b) Tree height
(c) Crown density
(d) Number of stems per acre

57. Selling standing timber is called ____ sale
(a) Live (b) Green
(c) Stumpage (d) Standing

58. A tree stand regeneration almost totally from sprouts is called a ____.
(a) Sprout forest
(b) Seedless forest
(c) Stump forest
(d) Coppice forest

59. Girdling is a method used to:
(a) Kill unwanted trees
(b) Harvest mature trees
(c) Properly space veneer quality trees
(d) Produce knot free lumber

60. Pruning is done to produce logs with _______.
(a) Knot free lumber
(b) A minimum of 24 inches in diameter
(c) No insects or diseases
(d) All of the above

61. The periodic removal of trees in order to produce better yields is:
(a) Clear cutting
(b) Seed-tree cutting
(c) Selective cutting
(d) Shelterwood cutting

62. Cuttings made at the sapling stage for the purpose of improving the composition or the quality of the growing timber stand is called:

(a) Cleaning cuttings
(b) Improvement cuttings
(c) Liberation cuttings
(d) Weeding cuttings

63. A non grass like herbaceous plant is called

(a) Forb (b) Forage
(c) Forest (d) Tree

64. The primary purpose of the flower is ______.

(a) Reproduction
(b) Respiration
(c) To attract beneficial insects
(d) Transpiration

65. Trees which are characterized by having flowers of both sexes on the same tree are:

(a) Coniferous (b) Deciduous
(c) Dioecious (d) Monoecious

66. The light colored part of a cross section of a tree that is responsible for conducting water and minerals to the crown is:

(a) Cambium (b) Heartwood
(c) Pith (d) Sapwood

67. The symbol for established regeneration whose height should be 2·5 m or more and dbh 10 cm is

(a) e (b) u+
(c) S+ (d) W+

68. A fire that has been started intentionally by a forester to be used as a management tool is a ______.

(a) Crown fire (b) Prescribed burn
(c) Surface fire (d) Wildlife

69. The part of a tree that grows into a seed is called as:

(a) Ovule (b) Stamen
(c) Ovary (d) Polyp

70. Flower part modified as a male reproductive structure

(a) Stamen (b) Stigma
(c) Embryo sac (d) Sepal

71. The science of classification of trees or any other living matter is:

(a) Identification
(b) Agronomy
(c) Horticulture
(d) Taxonomy

72. In measuring the site index of a tree, which criteria must be measured?

(a) Age of tree and merchantable height
(b) Age of a tree and total tree height
(c) Dbh and total tree height
(d) Merchantable height and DBH

73. _____ is the weight of a complete tree including roots, stump, stem, limbs, branches and leaves.

(a) Basal area
(b) Biomass
(c) Cubic volume
(d) Stumpage

74. A conifer with scales rather than needles is ______.

(a) Cedar (b) Fir
(c) Spruce (d) Yew

75. The system of binomial nomenclature, is credited to:

(a) Carolus Linnaeus
(b) Charles Darwin
(c) Charles E. Bessey
(d) George Bentham

76. Silvics is the science of tree ______ and species distribution in the forest.

(a) Cutting (b) Disease
(c) Growth (d) Harvest

77. The most expensive timber management activity is ______.

(a) Estimating or cruising
(b) Hand planting
(c) Marking
(d) Site preparation

78. A timber harvesting method used to allow faster growth of remaining trees, to remove diseased trees or to remove undesirable trees before the final harvest is called ______.

(a) Initial forestation

(b) Intermediate cuttings

(c) Selective cuttings

(d) Shelter wood

79. Thinning is designed to ______.

(a) Concentrate growth on the best trees

(b) Reduce crown density

(c) Reduce the number of trees in an immature stand

(d) Reduce the time period to the final harvest

80. An economical way to remove excess logging debris and undesirable vegetation prior to planting is by.

(a) Bulldozer

(b) Herbicide

(c) Prescribed fire

(d) Wildlife

81. Tree injectors are used to:

(a) Stimulate growth

(b) Mark trees

(c) Kill trees

(d) Inoculate trees

82. An example of a biotic factor in a forest is ______.

(a) Climate (b) Rainfall

(c) Soil depth (d) A tree

83. A cottonwood tree has a ____ leaf shape.

(a) Cordate (b) Deltoid

(c) Obovate (d) Oval

84. ______ is not a silvical characteristic.

(a) Growth habit

(b) Seeding habit

(c) Site and moisture requirements

(d) None

85. The term commercial forest means.

(a) Land is capable of producing economically useful trees

(b) Land produces only commercial size harvestable trees

(c) Government owns a majority of the saleable timber

(d) Timber companies own a majority of the forestland

86. A tree with a small crown in the general canopy level and receives little direct light from the side is called.

(a) Codominant (b) Dominant

(c) Intermediate (d) Suppressed

87. The class of trees angiosperms is made up of ___ trees.

(a) Broadleaf (b) conifer

(c) evergreen (d) hardwood

88. Roots which grow vertically downward and anchor the tree are ______.

(a) Feeder roots and hair roots

(b) Lateral roots

(c) Surface roots

(d) Tap roots

89. The trunk branches and roots grow in diameter by means of cell division in the

(a) Cambium layer (b) Heartwood

(c) Sapwood (d) Outer bark

90. The inner core of a woody stem, wholly composed of non-living cells, darker in color than the outer layer of the stem is ______

(a) Cambium layer

(b) Heartwood

(c) Sapwood

(d) Outer bark

91. Alternate, opposite and whorled describes.

(a) Leaf types

(b) Needle types

(c) Twig branching

(d) Tree classifications

92. A process which removes injured, disease or insect-infested trees is called ____.

(a) Liberation cutting
(b) Salvage cutting
(c) Sanitation cuttings
(d) Thinnings

93. In the ____________ harvest method, the entire stand of trees are cut except for a few which are selected for their quality and seed producing ability.

(a) Coppice
(b) Clearcutting
(c) Selective cutting
(d) Seed tree cutting

94. The purpose of this harvest method is to: produce a large immediate income, provide for the natural reseeding of the area and provide partial shelter for the seedlings as they begin to germinate.

(a) Coppice
(b) Selective cutting
(c) Seed tree cutting
(d) Shelter wood cutting

95. Trees and forests are considered to be ____.

(a) Producers
(b) First level consumers
(c) Second level consumers
(d) Decomposers

96. A very complex community of associated trees, shrubs, other plants and animals is a/an.

(a) Arboretum (b) Forest
(c) Greenhouse (d) Park

97. The flowering period of Teak is.....

(a) Apr-May
(b) May-June
(c) Nov-Jan
(d) Sep-Oct

98. A tree that bears seeds in cones and that have long, narrow (needle like) leaves is a

(a) Broad leaf (b) Conifer
(c) Deciduous tree (d) Foliage plant

99. The class of trees, Gymnosperm, is made up of _____ trees.

(a) Broad leaf (b) Coniferous
(c) Deciduous (d) Hardwood

100. Roots which grow horizontally and help keep the tree upright are _______.

(a) Feeder roots and root hairs
(b) Lateral roots
(c) Surface roots
(d) Tap roots

101. The process by which excess water from the tree is allowed to evaporate through openings in the leaves is

(a) Hydrolysis
(b) Osmosis
(c) Photosynthesis
(d) Transpiration

102. Applying tree seeds to the harvested area or prepared beds is called________.

(a) Seed tree method
(b) Direct seeding
(c) Natural seeding
(d) All of the above

103. Neither natural seeding or_____ will result in trees being planted in rows unless a grain drill is used.

(a) Seed tree method
(b) Direct seedings
(c) Natural seeding
(d) Shelterwood seeding

104. Dendrology is commonly defined as the ____.

(a) Study of forest measurement
(b) Study of tree growth & management
(c) Taxonomy of all woody plants
(d) Taxonomy of trees

105. As a general rule:
(a) Conifers are usually less tolerant of flooding than hardwoods
(b) Hardwood are usually less tolerant of flooding than conifers
(c) Most important forest species are flood tolerant
(d) No important forest species are flood tolerant

106. A silviculture prescription:
(a) Changes in forest density
(b) Is a type of forest inventory?
(c) Plans for pesticide application to control insects
(d) Usually contains cultural operations relating to objectives

107. Windrowing, herbicide application and prescribed burning are methods of:
(a) Artificial regeneration
(b) Natural regeneration
(c) Site preparation
(d) Uneven-aged management

108. A disadvantage to clearcutting is ____
(a) Nothing can grow in clear-cut area for 3 years
(b) Soil erosion can be a problem if there is no advance tree regeneration
(c) It is the most costly harvest method
(d) There are no disadvantages to clearcutting

109. The term "alkaline" refers to a substance with a pH of _____
(a) 7.0 (b) 7.1 or above
(c) 4.9 or below (d) 3.0

110. A soil with a pH of 3.2 is _____
(a) Neutral (b) Acidic
(c) Alkaline (d) All of the above

111. An herbicide that will not move or translocate within a plant is called.
(a) Surfactant (b) Contact
(c) Systemic (d) Cuticle

112. Stumpage refers to the ____________.
(a) Number of stumps left after a timber harvest
(b) Number of stems per ha
(c) Not knowing what to do next
(d) The value of trees as they stand uncut in the woods

113. A forest fire which is out of control is called.
(a) Ground fire
(b) Spot fire
(c) Suppression fire
(d) Wild fire

114. A mixed stand is one in which less than ____ % of the trees are of same species.
(a) 75 to 80 (b) 60 to 70
(c) 50 (d) 30 to 40

115. Litter from leaves, especially from needles, at various stages of decomposition is called ______.
(a) Slash (b) Litter
(c) Trash (d) Duff

116. Which of the following factors would affect the success or failure of direct seeding?
(a) Climatic conditions
(b) Consumption by animals
(c) Previous harvest method
(d) Both a and B

117. The debris left after logging, such as limbs, tops, culls and knockdown trees is called:
(a) Conk (b) Kerf
(c) Stumpage (d) Slash

118. The most important characteristics to use in tree identification are ___.
(a) Bark, fruit, twigs and leaves
(b) Bark, growth habit, leaves
(c) Fruit, twigs, leaves, color and taste
(d) Twigs, bark, fruit, site

119. The function of the xylem in the tree is.
(a) Develop leaves from nutrients
(b) Conduct water and minerals to the crown
(c) Grow wood and bark cells
(d) Produce carbohydrates through photosynthesis

120. A totally occupied forest stand that allows plenty of room for development of crop trees is called a (an)
(a) Fully stocked stand
(b) Mixed stand
(c) Overstocked stand
(d) Pure stand

121. The best starter fertilizer for growth of trees:
(a) Nitrogen (b) Phosphorous
(c) Potassium (d) All of the above

122. Which of the following is not a factor in using controlled burns?
(a) Area to be burned
(b) Intensity of the fire
(c) Humidity
(d) Site index

123. The harvest method that results in sustained production forestry by requiring foresters to choose trees for harvest based on maturity, size, species, growth rate or other factors is ________.
(a) Coppice cutting
(b) Clear cutting
(c) Selective cutting
(d) Thinning

124. Stumps should be cut ______.
(a) 1 foot above the ground
(b) At dbh
(c) At waist height
(d) Close to the ground to avoid waste

125. Thinning overstocked stands in order to manage or control pests is a _______ pest management strategy.
(a) Biological (b) Chemical
(c) Regulatory (d) Silvicultural

126. A snag is a ____________.
(a) Blocked firebreak
(b) A branch from one tree caught up in branches from another tree
(c) A standing dead tree
(d) A pinched chain saw bar

127. A hypo-hatchet is used for ________.
(a) Applying a blaze to trees to mark a boundary
(b) Cutting small trees quickly
(c) Girdling a tree
(d) Injecting a tree with herbicide

128. The mid-summer water temperature of a stream flowing through a clear-cut with no stream buffer would be ______ the temperature of the water prior to the clearcut.
(a) Lower than
(b) Higher than
(c) The same as
(d) Clear cutting would not affect the water temperature

129. Which of the following would be considered mechanical tree damage?
(a) Frost cracks (b) Girdling roots
(c) Powdery mildew (d) Scorching

130. Which of the following processes uses shelterwood methods of harvesting to promote regeneration?
(a) Cuttings (b) Direct seeding
(c) Natural seeding (d) Seedlings

131. The practice most commonly used to produce high grade stands is ________.
(a) Cleaning (b) Improvement
(c) Release (d) Thinning

132. The process to remove damaging dominant trees in an all-aged stand is called ________.
(a) Liberations
(b) Salvage cuttings
(c) Sanitation cuttings
(d) Thinnings

133. Tree cuttings made to allow faster growth of remaining trees, to remove diseased trees or to remove undesirable trees before the final harvest is called ___.___

(a) Early forestation
(b) Intermediate cuttings
(c) Seed tree cuttings
(d) Shelterwood

134. The most accurate method of cruising is the _________.

(a) Average cruise
(b) Ocular cruise
(c) One hundred percent cruise
(d) Point sampling

135. Which of the following produces reproduction from stump sprouts?

(a) Coppice cutting
(b) Seed tree cutting
(c) Selective cutting
(d) Shelterwood cutting

136. Coppice shoots and root suckers are examples of __________.

(a) Sexual reproduction
(b) Vegetative reproduction
(c) Cloning
(d) Wood production

137. Fallen tree tops are generally referred to as _________.

(a) Biomass (b) Brush piles
(c) Firewood (d) Slash

138. Incendiary fires are ______________.

(a) Ignited accidentally through use of power logging equipment
(b) Set maliciously
(c) Started by burning debris after logging
(d) The result of carelessly discarded ciagarettes

139. The species which occurs under sclerophyll forests

(a) *Quercus ilex*
(b) *Abies pindrow*
(c) *Cedrus deodara*
(d) *Populus tremuoides*

140. Contour lines show ______________.

(a) Elevation and shape of the land
(b) Land quality
(c) Land use potential
(d) Resource distribution

141. On a topographic map which of the following is a dark, heavy line with an elevation figure?

(a) Contour index
(b) Contour interval
(c) Contour line
(d) Horizontal control station

142. Using the natural shape and texture of a tree to complement the architecture of a building or to enhance the vertical elements of a structure is an example of a __________ benefit of urban trees.

(a) Aesthetic (b) Environmental
(c) Economic (d) Wildlife

143. The function of root hairs is to ________.

(a) Anchor the tree
(b) Absorb water and nutrients
(c) Produce chlorophyll
(d) All of the above

144. __________ lack root hairs.

(a) Aspen (b) Casuarina
(c) Birch (d) None

145. Instead of root hairs, pines have _________ attached to the rootlets which absob water and nutrients.

(a) Tap roots
(b) Branch roots
(c) Mycorrhizae
(d) All of the above

146. Windfalls are most common on ________.

(a) Forest stands that have been heavily cut over
(b) On shallow soils
(c) On wet soils
(d) All of the above

147. The process whereby chloroplasts convert radiant energy into chemical energy is called ______.
(a) Transpiration & respiration
(b) Phototranspiration
(c) Chlorophyll
(d) Photosynthesis

148. In photosynthesis, carbon dioxide from the air and water and minerals from the soil into ______ with oxygen as a by-product.
(a) Fats and proteins (b) Amino acids
(c) Sugars and starches
(d) Vitamins and minerals

149. The process of consuming carbohydrates and oxygen to obtain energy for the biological processes of life support, growth and reproduction is called ______.
(a) Transpiration (b) Respiration
(c) Perspiration (d) Chlorophyll

150. The loss of water vapor from the leaf of a tree is called ______.
(a) Transpiration (b) Respirations
(c) Perspiration (d) Chlorophyll

151. Most of the water vapor given off by a plant escapes through structures called______.
(a) Pores (b) Stomata
(c) Exit cells (d) Cutin

152. Excess carbohydrates not used in respiration are converted to ______ and ______ and is then stored in the tree.
(a) Fats and protein
(b) Vitamins and minerals
(c) Cellulose and lignin
(d) Chlorophyll and chloroplasts

153. ______ and ______ are the main constituents of wood.
(a) Fats and protein
(b) Vitamins and minerals
(c) Cellulose and lignin
(d) Chlorophyll and chloroplasts

154. Sprouting, suckering and layering are examples of ______ reproduction in trees.
(a) Sexual (b) Asexual
(c) Tissue Culture
(d) None of the above

155. During autumn and winter your best way to identify trees will be to compare ______ and ______ characteristics.
(a) Leaf and flower
(b) Fruit and stem
(c) Twig and bark
(d) All of the above

156. The leaf colour change before the leaf fall is called
(a) Autumn tint
(b) Discolouration of leaves
(c) Malformation of leaves
(d) All of the above

157. In deciduous trees, ____ is the most important identification feature in the dormant winter period.
(a) Fruit (b) Leaves
(c) Twigs (d) Bark

158. Poles and pilings are measured in ____.
(a) Linear feet (b) Cords
(c) Cubic feet (d) Weight

159. In a dormant condition, trees can withstand ______ temperature extremes than during periods of active growth.
(a) Smaller
(b) Greater
(c) The same
(d) None of the above

160. In the temperate zone, summer growth is often completed ______ high maximum summer temperatures are reached.
(a) Before
(b) Long after
(c) At the same time as
(d) At least 3 months before

161. Sunlight is important in regulating the following processes
(a) Growth
(b) Leaf fall and dormancy
(c) Flowering and reproduction
(d) All of the above

162. Day length or ____________ influences diameter growth in trees.
(a) Photoperiod
(b) Perspiration
(c) Photosynthesis
(d) Photochemical

163. Periods of long daylight in spring and early summer result in the production of ____________ cells.
(a) Large diameter, thick walled, spring wood
(b) Large diameter, thin walled, spring wood
(c) Small diameter, thick walled, summer wood
(d) Small diameter, thin walled, summer wood

164. Length of day light affects the time of flushing or ____________ in the spring and the onset of dormancy in the autumn.
(a) Phototrophism
(b) Defoliation
(c) New growth
(d) Bud formation

165. Light distribution can affect tree form, resulting ____________.
(a) Greater development of the tree on the side exposed to the most sunlight
(b) Less growth on the side of the tree ex posed to the most sunlight
(c) A tall, spindly tree because of exposure to the sun
(d) A short, overtopped tree because of ex posure to the sun

166. The quality of light refers to -------.
(a) Wavelength of the incoming radiation
(b) Intensity of light
(c) Temperature
(d) a and b

167. Seed formation depends on the amount of reserve ____________ and nitrogenous materials in a tree.
(a) Fatty acids
(b) Carbohydrates
(c) Proteins and amino acids
(d) Oxygen

168. The term ____________ is used to describe the physical expression of an inherited tree trait that can be measured and observed.
(a) Genotype (b) Genetic map
(c) Physical map (d) Phenotype

169. Winterkill is the ____________ of young trees and seedlings.
(a) Stunting (b) Deforming
(c) Killing (d) Weakening

170. In a forest ecosystem a tree with a small crown receiving sunlight from above and almost none from the sides would be considered a ____________ tree.
(a) Co-dominant (b) Dominant
(c) Inferior (d) Intermediate

171. To determine site index the two measurements needed are ____________.
(a) Age and diameter
(b) Age and height
(c) Diameter and distance from the closest tree
(d) Diameter and height

172. Growth at the end of the stem is called ____________.
(a) Preliminary growth
(b) Primary growth
(c) Terminal growth
(d) Turgid growth

173. Tree species that have male and female flowers on separate trees are called ___.

(a) Bioecious (b) Dioecious
(c) Monoecious (d) Triecious

174. Trees are planted along city streets to help increase ___.

(a) Gas removal & particulate collection
(b) Summer air temperatures
(c) Temperatures of sidewalks and road surfaces
(d) Wind velocity

175. Using containerized seedlings rather than bareroot seedlings usually results in

(a) A much shorter planting season
(b) Easier handling and transportation
(c) Improved rates of survival & growth
(d) Lower initial cost for seedlings

176. The most widely known and used hardwood is ___.

(a) Ash (b) Hickory
(c) Maple (d) Oak

177. The type of pit which is suitable for slopy regions?

(a) Ordinary pit (b) Saucer pit
(c) Ridge ditch pit (d) Ring pit

178. Which of the following is NOT a common method of testing improved seeds or seedlings?

(a) Clone testing
(b) Progeny testing
(c) Provenance testing
(d) Regulation testing

179. Which of the following silvicultural systems is most likely to result in uneven-aged forests?

(a) Clear cutting
(b) Shelterwood system
(c) Seed tree system
(d) Selection system

180. Beating up is a silvicultural operation that refers to

(a) Pollarding (b) Pruning
(c) Restocking (d) Thinning

181. Deciduous means :

(a) Falling off at maturity
(b) Tending to fall off
(c) Leaf colour change
(d) Both a & b

182. Which type of frost affects the forest the most?

(a) Radiation frost (b) Pool frost
(c) Advective frost (d) All of the above

183. Thinnings in young timber allow the timber grower to:

(a) Maintain high rates of growth
(b) Concentrate growth on the better trees
(c) Maintain a properly stocked forest
(d) All of the above

184. Removing trees that have been or may be killed or damaged by insects, ice, fire and other agents is called

(a) Liberation (b) Pruning
(c) Salvage (d) Sanitation

185. Site index is a term used by foresters to explain the relationship of ___ in forest stands.

(a) Height to age
(b) Diameter to age
(c) Diameter to height
(d) Volume to age

186. In India, Plantation forestry was started in

(a) 1942 (b) 1842
(c) 1947 (d) 1847

187. Quantity of seed required for a nursery bed can be worked out by

(a) W = (A X D X 100) / (P X N)
(b) W = (P X D X 100) / (A X N)
(c) A = (W X D X 100) / (P X N)
(d) W = (A X N X 100) / (P X D)

188. Tick the odd man out: Growth stages of tree

(a) Pole (b) Seedling stage

(c) Young timber (d) Sapling stage

189. Choose the best tree species for coastal afforestation

(a) *Albizia lebbeck*

(b) *Azadirachta indica*

(c) *Casuarina equisetifolia*

(d) *Eucalyptus tereticornis*

190. CVP index refers to

(a) Crop volume per cent index

(b) Cumulative volume per cent index

(c) Climate vegetation and productivity index

(d) Climate volume productivity index

191. The term provenance means

(a) Land race

(b) Geographical source

(c) Seed source

(d) None of the above

192. Tropical dry evergreen forests are found in

(a) Carnatic coast

(b) Andaman and Nicobar island

(c) Coastal areas of Kerala

(d) North east India

193. Shelterbelts have (pick the wrong statement)

(a) Single tree species

(b) Multiple tree species

(c) Used to arrest wind

(d) Prevent sand dune formation

194. Ecotype refers to

(a) Variation in the phenotypic characters

(b) Phenotypic variation of different species

(c) Differential response of different species to environmental factors

(d) Differential genetical response of the same species to environmental factors

195. Which is not a characteristic of gregarious flowering in Bamboo?

(a) Flowering occurs almost over the entire area

(b) It involves only few proportion of the culms

(c) Flowering is followed by the death of clump

(d) It follows a cycle which occur after long interval

196. High forest refers to

(a) Forests regenerated through cutting

(b) Forests regenerated through seeds

(c) Forest regenerated through coppice

(d) Forest regenerated through tissue culture plants

197. Pure forest are those which cover

(a) 40% of the crop consist of a single species

(b) 50% of the crop consist of a single species

(c) 60% of the crop consist of a single species

(d) 80% of the crop consist of a single species

198. Champion and Seth classified the Indian Forests into major

(a) Six groups (b) Sixteen groups

(c) Sixty groups (d) None

199. The leader of the 'Chipko' movement is

(a) Westoby

(b) Sundarlal Bahuguna

(c) M.S. Swaminathan

(d) Tewari

200. Trees which do not form part of the upper most leaf canopy but the leading of shoots of which are not definitely overtopped by the neighbouring trees are called

(a) Dominant (b) Co-dominant

(c) Dominated (d) Pre-dominants

201. The mean annual temperature of montane sub-tropical forest ranged between
(a) 10°C and 18°C (b) 7°C and 17°C
(c) 17°C and 24°C (d) 18°C and 24°C

202. The mean annual temperature of montane temperate forest ranged between
(a) 10°C and 18°C (b) 7°C and 17°C
(c) 17°C and 24°C (d) 18°C and 24°C

203. The term 'forest succession' was first used by
(a) Clement (b) Cawles
(c) Thoreau (d) Dawson

204. An organism on or within which another organism feeds and develops is called
(a) Parasites (b) Epiphytes
(c) Host (d) None of the above

205. ICFRE was established during the year
(a) 1986 (b) 1980
(c) 1975 (d) 1990

206. The climax stage of estuarine succession
(a) Mangrove scrub
(b) Mixed forests of spruce and fir
(c) Evergreen forest
(d) Dry deciduous forest

207. Tropical Forest Research Institute is located at
(a) Jodhpur (b) Jorhat
(c) Jabalpur (d) Jhansi

208. Study of the biotic community and its abiotic environment is called
(a) Population ecology
(b) Community ecology
(c) Ecosystem ecology
(d) None

209. Population ecology is otherwise called
(a) Synecology
(b) Autecology
(c) Ecosystem ecology
(d) None

210. The poly climax theory in plant succession was developed by
(a) Clement (b) Tansley
(c) Fosberg (d) Whittaker

211. Fungal roots are otherwise called
(a) Frankia (b) Mycorrhiza
(c) Rhizobium (d) Azospirillum

212. The Father of Indian Forestry is
(a) O'Conolly
(b) Dietrich Brandis
(c) Voelckar
(d) Gadgil

213. The Moist Deciduous and Rain Forest Research Institute is located in
(a) Jorghat (b) Simla
(c) Jabalpur (d) Jhansi

214. The first Conservator of Forests.
(a) Trevor (b) Troupe
(c) Watson (d) Hartig

215. The relative extent to which a crop is opened up in thinning is called
(a) Thinning cycle
(b) Thinning intensity
(c) Thinning regime
(d) Thinning grade

216. The per cent forest area in the country as per 2011 FSI report
(a) 19.55 (b) 23.81
(c) 21.55 (d) 22.55

217. The OFI is located
(a) London (b) Edinburg
(c) Brussels (d) Rome

218. Which of the following plants might be found in a primary state of succession?
(a) Beech (b) White pine
(c) Aspen (d) Lichen

219. Foresters recognize which of the following as the basic management unit in a forest?
(a) Stand (b) Lot
(c) Section (d) District

220. Secondary succession might be considered to start after which of the following?
(a) A landslide
(b) Back filling of a strip mine
(c) A glacier
(d) All of the above

221. Site index is a term used by foresters to explain the relationship of ______ in forest stands.
(a) Height to age
(b) Diameter to age
(c) Diameter to height
(d) Volume to age

222. Within the forest, trees which do not part of the upper most leaf canopy but the leading shoots of which are not definitely by the neighbouring trees and the height is about ¾ of the tallest trees:
(a) Suppressed (b) Co-dominant
(c) Dominated (d) Intermediate

223. Under which grade very ordinary thinning is carried out?
(a) B grade (b) C grade
(c) D grade (d) E grade

224. Which is the closest to a hectare?
(a) 1 acre (b) 2.5 acres
(c) 5.2 acres (d) 7 acres

225. A typical example of the northern tropical wet evergreen forest is
(a) *Lagerstroemia lanceolata*
(b) *Pterocarpus dalbergioides*
(c) *Diperterocarpus turbinatus*
(d) *Rhododendron nilgiricum*

226. An abandoned field changes into a mature forest over an extended period of time through a process called:
(a) Succession (b) Dominance
(c) Recycling (d) All of the above

227. Stump planting is very common in
(a) Teak (b) Sal
(c) Deodar (d) Red sanders

228. Deodar is managed under
(a) Clear felling system
(b) Selection system
(c) Shelter wood system
(d) None of these

229. Scientific name of Toon
(a) *Tectona grandis*
(b) *Toona ciliata*
(c) *Shorea robusta*
(d) *Cedruš deodara*

230. *Eucalyptus* is a
(a) Bad coppicer
(b) Moderately good coppice
(c) Good Coppicer
(d) Very bad coppicer

231. Example for Nitrogen fixing tree
(a) Casuarina (b) Acacia
(c) Prosopis (d) All of the them

232. Tendu is a source of
(a) Tannin (b) Dye
(c) Gum (d) Beedi Leaf

233. Sandal is commonly affected by
(a) Spike disease (b) Pink Disease
(c) Stem wilt (d) Leaf galls

234. *Terminalias* belong to the family
(a) Dipterocarpaceae
(b) Verbenaceae
(c) Papilionaceae
(d) Combretaceae

235. Calculate number of plants /ha when teak is planted at 2m x 3m spacing
(a) 666 (b) 1,666
(c) 2,000 (d) 2,666

236. Ligno tuber is present in
(a) Teak (b) Casuarina
(c) Rose wood (d) Eucalyptus

237. *Prosopis juliflora* is native of
(a) Brazil (b) Mexico
(c) Thailand (d) Indonesia

238. Viability of the following tree seed slow down quickly
(a) Sal (b) Teak
(c) Eucalyptus (d) Terminalia

239. "Mimosin" is present in
(a) Leucaena (b) Acacia
(c) Neem (d) Sal

240. An important Multi-purpose tree in coastal area
(a) Casuarina (b) Neem
(c) Karanj (d) Albizia

241. Which is the family of Bamboo
(a) Ebenaceae (b) Verbanaceae
(c) Combretaceae (d) Poaceae

242. Casuarina root nodules are formed in association with
(a) Frankia (b) Rhizobium
(c) VAM fungi (d) AM fungi

243. *Shorea robusta* is
(a) Susceptible to frost
(b) Frost resistant
(c) Drought resistant
(d) Susceptible to drought.

244. Santalum album is a
(a) Partial stem parasite
(b) Complete stem parasite
(c) Partial root parasite
(d) Complete root parasite

245. *Mesua ferrea* belongs to
(a) Malvaceae (b) Combretaceae
(c) Guttiferae (d) Meliaceae

246. Most suitable Silvicultural system of deodar in hilly terrain for control of soil erosion is
(a) Selection system (b) Clear felling
(c) Coppice system (d) All

247. The Silviculture of Indian Trees was written by
(a) L.S.Khanna (b) Brandis
(c) R.S.Troup (d) Banerjee

248. Neem is found to occur wild in India in
(a) Irawadi valley
(b) Western ghats
(c) Anamalai hills
(d) All

249. The scientific name of nagasampigo
(a) *Hopea parviflora*
(b) *Mesua ferrea*
(c) *Acacia catechu*
(d) *Vateria indica*

250. Fragrant flowers characterized for
(a) *Michelia champaca*
(b) *Gmelina arborea*
(c) *Santalum album*
(d) *Acacia catechu*

251. Very strong durable wood
(a) *Casuarina equisetifolia*
(b) *Anogeissus latifolia*
(c) *Mesua ferrea*
(d) *Kydia calycina*

252. Commonest companion tree of Sal
(a) *Eugenia jambolana*
(b) *Buchanania latifolia*
(c) *Trevia nudiflora*
(d) *Anogeissus latifolia*

253. Iron wood of Malabar
(a) *Mesua faerrea*
(b) *Hopea parviflora*
(c) *Pterocarpus marsupium*
(d) *Tectona grandis*

254. Native of *Swietenia mahagoni*
(a) West Indies (b) India
(c) Burma (d) Malaysia

255. *Dalbergia sissoo* grows most typically in
(a) Alluvial (b) Laterite
(c) Loamy (d) Sandy loam

256. ________ is a special kind of pruning
(a) Dry pruning (b) Green pruning
(c) Bud pruning (d) Artifical pruning

257. Forests of *Acacia arabica* are frequently termed as
(a) Babul meadow
(b) Mesquite bana
(c) Sundra bana
(d) None

258. Calorific value of wood is highest in
(a) *Casuarina equisetifolla*
(b) *Acacia nilotica*
(c) *Tectona grandis*
(d) Betula

259. Artificially, *Terminalia belerica* can be raised by
(a) Direct sowing
(b) Entire plants
(c) Container seedlings
(d) All

260. Kail is
(a) *Acacia nilotica*
(b) *Acacia catechu*
(c) *Pinus wallichiana*
(d) *Pinus excelsa*

261. Example for presence of root nodules in trees
(a) *Casuarina equistifolia*
(b) *Terminalia tomentosa*
(c) *Adina cordifolia*
(d) *Tectona grandis*

262. *Cedrus deodara* is a
(a) Good coppicer
(b) Bad coppicer
(c) Do not coppice
(d) Moderate coppicer

263. *Vitex negundo* belongs to the family
(a) Verbenaceae (b) Rhizophoraceae
(c) Leguminsoe (d) Rubiacea

264. *Hopea parviflora* belongs to the family
(a) Dipterocarpaceae
(b) Leguminosae
(c) Guttiferae
(d) Mslvaceae

265. Scientific name of yellow bamboo is
(a) *Bambusa tulda*
(b) *Bambusa vulgaris*
(c) *Bambusa arundinaria*
(d) *Dendrocalamus strictus*

266. Sal is a
(a) Poor coppicer
(b) Good coppicer
(c) Moderate coppicer
(d) Non-coppicer

267. *Dalbergia sissoo* is
(a) Frost hardy
(b) Moderately frost hardy
(c) Frost tender
(d) None of these

268. *Atteva fabriciella* is a common pest of
(a) *Ailanthus excelsa*
(b) *Acacia auriculiformis*
(c) *Tectona grandis*
(d) All of these

269. Silk cotton tree belongs to the family
(a) Malvaceae (b) Guttiferae
(c) Meliaceae (d) Caesalpinaceae

270. Good pollarding ability tree
(a) *Mesua ferrea*
(b) *Hopea parviflora*
(c) *Pterocarpus marsupium*
(d) *Hardwickia binata*

271. Santalin is present in
(a) *Santalum album*
(b) *Pterocarpus marsupium*
(c) *Pterocarpus santalinus*
(d) *Tectona grandis*

272. Mesquite bean is the common name for
(a) *Prosopis spicigera*
(b) *Prosopis juliflora*
(c) *Acacia tortilis*
(d) None

273. *Azadirachta indica* belongs to
(a) Myrtaceae (b) Meliaceae
(c) Malvaceae (d) Mimoceae

274. *Eucalyptus tereticornis* belongs to the family
(a) Malavaceae (b) Mimoceae
(c) Myrtaceae (d) Meliaceae

275. Dry teak forest association
(a) *Terminalia tomentosa*
(b) *Acacia sundra*
(c) *Xylia xylocarpa*
(d) All

276. *Gmelina arborea* belongs to the family
(a) Verbenaceae (b) Combretaceae
(c) Myrtaceae (d) Meliaceae

277. *Pinus wallichlana* is a
(a) 2 needle pine (b) 3 needle pine
(c) 5 needle pine (d) 7 needle pine

278. Origin of *Mangifera indica* is
(a) Pakistan (b) India
(c) Srilanka (d) Burma

279. Indian Copal tree
(a) *Vateria indica*
(b) *Mesua ferrea*
(c) *Artocarpus hirsuta*
(d) *Michelia champaca*

280. *Terminalia arjuna* belongs to the family
(a) Verbenaceae (b) Combretaceae
(c) Malvaceae (d) Meliaceae

281. Karanj oil is extracted from
(a) *Pongamia pinnata*
(b) *Terminalia tomentosa*
(c) *Azadirachta indica*
(d) *Santalum album*

282. *Pinus roxburghii* is a
(a) 2 needle pine
(b) 3 needle pine
(c) 5 needle pine
(d) 7 needle pine

283. ______ is the most common pattern of planting method
(a) Square planting
(b) Line planting
(c) Triangular planting
(d) Quincunx planting

284. A low forest is one that has developed from
(a) Coppice (b) Cuttings
(c) Root sucker (d) Seed

285. A high forest is one that has developed from
(a) Coppice (b) Cuttings
(c) Root sucker (d) Seed

286. Examples of light demander species in tropical wet evergreen forest is:
(a) *Artocarpus* spp.
(b) *Dipterocarpus* spp.
(c) *Mesua ferrea*
(d) *Pinus wallichiana*

287. Which one of the following is not a stage in the growth of a tree
(a) Sapling (b) Juvenile
(c) Pole (d) Saw timber

288. Canopy density is said to be dense when the density is
(a) 1.0 (b) 0.75- 1.0
(c) 0.50 -0.75 (d) < 0.50

289. Cane brakes in the Northern tropical wet evergreen forest is an example of
(a) Climatic climax
(b) Edaphic climax
(c) Pre climax
(d) Post climax

290. A typical example of the southern montane wet temperate forest is
(a) *Diperterocarpus turbinatus*
(b) *Pertocarpus dalbergioides*
(c) *Lagerstroemia lanceolata*
(d) *Rhododendron nilgiricum*

291. Seed dispersal in Dipterocarpus is by the agency of
(a) Water (b) wind
(c) gravity (d) birds

292. Tick the odd man out amount the following species that coppice strongly
(a) *Acacia catechu*
(b) *Adina cordifolia*
(c) *Eucalyptus tereticornis*
(d) *Albizia lebbek*

293. Tick the odd man out among the following kinds of sowing
(a) Continuous line sowing
(b) Parallel sowing
(c) Strip sowing
(d) Patch sowing

294. The number of plants required for 10 ha of plantation in which the plants are planted in a triangular pattern 2.5 m apart is
(a) 4620 (b) 9840
(c) 13860 (d) 18480

295. Calculate the spacing of plants if there are 400 plants in a ha. of plantation raised on square planting pattern
(a) 4 m (b) 5 m
(c) 8 m (d) 10 m

296. The crop grown to foster the growth of important tree crop in its early stages called
(a) Cover crop (b) Beating up
(c) Nurse crop (d) Restocking

297. Tick the odd man out of the following terms for ordinary thinning
(a) Crown thinning
(b) Low thinning
(c) Thinning from below
(d) German thinning

298. Maximum thinning is attributable to
(a) Heck (b) Craib
(c) Gehrhardt (d) Glover

299. The rotation followed for deodar in U.P. is
(a) 80 years (b) 100 years
(c) 120 years (d) 160 years

300. The practice of planting other valuable species under the main crop when the natural regeneration of the valuable species in the upper storey cannot be assured is called
(a) Supplementary planting
(b) Cover planting
(c) Enrichment planting
(d) Underplanting

301. Crop diameter
(a) Diameter of a representative tree
(b) Diameter corresponding to the mean basal area of a uniform, generally pure crop
(c) Average of diameter measurements of 250 trees
(d) None

302. A symbiotic association of fungus with the subterranean part of a plant is said to be
(a) Mycorrhiza (b) Endophyte
(c) Exophyte (d) Parasite

303. Edaphic factor is related to
(a) Temperature
(b) Wind
(c) Soil
(d) Rainfall

304. A pure forest
(a) Consists of trees of exactly the same age
(b) Consists of only softwood species
(c) Consists of only hardwood species
(d) Is in which over 80% of the crop consists of a single species

305. A mixed crop is
(a) One in which a number of species may be present
(b) In which trees of all ages are growing together
(c) An inventory of logs of various commercial timbers
(d) None

306. Wind throw
(a) Is trees which fall easily due to wind
(b) Is when all the pollen from trees is blow in one direction
(c) Is when all the dry leaves suddenly drop due to heavy wind
(d) None

307. Epiphytes are
(a) Plants severely affected by an epidemic
(b) Plants restricted to a particular area
(c) Plants growing on the boundaries of a thick forest
(d) Plants growing on another plant, but making their own food

308. Dry nursery is
(a) One which is established in a desert area
(b) One which is maintained on a rocky site
(c) One which is maintained without artificial irrigation
(d) The planting stock stored under low temperature during winter

309. Sunken bed is required for ——— area
(a) Dry (b) Wet
(c) Shallow sites (d) None

310. Living fossil is
(a) *Rhododendron* spp.
(b) *Gingko biloba*
(c) *Sequioia semeperverens*
(d) *None*

311. Total number of plants that can be accommodated in 1.0 ha area with the spacing of 3 m X 3 m spacing is
(a) 1500 (b) 1800
(c) 1111 (d) 2500

312. Difference between western and eastern part of Himalayan region is
(a) Pinus spp is more
(b) Denuded land
(c) Most warmer
(d) None

313. Pnuematophore has function of
(a) Respiration
(b) Digestion
(c) Photosynthesis
(d) All of these

314. Tidal vegetation is found in
(a) Coastal region
(b) Gangetic plain
(c) Himalayan region
(d) None of these

315. For full stock, crown density should be
(a) >1 (b) <1
(c) 0.5 (d) 0.75

316. Forest types of India was given by
(a) Khanna
(b) Champion & Seth
(c) Pascal
(d) None

317. Tropical wet evergreen forests are found in
(a) Leh and Ladakh area
(b) Gangetic plains
(c) Rajasthan
(d) Andamans and Southern India

318. Pneumatophores are commonly seen in
(a) Dry deciduous forests
(b) Mangrove forests
(c) Open forests
(d) Sub-tropical montane temperate forests

319. Non leguminous N_2 fixing tree is
(a) *Alnus nepalensis*
(b) *Leuceanea leucocephala*
(c) *Acacia nilotica*
(d) *Calliandra callothyrsus*

320. Sal is naturally found in
(a) Littoral moist deciduous forests
(b) Tropical moist deciduous forests
(c) Temperate forests
(d) Alpine forests

321. Teak is naturally found in
(a) Littoral moist deciduous forests
(b) Tropical evergreen forests
(c) Tropical dry deciduous forests
(d) Alpine forests

322. The polyclimax theory of succession is attributable to
(a) Whittaker (b) Element
(c) Tansley (d) Folsberg

323. Presence of chloroxylon in moist deciduous forest is
(a) Pre climax (b) Post climax
(c) Sub climax (d) Zootic climax

324. ______________ is the upper limit of the tree growth beyond which the growth of the trees is not seen
(a) Tree line
(b) Tropic of cancer
(c) Tropic of Capricorn
(d) None of these

325. Buttresses are commonly found in
(a) Temperate evergreen forests
(b) Tropical ever green forests
(c) Tropical Rain forest
(d) Tropical thorn forests

326. Rain forests occurring above the cloud level is known as
(a) Cloud forests (b) Heath forests
(c) Mequis (d) Mangrove forests

327. In Mediterranean region ________ forests are seen
(a) Sclerophllous forests
(b) Tropical evergreen forests
(c) Mangrove forests
(d) None

328. Tallest tree species of the world is seen at
(a) Tropical seasonal forests
(b) Temperate forests
(c) Boreal Forests
(d) Maquis

329. Trees tend to more biomass for girth increment when they are planted at
(a) Wider spacing
(b) Closer spacing
(c) Both
(d) None

330. What does the type 1A/C4 indicate?
(a) West coast tropical evergreen forest
(b) Southern tropical semi evergreen forest
(c) Southern hill top tropical evergreen forest
(d) Giant evergreen forest

331. Practical application of silvics is
(a) Silviculture (b) Silvics
(c) All of the above (d) None

332. The study of life history and general characteristics of forest trees and crops with reference to environment is
(a) Silviculture (b) Silvics
(c) All of the above (d) None

333. Which among the following species is both deciduous and Evergreen species
(a) *Melia azaderach*
(b) *Santalum album*
(c) *Mangifera indica*
(d) *None*

334. The year in which a given species produces abundant first year seedling
(a) Mast year
(b) Seedling year
(c) both a and b
(d) All the above

335. What poisonous alkaloid is extracted from tobacco leaves and widely used as an insecticide?
(a) Cinerin (b) Nicotine
(c) Rotenone (d) Pyrethrum

336. The study of phenomena at very low temperatures is called
(a) Cryogenics (b) Cytogenetics
(c) a and b (d) None

337. Pollination by wind is called
(a) Anemophily (b) Zoophily
(c) Entomophily (d) All the above

338. Pollination by birds is called
(a) Anemophily (b) Zoophily
(c) Entomophily (d) All the above

339. Pollination by insects is called
(a) Anemophily (b) Zoophily
(c) Entomophily (d) All the above

340. The portion of the solar energy which is visible to human eye is known as
(a) Light (b) Temperature
(c) Moisture (d) Wind

341. In India the tree line is found around —————— from MSL
(a) 3500 M (b) 3000 M
(c) 2500 M (d) 5000 M

342. Which is a topographic component of environment?
(a) Aspect (b) Slope
(c) Exposure (d) None

343. The direction towards which the slope face is known as
(a) Aspect (b) Slope
(c) Exposure (d) None

344. Tropical rain forest is not found in
(a) Central and South Australia
(b) North and Central south America
(c) Central and West Africa
(d) Indo-Malayan-Burmeo and New Guinea

345. Deodar is managed under
(a) Clear felling systems
(b) Simple Coppice Systems
(c) Shelter wood system
(d) None

346. Epicormic branches originate from
(a) Adventious buds
(b) Auxiliary buds
(c) Meristematic regions
(d) Internodal buds

347. Example for Nitrogen fixing tree
(a) Dalbergia (b) Acacia
(c) Leucaena (d) All the above

348. Vana Mahotsva was started in the year
(a) 1952 (b) 1950
(c) 1988 (d) None

349. Knee shaped or spike like projection of the roots of swamp tree
(a) Silt roots (b) Prop roots
(c) Pneumatophore (d) All the above

350. Examples of pneumatophore
(a) Heretiera (b) Bruguiera
(c) a and b (d) None

351. Examples of Lignotubers
(a) Eucalyptus (b) Teak
(c) Cassia spp. (d) Dalbergia spp.

352. The science that deals with the time of appearance of characteristic periodic events
(a) Silviculture (b) Phenology
(c) Vegetative propagation
(d) None

353. The forest composed of even age woods called
(a) Even aged (b) Regular forest
(c) a and b (d) None

354. If the male and female flower parts occur in separate flowers on the same tree, the species is said to be
(a) Monoecious (b) Dioecious
(c) Polygamous (d) None

355. The most important climate indices for estimating potential productivity of a site
(a) CVP index
(b) Ecological index
(c) Precipitation effective index
(d) Dryness factor

356. Prop roots is a common phenomenon in
(a) Ficus (b) Bamboo
(c) Rhizophora 1(d) Heritieras

357. A stand of trees produced after a clear cut is most likely to be
(a) Even aged (b) Uneven aged
(c) a and b (d) None of these

358. The profession dealing with maintenance of ornamental tress is called
(a) Arboriculture (b) Silviculture
(c) Floriculture (d) None

359. Saucer pits is suitable for _____ soil and _____ region
(a) Loamy ; Dry (b) Loamy ; Moist
(c) Saline; Dry (d) Saline; Moist

360. The leaf stem is called the
(a) Pedicel (b) Peduncle
(c) Petiole (d) None

361. Identification of trees according to species is called
(a) Dendrology (b) Wood Technology
(c) Silviculture (d) All the above

362. A tree sprout arising from the root or stump is called a
(a) Coppice (b) Sucker
(c) Cuttings (d) All the above

363. New terminal growth on pines is called
(a) Candles (b) Apical shoot
(c) Meristem (d) None

364. ______ species is pollinated by bats
(a) Bombax
(b) Butea
(c) Anthocephalus
(d) Coniferrous

365. Maximum forest area in India comes under
(a) Tropical wet evergreen
(b) Tropical moist deciduous
(c) Tropical dry deciduous
(d) Tropical dry evergreen

366. The height of *Tectona grandis* in the first quality site
(a) 30-36 m (b) 24-30 m
(c) 18-24 m (d) 12-18 m

367. *Shorea robusta* is
(a) Susceptible to drought.
(b) Drought Hardy
(c) Moderately drought hardy
(d) None

368. *Mesua ferrea* belongs to
(a) Guttiferae
(b) Combertaceae
(c) Myrtaceae
(d) Verbanaceae

369. Most suitable Silvicultural system of deodar in hilly terrain for control of soil erosion is
(a) Selection system
(b) Clear felling systems
(c) Shelter wood systems
(d) Coppicing systems

370. Forest of *Acacia arabica* are frequently termed as
(a) Babul meadow
(b) Carnatic thorn
(c) Black thorn
(d) All the above

371. Dry teak forest associated species.
(a) *Terminalia tomentosa*
(b) *Indigofera tinctoria*
(c) *Mallotus* spp.
(d) *Sterculia* spp.

372. *Melia dubia* belongs to the family
(a) Verbenaceae (b) Meliaceae
(c) Laminaceae (d) Myrtaceae

373. The principles and practice of silviculture was written by
(a) Banerjee (b) Khanna
(c) Troup (d) Negi

374. Site quality can be assured by
(a) Indicative species
(b) Rate of growth of trees
(c) Total volume of trees
(d) All of the above

375. It is a measure of the relative productive capacity of a site for a particular species is called

(a) Site factors (b) Site quality

(c) a and b (d) None

376. The average height of the dominant trees in a stand

(a) Top height

(b) Crop height

(c) Crop diameter

(d) Crop volume

377. The first Inspector General of Forests of Government of India was

(a) Dietrich Brandis

(b) Watson

(c) Chathu Menon

(d) Conolly

378. Size distribution of particles in the soil is called

(a) Soil Texture (b) Soil Structure

(c) a and b (d) None

379. Oxides of Iron and Aluminum is called

(a) Sesquioxides (b) Silica

(c) a and b (d) None

380. The plants which grow on the trunk and branches of other plants for nourishment is called

(a) Epiphytes (b) Climbers

(c) Parasites (d) Abnoxious weeds

381. Rhizobium is .... fixing bacteria

(a) Nitrogen (b) Phosphorus

(c) Potassium (d) All the above

382. The best indicator for the site quality is .... Annual Increment

(a) Mean (b) Periodic

(c) Total (d) None

383. When regeneration is obtained from the seed, the forest may be named as

(a) High Forest (b) Low Forest

(c) Coppice forest (d) All the above

384. Removing dead, dying, diseased and suppressed stems during thinning process is

(a) B grade (Moderate) thinning

(b) C grade (Heavy) thinning

(c) A grade (Light) thinning

(d) D grade (Very heavy) thinning

385. Central Board of Forestry was constituted in the year

(a) 1947 (b) 1948

(c) 1952 (d) 1988

386. The effect of light on the form of trees is called

(a) Phototropism

(b) Photoperiod

(c) Photoperiodism

(d) None

387. The reflectivity of a surface or substance to the solar radiation is called

(a) Albedo (b) Solar radiation

(c) Ultra violet (d) Infra red

388. Local climate of an area which differ from the general climate is

(a) Microclimate (b) Weather

(c) Bio climate (d) None

389. The study of an individual organism in relation to its environment is

(a) Autecology

(b) Synecology

(c) Ecosystem ecology

(d) All the above

390. The series of stages in succession is called

(a) Climax (b) Lithosere

(c) Psammosere (d) Seres

391. When natural succession is pushed back to several stages due to natural or biotic factors is called

(a) Retrogression

(b) Regression

(c) a and b

(d) None

392. Succession comes under the category of .... ecology
(a) Autecology
(b) Synecology
(c) Ecosystem ecology
(d) All the above

393. The plants which grow on the trunk and branches of other plants for support is called
(a) Epiphytes (b) Climbers
(c) Parasites (d) Abnoxious weeds

394. The negative logarithm of hydrogen ion concentration is called
(a) pH
(b) EC
(c) Cation exchange capacity
(d) All the above

395. Casuarina root nodules are formed in association with
(a) Mycorrhiza (b) Frankia
(c) a and b (d) None

396. The largest family among the Indian flora is
(a) Asteraceae (b) Poaceae
(c) Fabaceae (d) Myrtaceae

397. Sucking roots of parasites are
(a) Haustoria (b) Hypophysis
(c) Radicle (d) Transverse cell

398. The origin of teak is
(a) Burma (b) Nilambur
(c) West Bengal (d) None

399. In India, silvicultural systems are classified based on
(a) Type of crop
(b) Tending operations
(c) Mode of regeneration
(d) All the above

400. Low forests originated from
(a) Regeneration through coppice
(b) Regeneration through seeds
(c) a and b
(d) None

401. The Latin name for Andaman Padauk tree is
(a) *Pterocarpus dalbergioides*
(b) *Dipterocarpus macrocarpus*
(c) *Messua ferrea*
(d) *Dipterocarpus tubinatus*

402. The habitat of deodar is
(a) Evergreen forest
(b) Temperate forest
(c) Deciduous forest
(d) Alpine forest

403. The tree species belonging to the family Myrtaceae
(a) Eucalyptus (b) Casuarina
(c) Teak (d) Terminalia

404. Development of embryo without fertilization is termed
(a) Apomixis (b) Parthnogensis
(c) Apogamy (d) Apospory

405. The Latin name of golden bamboo is
(a) *Bambusa vulgaris*
(b) *Bambusa bamboo*
(c) *Bambusa balcooa*
(d) *Bambusa lineata*

406. Nitrogen fixing non-leguminous tree
(a) Dalbergia (b) Acacia
(c) Leucaena (d) Casuarina

407. The upper branchy part of a tree above the bole called
(a) Stem (b) Crown
(c) Root (d) None

408. The expansion of crown measured as crown length and crown width called as
(a) Crown development
(b) Crown expansion
(c) a and b (d) None

409. Example for Deciduous tree
(a) *Acacia catechu*
(b) *Melia azedarach*
(c) *Ailanthus excelsa*
(d) *All the above*

410. Example for Evergreen tree
(a) *Abies pindrow*
(b) *Michelia champaca*
(c) *Hopea parviflora*
(d) *All the above*

411. The decrease in diameter of the stem of a tree from the base upwards is known as
(a) Taper (b) Clean bole
(c) Commercial Bole (d) None

412. The part of the bole free of branches is called as
(a) Stem (b) Crown
(c) Root (d) Both a and b

413. The irregular involutions and swelling on the bole called
(a) Forked (b) Fluting
(c) Buttressed (d) None

414. Example of buttressed stem
(a) *Bombax ceiba*
(b) *Terminalia myriocarpa*
(c) *Acrocarpus fraxinifolius*
(d) All the above

415. The roots produced from parts of the plants other than the radicle is called
(a) *Adventitious roots*
(b) *Taproot*
(c) *Lateral roots*
(d) *All the above*

416. Example for Prop- roots
(a) *Ficus bengalensis*
(b) Bamboo
(c) Rhizophora
(d) Heritieras

417. The tree appears on flying buttresses is called as
(a) Pneumatophore
(b) Silt roots
(c) Prop roots
(d) All the above

418. A knee-shaped or spike like projection of the roots
(a) Pneumatophore
(b) Silt roots
(c) Prop roots
(d) All the above

419. The structure produced from the combination of the modified rootlet with fungal tissue called as
(a) Fungi (b) Mycorrhiza
(c) Frankia (d) None

420. ________ increase the surface area of roots.
(a) Plasmodesmata
(b) Mycorrhizae
(c) Root hairs
(d) Root hairs and mycorrhizae

421. Flame of the forest
(a) *Butea monosperma*
(b) *Tectona grandis*
(c) *Shorea robusta*
(d) *Cassia fistula*

422. The condition of the tree with dead upper branches projecting above the green portion of the crown is called
(a) Thinning (b) Pruning
(c) Climber cutting (d) Stagheaded

423. Natural forest uninfluenced by human activity
(a) Natural Forest (b) Rain Forest
(c) Virgin forest (d) All the above

424. Trees have mostly ---------- flowers
(a) Bisexual (b) Unisexual
(c) Polygamous (d) None

425. The unit of Silviculture is
(a) Silvix (b) Stand
(c) Forest (d) Forestry

426. The Unisexual and bisexual flowers are found on the same individual called
(a) Monoecious (b) Dioecious
(c) Polygamous (d) None of these

427. Example of monoecious
(a) Pinus (b) *Abies pindrow*
(c) Salix (d) a and b

428. Examples of dioecious
(a) Salix (b) Diospyros
(c) *Cedrus deodara*
(d) All the above

429. Examples of polygamous flowering
(a) Sterculia (b) Rhus
(c) Garuga (d) All the above

430. The height corresponding to the mean diameter of the 250 biggest diameters per hectare
(a) Crop Height (b) Top height
(c) Crop diameter (d) Crop Volume

431. Average height of regular crop is called as
(a) Crop diameter (b) Crop Height
(c) Crop Volume (d) Top height

432. Average height of regular crop is determined by
(a) Lorey's formula
(b) Smalian formula
(c) Heyer's formula
(d) None

433. The diameter corresponding to the mean basal area of a uniform crop is called as
(a) Crop diameter
(b) Crop Height
(c) Crop Volume
(d) Top height

434. The relative completeness of the stocking expressed as a decimal coefficient, taking normal number of trees, basal area or volume as unity is called as
(a) Crop density (b) Canopy density
(c) a and b (d) None

435. The relative completeness of canopy usually expressed as a decimal coefficient, taking closed as unity is called as
(a) Crop density (b) Canopy density
(c) a and b (d) None

436. The canopy density is 0.75 to 1.0
(a) Closed (b) Thin
(c) Dense (d) Open

437. The canopy density is below 0.5
(a) Closed (b) Thin
(c) Dense (d) Open

438. Examples of light demanders in Moist temperate forest
(a) *Pinus wallichiana*
(b) *Populus ciliata*
(c) Both a and b
(d) None

439. Examples of shade demanders for tropical Moist deciduous forest
(a) *Syzygium cumini*
(b) *Schleichera oleosa*
(c) *Mallotus philipinensis*
(d) *All the above*

440. Examples of Shade bearers in tropical wet evergreen forest
(a) Artocarpus spp.
(b) Dipterocarpus spp.
(c) Calophyllum
(d) *Mesua ferrea*

441. Based on the methods of regeneration, forests are classified as
(a) Normal and Abnormal forest
(b) Evenaged and unevenaged
(c) Pure and Mixed forest
(d) High and low forest

442. The mean annual temperature of montane sub-tropical forest ranged between
(a) 24°C (b) 17°C and 24°C
(c) 7°C to 17°C (d) <7 °C

443. The Forest Survey of India (FSI) conducts Forest Cover Survey once in every
(a) Two years (b) Five Years
(c) Ten Years (d) None

444. Climate Vegetation Productivity Index was given by
(a) Thoreau (b) Clements
(c) Dawson (d) Paterson

445. First Teak plantation in india
(a) Nilampur (b) West Bengal
(c) Siwalik (d) Gorakhpur

446. Examples of frost hardy species
(a) *Acacia catechu*
(b) *Anogeissus pendula*
(c) *Toona ciliate*
(d) *All the above*

447. Examples of frost tender species
(a) *Azadirachta indica*
(b) *Tectona grandis*
(c) *Terminalia arjuna*
(d) *All the above*

448. Examples of drought hardy species
(a) *Acacia arabica*
(b) *Diospyros melanoxylon*
(c) *Syzygium cumini*
(d) *All the above*

449. A month which has < 50 mm of rainfall is referred to as
(a) Dry month (b) Rainy days
(c) a and b (d) None

450. A day which has minimum of 2.5 mm of rainfall is referred to as
(a) Dry month (b) Rainy days
(c) a and b (d) None

451. The commercial beedi leaves are plucked from
(a) *Diospyros melanoxylon*
(b) *Shorea robusta*
(c) *Pterosperma* spp.
(d) *Gmelina arborea*

452. The dying back is a common disease in
(a) Teak
(b) Deodar
(c) Sal
(d) None of these

453. Example of indigenous populus species
(a) *Populus ciliate*
(b) *Populus robusta*
(c) *Populus deltoides*
(d) *Populus rubrapoiret*

454. Agar wood tree is
(a) *Aquilaria malaccensis*
(b) *Teactona grandis*
(c) *Dalbergia sissoo*
(d) *Acacia catechu*

455. The blowing of strong noisy wind up and down the valley locally called as
(a) Dadu
(b) Nor westers
(c) Mango showers
(d) All the above

456. A plant which grows in dry situation is called
(a) Hydrophyte (b) Mesophyte
(c) None (d) Xerophyte

457. A plant which grows in moist situation is called
(a) Hydrophyte (b) Hygrophyte
(c) Mesophyte (d) None

458. A plant which grows wholly or partly immersed in water
(a) Hydrophyte (b) Hygrophyte
(c) Mesophyte (d) None

459. A plant whose normal habitat is neither very wet nor very dry
(a) Hydrophyte (b) Hygrophyte
(c) Mesophyte (d) None

460. Plants growing on, but not nourished by another plant is referred to as
(a) Epiphytes (b) Climbers
(c) Parasites (d) Abnoxious weeds

461. Epiphytes commonly present in
(a) *Ficus bengalensis*
(b) *Ficus religiosa*
(c) Orchids
(d) All the above

462. Example of Abnoxious weeds
(a) *Lantana camara*
(b) *Eupatorium odoratum*
(c) a and b
(d) None

463. Succession which have previously not born vegetation
(a) Primary succession
(b) Secondary succession
(c) a and b
(d) None

464. The series of plant communities resulting from processes of succession is called
(a) Seres (b) Climax
(c) Lithosere (d) Psammosere

465. Who developed the idea and theory of plant succession
(a) Thoreau (b) Clement
(c) Cawles (d) Dawson

466. Making high demands as regard soil and other factors is called
(a) Exacting (b) Reaction
(c) Competition (d) None

467. The succession initiated in extremely dry situation is called
(a) Xerarch succession
(b) Hydrarch succession
(c) a and b
(d) None

468. The successional stages of Xerarch succession
(a) Xerarch succession
(b) Xeroseres
(c) a and b
(d) None

469. Xerosere which originates on rock surface is called
(a) Lithosere
(b) Psammosere
(c) Xerarch succession
(d) None

470. Xerosere which originates on sand is called
(a) Lithosere
(b) Psammosere
(c) Xerarch succession
(d) None of these

471. What is the scientific term for a leaf blade?
(a) Lamina (b) Pedicel
(c) Peduncle (d) None

472. An ash leaf generally contains ———————— leaflets
(a) Nine (b) Five
(c) Seven (d) None

473. What is the scientific term for a leaf stalk?
(a) Petiole (b) Pedicel
(c) Peduncle (d) None

474. Example of edible 'Fungi'
(a) Mucor (b) Pencillium
(c) Agaricus (d) Rhizopus

475. The intensity of regeneration survey generally used to ______
(a) 2-4 % (b) 10-20 %
(c) 20-40% (d) 80-100 %

476. Coir which is obtained from fruit of coconut is
(a) Endocarp (b) Periderm
(c) Epicarp (d) Mesocarp

477. Solarisation is
(a) Formation of chlorophyll
(b) Utilisation of sunlight
(c) Effects of solar light
(d) Destruction of chlorophyll

478. 'Pneumatophores' are present /common in
(a) Xerophytes
(b) Hygrophytes
(c) Halophytes
(d) Mesophytes

479. The plant does not produce coppice
(a) *Abies pindrow*
(b) *Cedrus deodar*
(c) *Pinus wallichiana*
(d) all of the above

480. In which form, does the food transported in plants
(a) Sucrose (b) Fructose
(c) Glucose (d) Lactose

481. Clear strip silviculture system is applied for
(a) *Pinus keysia*
(b) *Acacia mearnsii*
(c) *Tectona grandis*
(d) *Populus deltoides*

482. In angiosperms, the strucutres that produce male gametophyte are called the
(a) Stigmas (b) Anthers
(c) Pollen tubes (d) Sepals

483. The outmermost circle of flower parts consists of several
(a) Sepals (b) Petals
(c) Carpels (d) Corollas

484. Pollination occurs when pollen lands on the
(a) Stigma (b) Style
(c) Filament (d) Anther

485. The seed leaves of flowering plants are known as
(a) Endosperm (b) Carpels
(c) Radicals (d) Cotyledons

486. Grasses have ______ meristems.
(a) Intercalary (b) Apical
(c) Lateral (d) None

487. The scientist credited with being the Father of Botany is
(a) Darwin (b) Theophrastus
(c) Hinterthuer (d) Linnaeus

488. The part of cotton producing pure cellulose is
(a) Root hair (b) Leaf hair
(c) Seed hair (d) Stem hair

489. Which is considered to be the best chemical method of fixing atmospheric nitrogen?
(a) Fisher method
(b) Decan method
(c) Haber-Bosch method
(d) Paranas-Meyerhoff method

490. ______ planting techniques is adapted in mangroves
(a) Fishbore (b) Line
(c) Square (d) Quincuex

491. Group silviculture system is applied for ______ species
(a) Deodar (b) Acacia
(c) Eucalyptus (d) Casuarina

492. Active division takes place in the cells of ______
(a) Xylem (b) Pholem
(c) Sclerenchyma (d) Cambium

493. The larger proportion of mature trees in the forests, the length of conversion is
(a) Shorter (b) Longer
(c) Moderate (d) Extreme

494. Plants growing in extremely cold soils are
(a) Xerophyte (b) Psychrophyte
(c) Mesophyte (d) Hygrophyte

495. Xerophytes have long roots
(a) Lower rate of transpiration than mesophytes
(b) Transpiration rate equal to mesophytes
(c) Transpiration rate higher than mesophytes
(d) No transpiration

496. The orchid plants, which are found growing on the trees, are
(a) Epiphytes
(b) Climbers
(c) Parasites
(d) Abnoxious weeds

497. The first plants to reappear in a badly burned forest area will most probably be
(a) Liverworts (b) Ferns
(c) Grasses (d) Mosses

498. The transfer of energy from organism to organism in a natural community establishes
(a) Food chains (b) Food web
(c) Ecosystems (d) Population

499. Which is a primary consumer
(a) Carnivore (b) Herbivore
(c) Omnivore (d) Detritivore

500. Trophic levels are formed by
(a) Plants only
(b) Animals only
(c) Carnivores only
(d) Organisms linked in food chain

501. Driving force of ecosystem is
(a) Carbohydrate in plants
(b) Biomass
(c) Solar energy
(d) Producer

502. Which group of plants possesses both animal and plant characteristics?
(a) Myxomycetes (b) Basidiomycetes
(c) a and b (d) None

503. Cleistogamous flowers are always
(a) Self pollinated (b) Cross Pollinated
(c) a and b (d) None

504. Flowers are small and unattractive in
(a) Anemophily (b) Ornithophily
(c) Entomophily (d) None

505. Examples of annual flowering bamboo species is
(a) *Arundinarina wightiana*
(b) *Bambusa lineata*
(c) Ochlandra stridula
(d) All of the above

506. Palynology deals with study of
(a) Pollen grains (b) Fossils
(c) Flowers (d) All the above

507. The color of light least effective in driving photosynthesis is
(a) Yellow (b) Green
(c) Blue (d) Orange

508. Sandal wood tree generally flowers in the month of
(a) May to June
(b) February to April
(c) September to February
(d) March to June

509. The collection period for Neem *(Azadiracta indica)* seed is
(a) February to April
(b) September to February
(c) March to June
(d) May-june

510. *Dendrocalamus strictus* is a good associate of
(a) Trema (b) Callicarpa
(c) Macaranga (d) Teak

511. The beef wood is also known as
(a) Casuarina (b) Teak
(c) Dalbergia (d) Bamboo

512. Taiga form of vegetation are found in
(a) Russia (b) Brazil
(c) Canada (d) Italy

513. The tropical and sub-tropical grassland with either sparsely scattered trees or bushes are called
(a) Savannas (b) Prairies
(c) Steppe (d) Range land

514. Clear felling system has also been termed as
(a) Coupe blanche
(b) Coupe rare
(c) Clear cutting system
(d) All the above

515. Trade/common name of *Dalbergia latifolia* is
(a) Rose Wood (b) Beef wood
(c) Soft wood (d) Hard wood

516. Haustorial roots are mostly found in
(a) *Viscum* (b) Teak
(c) Sal (d) Salix

517. Coarse textured soils such as sands are _____ in water holding capacity and nutrient content.
(a) Low (b) High
(c) Medium (d) None

518. The coppice arising from the shoot on a living stump is called.....
(a) Stem coppice
(b) Coppice shoots
(c) Seedling coppice
(d) Stool coppice

519. Based on crown classification which is defined as a smaller tree with a crown below the upper level of the forest canopy and receiving only scattered sunlight from above?
(a) Suppressed (b) Co-dominant
(c) Dominant (d) Intermediate

520. Bud pruning is done for --------species
(a) *Salix alba* (b) Coerulea
(c) Poplars
(d) All of the above

521. The outgrowth from the base of the trees connecting it with the roots, especially common in tropical rain forest species is
(a) Fluting (b) Forking
(c) Tapering (d) Buttressing

522. Natural pruning in trees can be accelerated by
(a) Density of planting
(b) Species composition
(c) Retention of understory trees
(d) All of the above

523. Which one of the following is correctly matched?
(a) *Picea smithiana- Spruce*
(b) *Dysoxylum malabaricum- Elm*
(c) *Ulmus wallichiana- White cedar*
(d) *Tectona grandis- Ven Teak*

524. Which of the following is a moisture loving plant?
(a) *Tectona grandis*
(b) *Prosopis juliflora*
(c) *Shorea robusa*
(d) *Terminalia arjuna*

525. Which sub-type of forest will not be under Tidal Swamp Forests?
(a) Myristica swamp forest
(b) Mangrove scrub
(c) Mangrove forest
(d) Palm swamp

526. Dry red sanders forests occur in which state?
(a) Andhra Pradesh
(b) Tamil Nadu
(c) Karanataka
(d) Madhya Pradesh

527. The climate that modified by the adjusted climatic factors is called
(a) Bioclimate (b) Climate
(c) Microclimate (d) All the above

528. Koppen classified world climate into ___________ groups
(a) Five (b) Six
(c) Seven (d) Sixteen

529. Solar radiation, temperature and rainfall is mainly affected by
(a) Altitude (b) Slope
(c) Exposure (d) All the above

530. Moisture regime of the soil is highly influenced by the ________
(a) Altitude (b) Slope
(c) Exposure (d) All the above

531. ------------- of place refers its elevation from the mean sea level
(a) Altitude (b) Slope
(c) Exposure (d) Aspect

532. The relation of a site in relation to sun and wind is called
(a) Altitude (b) Slope
(c) Exposure (d) Aspect

533. Southern aspect are __________ than northern aspect

(a) Drier (b) Wetter

(c) Moisture (d) None

534. The climate of small area that differ generally from the general climate is called

(a) Microclimate (b) Bioclimate

(c) a and b (d) None

535. Microclimate is important in

(a) Artificial regeneration

(b) Natural regeneration

(c) a and b

(d) None

536. The upper most weathered layer of the earth crust is called

(a) Soil (b) Bed rock

(c) Regolith (d) Rock

537. The geological process through which the soil is formed is called

(a) Weathering (b) Soil formation

(c) Soil tilth (d) None

538. Physical disintegration of rock is called

(a) Physical weathering

(b) Chemical weathering

(c) Biological weathering

(d) All the above

539. Disintegration of rocks chemically is called

(a) Physical weathering

(b) Chemical weathering

(c) Biological weathering

(d) All the above

540. The structure of the surface layer of the soil is called

(a) Weathering (b) Soil formation

(c) Soil tilth (d) None of these

541. A sequence of soil derived from the same age and same parent material is called

(a) Catena (b) Mature soils

(c) Immature soils (d) None

542. Basis of soil forming factor, the soils are classified into

(a) Endodynamorphic

(b) Ectodynamorphic

(c) Endoectododynamorphic

(d) All the above

543. Soil host properties are influenced mainly by parental materials

(a) Endodynamorphic

(b) Ectododynamorphic

(c) Endoectododynamorphic

(d) None of these

544. Soil formed from parent material in situ

(a) Sedentary soil

(b) Alluvial soil

(c) Secondary soil

(d) Aeolian soil

545. Soil derived from pre-existing soils by translocation and redeposition

(a) Secondary soil

(b) Alluvial soil

(c) Aeolian soil

(d) Sedentary soil

546. Soil whose properties are influenced by the parent material is called

(a) Endodynamorphic

(b) Ectododynamorphic

(c) Endoectododynamorphic

(d) None

547. Soil whose properties are influenced by other than the parent material is called

(a) Endodynamorphic

(b) Ectododynamorphic

(c) Endoectododynamorphic

(d) None

548. A secondary immature soil derived from soil material of older origin deposited by local erosion on the lower slope of gentle hills is called

(a) Colluvial soils

(b) Alluvial soil

(c) Aeolian soil

(d) Sedentary soil

549. The appearance of soils in vertical section with the particular reference to the sequence of layers that may be differentiated is termed as
(a) Soil profile (b) Soil horizon
(c) Regolith (d) Bed rocks

550. Soils which contains an excess of soluble salts with low exchangeable Sodium is
(a) Saline soils (b) Alkali soils
(c) Acid soils (d) All the above

551. Kankar pan is commonly found in
(a) Temperate region
(b) Tropical region
(c) Alluvial soils
(d) Red soils

552. Silican pan is formed in
(a) Podsolic iron pan
(b) Laterite and lateritic soils
(c) Kankar pan
(d) Clay pan

553. Podsolic iron pan is formed in podsol soils of
(a) Temperate region
(b) Tropical region
(c) Alluvial soils
(d) Red soils

554. A plant which tend to be restricted to calcarious soil is
(a) Calciphile (b) Calcifuge
(c) Calcigaurd (d) None

555. Laterization is a process of accumulation in soil
(a) Iron and Aluminium oxides
(b) Silica
(c) $CaCo_3$
(d) $CaSo_4$

556. Podzolization is a process of accumulation in soil
(a) Iron and Aluminium oxides
(b) Silica
(c) $CaCo_3$
(d) $CaSo_4$

557. Concentration of nitrogen in soil air is
(a) 70 % (b) 74 %
(c) 78 % (d) 82 %

558. Vertical section of a the soil is known as
(a) Solum (b) Regolith
(c) Soil profile (d) A horizon

559. Unconsolidated material overlying rocks
(a) Solum (b) Regolith
(c) Bed rocks (d) A horizon

560. The solum is comprised of
(a) A, B and E horizon
(b) A and B horizon
(c) A, B and C horizon
(d) A + regolith

561. The B horizon is generally absent in
(a) Black soils (b) Red soils
(c) Desert soils (d) Brown hill Soils

562. Passive factor of soil formation includes
(a) Climate (b) Organisms
(c) Parent material (d) None

563. The relative proposition of various size groups of individual soil particles is termed as
(a) Soil texture (b) Soil Structure
(c) Soil class (d) None of these

564. Which type of structure is dominant in grasslands?
(a) Spheroidal (b) Platy
(c) Columnar (d) Blocky

565. White alkali soils are
(a) Acid soils (b) Saline soil
(c) Saline sodic soil (d) Sodic soil

566. A group of soils having same range in particle size and physical properties
(a) Soil class (b) Soil Structure
(c) Soil texture (d) None

567. Arrangement of individual soil particles into aggregates is called
(a) Soil class (b) Soil Structure
(c) Soil texture (d) None

568. The root development of plant is affected by
(a) Soil class (b) Soil Structure
(c) Soil texture (d) None

569. The percentage of finer particle governess ———— status of soil
(a) Nutrient (b) Soil Texture
(c) Soil structure (d) C : N ratio

570. In the nutrition of a plant, which element is classified as a macronutrient?
(a) Zinc (b) Chlorine
(c) Calcium (d) Molybdenum

571. The most suitable C:N ratio is
(a) 10 : 1 (b) 17 : 1
(c) 33 : 1 (d) 12 : 1

572. The influence of living organisms on the vegetation is called
(a) Climatic factors
(b) Edaphic factors
(c) Biotic factors
(d) Topographic factors

573. An organism that draws part or whole of its nourishments from another living organisms is called
(a) Parasite (b) Epiphytes
(c) Climbers (d) Abnoxious weeds

574. Viscum and Loranthus are the
(a) Partial stem parasite
(b) Total stem parasite
(c) Partial root parasite
(d) Total root parasite

575. Cuscuta and Cassytha are the
(a) Partial stem parasite
(b) Total stem parasite
(c) Partial root parasite
(d) Total root parasite

576. Sandal is a
(a) Partial root parasite
(b) Partial stem parasite
(c) a and b
(d) None

577. A plant growing on but not nourished by another plant is
(a) Parasite (b) Epiphytes
(c) Climbers (d) Abnoxious weeds

578. Ficus, orichids and tinospora are
(a) Parasite (b) Epiphytes
(c) Climbers (d) Abnoxious weeds

579. Large woody climbers in tropical areas are called
(a) Liana or Liane
(b) Parasite
(c) Epiphytes (d) Climbers

580. Dioscorea, Ipomaea are the example of
(a) Parasite (b) Epiphytes
(c) Climbers (d) Lianas

581. Eupotorium and Lantana are classified as
(a) Parasite (b) Epiphytes
(c) Climbers (d) Abnoxious weeds

582. A species that invades in a bare area such as newly exposed soil are rock surface is called
(a) Pioneer species
(b) Succession
(c) Xerarch succession
(d) Lithosere

583. The gradual replacement of one community by another in the development of the vegetation towards a climax is called
(a) Succession (b) Climax
(c) Pioneer species (d) None

584. Culmination stage in plant succession for a given environment is called
(a) Climax (b) Climatic climax
(c) Biotic climax (d) Pre climax

585. The series of plant communities resulting from process of succession is called
(a) Primary Succession
(b) Secondary succession
(c) Seres
(d) All the above

586. What is haustoria?
(a) Sucking root of plants
(b) Creeping root of climbers
(c) Sucking root of parasite
(d) Climbing root of plants Agroforestry

587. Widely accepted succession theory was
(a) Clements Theory
(b) Whittaker's theory
(c) Aubreville's theory
(d) All the above

588. The Succession initiated in extremely dry situation is called
(a) Xerarch succession
(b) Hydrarch succession
(c) Hydro succession
(d) None

589. The different stages in a Xerarch Succession is called
(a) Xeroseres (b) Hydrosere
(c) a and b (d) None

590. Succession begins in water is defined
(a) Xerarch succession
(b) Hydrarch succession
(c) Hydro succession
(d) None

591. The different stages in a hydrarch succession is called
(a) Lithosere (b) Psammosere
(c) Hydrosere (d) All the above

592. Succession take place where there is no previous vegetation is called
(a) Primary Succession
(b) Secondary succession
(c) a and b
(d) None

593. A succession in which vegetation affects the ecosystems to bring about consequently changes is called
(a) Allogenic succession
(b) Autogenic succession
(c) a and b (d) None

594. A succession which take place after the destruction of the whole or part of the original vegetation is called
(a) Secondary Succession
(b) Primary succession
(c) a and b (d) None

595. Secondary Succession otherwise termed as
(a) Allogenic succession
(b) Autogenic succession
(c) a and b (d) None

596. The whole process where a plant establishes itself in a new area from germination to reproduction is called
(a) Migration (b) Ecesis
(c) Aggregation (d) Competition

597. Grouping followed by establishment of scattered colonizing invaders as a result of propagation is called
(a) Migration (b) Ecesis
(c) Aggregation (d) Competition

598. The effect of vegetation on the site is called
(a) Reaction (b) Migration
(c) Aggregation (d) Competition

599. Riverine succession of Gangetic alluvium is ______________ Succession
(a) Primary Succession
(b) Secondary Succession
(c) a and b (d) None

600. A Major unit of vegetation comprising the climax communities of an area uniform in its major physiognomic features is termed as
(a) Formation (b) Migration
(c) Establishment (d) Grouping

601. Monoclimax Theory was proposed by
(a) Clements (b) Tansley
(c) Whittiaker (d) Aubreville

602. Mosaic theory was proposed by
(a) Clements (b) Tansley
(c) Whittiaker (d) Aubreville

603. Theory of vegetation gradient and climax patterns was proposed by
(a) Clements (b) Tansley
(c) Whittiaker (d) Aubreville

604. The climax which owes distinctive characters due to climatic factors is referred as
(a) Climatic climax
(b) Edaphic climax
(c) Pre climax
(d) Post climax

605. A plant which differs from the climatic climax due to the influence of special soil factors is defined
(a) Edaphic climax
(b) Pre climax
(c) Climatic climax
(d) Post climax

606. The plant community immediately preceding the climatic climax is called
(a) Edaphic climax
(b) Pre climax
(c) Climatic climax
(d) Post climax

607. A plant community more exacting than the climatic climax is defined
(a) Edaphic climax
(b) Pre climax
(c) Climatic climax
(d) Post climax

608. A climax which differs from the climatic climax of the area due to biotic factors is referred as
(a) Biotic Climax (b) Climatic climax
(c) Pre climax (d) Post climax

609. A vegetation development towards the climatic climax has been arrested due to some stage short of normal climax is called
(a) Biotic climax
(b) Sub climax
(c) Pre climax
(d) Post climax

610. Study of plant succession is very important for classification of forests into
(a) Forest types
(b) Artificial regeneration
(c) Successional stage
(d) All the above

611. A category of forest defined with reference to its geographical location, climatic and edaphic feature, composition and condition is called
(a) Forest type (b) Forestry
(c) Forest (d) None

612. Indian Forests are classified into forest types based on ——————classification
(a) Mayr
(b) Koppen
(c) Seth and Khan
(d) Champion and Seth

613. A general appearance of forest community is called
(a) Physiognomy (b) Physiology
(c) Phenology (d) None

614. Vegetation first classified according to temperate zones
(a) Schimper (b) Mayr
(c) Koppen (d) Seth and Khan

615. Classification of Vegetation according to temperate combined with latitude and altitude was done by
(a) Schimper (b) Mayr
(c) Koppen (d) Seth and Khan

616. Thornthwaite classified the vegetation based on
(a) Season of Rain fall
(b) Temperature efficiency
(c) Bioclimatic
(d) Eco systems

617. Classification of vegetation based on thermodynamic principles was done by
(a) Shanbhag (b) Thornthwaite
(c) Koppen (d) Seth and Khan

618. Indian forests are classified into forest types in the year
(a) 1947 (b) 1935
(c) 1952 (d) 1988

619. According to Champion and Seth, Indian forests have been divided into — groups
(a) Five (b) Sixteen
(c) 221 (d) None

620. According to Champion and Seth Indian forests have been divided into ——— Sub groups
(a) 5 (b) 16
(c) 221 (d) None

621. According to Champion and Seth Indian forest have been divided into ——— types
(a) 5 (b) 16
(c) 221 (d) None of these

622. A renewal of forest crop by natural or artificial means
(a) Regeneration
(b) Natural regeneration
(c) Artificial regeneration
(d) All the above

623. The renewal of a forest crop by self-sown seed or by coppice is called
(a) Regeneration
(b) Natural regeneration
(c) Artificial regeneration
(d) All the above

624. The regeneration obtained from seed forms a crop called
(a) Seedling crop (b) Coppice crop
(c) a and b (d) None

625. Girlding is usually done on trees which are more than ------------diameter
(a) 5 cm (b) 10 cm
(c) 20 cm (d) 30 cm

626. Crown thinning is applied to -------- crops
(a) Even aged regular
(b) Uneven aged
(c) Mixed uneven aged
(d) None

627. Coppice forest otherwise called as
(a) Low forest
(b) High forest
(c) a and b
(d) None

628. A year in which a given species bear seed abundantly called as
(a) Seed Year (b) Seedling Year
(c) Germination (d) None

629. Examples of seed dispersal by wind
(a) *Pterocarpus marsupium*
(b) *Acacia catechu*
(c) Casuarina spp.
(d) All the above

630. Examples of seed dispersal by water
(a) Mangrove species
(b) Dalbergia
(c) Teak
(d) All the above

631. Examples of seed dispersal by gravity
(a) Oaks (b) *Juglans regia*
(c) Aesculus (d) All the above

632. Examples of seed dispersal by birds
(a) Mulberry
(b) *Diospyros melanoxylon*
(c) Prunus spp.
(d) All the above

633. Examples of seed dispersal by animals
(a) *Anthocephalus cadamba*
(b) *Acacia Arabica*
(c) *Prosopis juliflora*
(d) *All the above*

634. The potential capacity of a seed to germinate is called
(a) Viability
(b) Germination
(c) Germinative capacity
(d) None

635. The percentage, by number of seeds in a given sample that actually germinate, irrespective of time is called
(a) Germinative Energy
(b) Germinative capacity
(c) Germination
(d) Plant percent

636. The percentage, by number of seeds in a given sample that have germinated up to the time when the rate of germination reaches its peak is called
(a) Germinative Energy
(b) Germinative capacity
(c) Germination
(d) Plant percent

637. The percentage of number of seeds in a sample that develop into seedlings at the first growing season is called
(a) Germinative Energy
(b) Germinative capacity
(c) Germination
(d) Plant percent

638. The species produces abundant first year seedlings is called
(a) Seedling year (b) Seed year
(c) a and b (d) None

639. Area where plants are raised for eventual planting out is called
(a) Nursery (b) Sowing
(c) Grading (d) All the above

640. Salvage cutting is done primarily for
(a) Improving the hygiene of the stand
(b) Salvaging the timber
(c) Salvaging the stand from deterioration
(d) Producing of large trees

641. Pick out the trees species with the correct spacing
(a) Eucalyptus - 3 X 3 m
(b) Casuarina - 2 X 3 m
(c) Bamboo - 6 X 6 m
(d) *Ceiba pentandra* - 2.5 X 2.5 m

642. Wide Nursery bed is called
(a) Permanent nursery
(b) Temporary nursery
(c) Kamara (d) None

643. Standard soil mixture used for the nursery is (Soil : Sand : FYM)
(a) 2 : 1 : 1 (b) 3 : 2 : 1
(c) 3 : 2 : 2 (d) 2 : 2 : 1

644. Standard size of nursery bed in forestry is
(a) 12.5 m X 1.2 m
(b) 10 m X 1.0 m
(c) 12.5 m X 1.5 m
(d) 10 m X 2 m

645. The technique of joining together parts of two plants to form one composite plant is
(a) Budding (b) Cutting
(c) Grafting (d) Sowing

646. Grafting can be used only in plants having
(a) Soft system (b) Hard tissue
(c) Buds (d) Cambium

647. In grafting scion forms
(a) Root system (b) Shoot system
(c) Hybrid plant (d) New plant

648. Scion is
(a) Graft of shoot (b) Graft of root
(c) A bud (d) Clone

649. In which of the following, diameter of stock is larger than scion
(a) Side grafting
(b) Tongue grafting
(c) Wedge grafting
(d) Crown grafting

650. More than one scion are grafted on single stock, the grafting is
(a) Crown grafting
(b) Wedge grafting
(c) Layering
(d) Bud grafting

651. Special type of grafting, in which scion unite with the stock while still attached to parent plant
(a) Approach grafting
(b) Wedge grafting
(c) Side grafting
(d) Tongue grafting

652. Grafting technique is successful between two plants having
(a) Viable seed
(b) Parenchymatous tissue
(c) Active buds
(d) Active vascular cambium

653. Which is the most common pattern of planting in forestry ?
(a) Rectangular Planting
(b) Line Planting
(c) Triangular Planting
(d) Quincunx Planting

654. Planting of cuttings is common in
(a) Ficus spp. (b) Bombax
(c) Salix (d) All of the above

655. Naked root planting has been successful in
(a) *Casuarina equisetifolia*
(b) *Alnus nitida*
(c) *Juglans regia*
(d) All the above

656. Which is not a tending operation ?
(a) Weeding (b) Soil Working
(c) Thinning (d) Cleaning

657. Which of the following is not a harvesting system?
(a) Coppice cutting
(b) Clear cutting
(c) Selective cutting
(d) Thinning

658. Removing a part of the stand to promote faster growth of the remaining stand is called ____________.
(a) Thinning
(b) Release cutting
(c) Improvement cutting
(d) Under story cutting

659. French thinning is otherwise called as
(a) Ordinary thinning
(b) Crown thinning
(c) Free thinning
(d) Advance thinning

660. Maximum thinning was developed by
(a) Heck (b) Gerhardt
(c) Craib (d) Glover

661. Germen thinning is otherwise called as
(a) Ordinary thinning
(b) Crown thinning
(c) Free thinning
(d) Advance thinning

662. The whole set of thinning carried out in a crop for the earliest to maturity
(a) Thinning cycle
(b) Thinning regime
(c) Thinning intensity
(d) Thinning

663. Thinnings in young timber allow the timber grower to:
(a) Maintain high rates of growth
(b) Concentrate growth on the better trees
(c) Maintain a properly stocked forest
(d) All of the above

664. The thumb rule method adopted for deodar as per Gloven formula is
(a) d = aD+b (b) d=D
(c) d=1.5D (d) d=3D

665. Laurie's formula for mechanical thinning for irregular sal is
(a) d = aD+b (b) d=D
(c) d=1.5D (d) d=3D

666. Free Thinning is otherwise called as
(a) Single Stem Silviculture
(b) Elite thinning
(c) a and b
(d) None

667. In which of these system, the regeneration fellings are distributed over the whole felling series
(a) Clear felling systems
(b) group systems
(c) shelter wood systems
(d) selection systems

668. In which of the tree species uniform systems is followed
(a) Teak
(b) Bamboo
(c) Eucalyptus
(d) Sal

669. Salix in Kashmir valley is worked under
(a) Coppice with standard
(b) Coppice with two rotation
(c) Pollard systems
(d) Selection systems

670. Silvicultural systems followed for Bamboo is
(a) Clear felling systems
(b) Shelter wood systems
(c) Group selection systems
(d) Culm selection systems

671. Silviculture system for road side plantations is
(a) Simple coppice systems
(b) Shelter wood systems
(c) Clear strip systems
(d) None

672. Teak was managed under
(a) Selection systems
(b) Clear felling systems
(c) Shelter wood systems
(d) Group selection systems

673. Natural selection sal is best achieved under
(a) Clear felling systems
(b) Coppice systems
(c) Shelter wood systems
(d) None

674. Poor regeneration of sal can be attributed to
(a) Seed Dormancy
(b) Dying back of seedlings
(c) Orthodox nature of the seeds
(d) None

675. A patch usually 20 X 20 m in size burnt and artificially regenerated in a felled coupe to supplement natural regeneration is known as
(a) Rab
(b) Block
(c) Compartment
(d) None

676. Teak is generally
(a) Shade bearer
(b) Light demander
(c) Shade demander
(d) None of these

677. Silvicultural system prevalent during early part of 20th century was
(a) Shelterwood system
(b) Clear felling system
(c) Selection system
(d) Coppice system

678. The clear felling or clear strip system is suitable for
(a) Shade demanders
(b) Shade bearers
(c) Light demanders
(d) Light bearers

679. Example of clear-felling system
(a) Clear-felling system
(b) Clear-strip system
(c) Alternate-strip system
(d) All the above

680. Example of shelter wood system
(a) Clear felling system
(b) Uniform system
(c) Selection system
(d) None

681. The first stage of regeneration felling under a shelterwood system
(a) Seeding felling
(b) Final felling
(c) Secondary felling
(d) Felling series

682. Example of clear-felling followed by artificial regeneration by taungyas
(a) Sukna technique for sal
(b) Allapalli technique for teak
(c) Cryptomeria plantations of WB
(d) None

683. The shelterwood coppice system is applicable where
(a) Frost is common occurrence
(b) Locality is poor
(c) There is frequent fire
(d) None

684. Coppice-with-standards system is applicable in
(a) Sal forests
(b) Teak forests
(c) Moist deciduous forests
(d) None

685. Reservation of old crops in C.W.R. is done
(a) By area (b) By species
(c) By trees (d) All

686. Pollard system is applied for
(a) Jamun (b) Hardwickia
(c) Teak (d) None

687. Example of accessory system
(a) Alternate strip system
(b) The Group system
(c) Selection system
(d) Improvement fellings

688. Example of coppice system
(a) Clear-strip system
(b) Group system
(c) Selection system
(d) Pollard system

689. Regeneration felling includes
(a) Seeding felling
(b) Secondary felling
(c) Final felling
(d) All

690. Important consideration for application of clear felling system is
(a) Factors of locality
(b) Potential productivity of site
(c) Species and composition of crop
(d) All of these

691. Example of clear felling followed by natural regeneration
(a) Natural regeneration by seed stored in the are
(b) Natural regeneration by seed received from outside
(c) Natural regeneration by seed from advance growth
(d) All of these

692. Factors affect the length of regeneration period in uniform system
(a) Soil conditions
(b) Fire
(c) Grazing and browsing
(d) All

693. Operations carried out in Indian irregular shetlerwood system in addition to prescribed cultural operations
(a) Burning
(b) Soil working
(c) Weeding and shrub cutting
(d) All of these

694. The major advantage of selection system over other systems is
(a) More growing stock
(b) Producing large sized trees
(c) Produces of forest of aesthetic & scenic values
(d) All

695. The height of stumps in simple coppice system for Eucalyptus hybrid is
(a) 5- 10 cm (b) 10-15 cm
(c) 15-25 cm (d) 25-35 cm

696. The main objective of the coppice of two rotations system
(a) Large sized timber
(b) Soil conservation
(c) To protect young regeneration from frost
(d) None

697. The shelterwood coppice system is applicable where there is
(a) Good locality
(b) High rainfall areas
(c) Poor soil
(d) None

698. Coppice-with-standards system is applicable in
(a) Jamun belts
(b) Bamboo plantations
(c) Teak plantations
(d) None

699. Example of Reservation of old crop by species in C.W.R. is
(a) Teak (b) Jamun
(c) Eucalypts (d) None

700. Pollard system is applicable to
(a) Salix (b) Teak
(c) Jamun (d) None

701. Length of conversion period depends on
(a) Sacrifice of immature crop
(b) Proportion of over mature growing stock with negative increment
(c) Both a and b
(d) None

702. The silvicultural stem in which regeneration fellings are distributed over the whole felling series and the crop is always irregular is known as
(a) Systems of diffused regeneration
(b) Accessory system
(c) System of concentrated regeneration
(d) Coppice system

703. The alternate strip system was experimentally applied to
(a) *Cedrus deodara*
(b) *Shorea robusta*
(c) *Acacia auriculiformis*
(d) *Pinus roxburghii*

704. The parts of forest set aside to be regenerated or otherwise treated during a specified period is known as
(a) Regeneration period
(b) Periodic block
(c) Felling series
(d) Annual coupe

705. The shelterwood coppice system is a modification of
(a) Simple coppice system
(b) The coppice of two rotation system
(c) Coppice selection system
(d) The coppice with reserve system

706. Most suitable Silvicultural system of deodar in hilly terrain for control of soil erosion is
(a) Selection system
(b) Clear felling
(c) Coppice system
(d) All

707. The natural death and fall of branches of standing trees caused by the agency of nature is called
(a) Artificial pruning
(b) Natural pruning
(c) Dry pruning
(d) Green pruning

708. The pattern of felling to be adopted in harvesting a matured crop to be regenerated depends on ______ requirements of the species
(a) Mode of regeneration
(b) Type of crop
(c) Tending operation
(d) Character of the crop

709. The form or character of the new crop produced in any Silvicultural systems depends on——————.

(a) Spacing (b) Age

(c) Light (d) Moisture

710. High forest systems are those Silvicultural systems in which the regeneration is normally of —————— origin

(a) Coppice (b) Cuttings

(c) Seed origin (d) Root suckers

711. In ———— systems equal or equi productive of matured crop are successively clear felled in one operation

(a) Clear felling system

(b) Uniform system

(c) Group system

(d) Shelter wood strip systems

712. High forest with reserves system is categorized under ———— systems

(a) Clear felling systems

(b) Shelter wood systems

(c) Selection systems

(d) Accessory Systems

713. Opening the canopy of a matured stand to provide conditions for securing regeneration from the retained seed trees is called ——————

(a) Seeding Felling

(b) Secondary Felling

(c) Final felling

(d) Felling cycle

714. Wagner's Blender Saumschlag is otherwise called as ———— systems

(a) Strip selection system

(b) Pujab shelter wood system

(c) Progressive clear strip system

(d) Compartment system

715. In ——— system tree stem has been cut off in order to obtain a flush of shoots

(a) Coppice systems

(b) Clear felling systems

(c) Selection systems

(d) Shelter wood systems

716. Change from one Silvicultural system or one species to another is called ————

(a) Conversion (b) Regeneration

(c) Rotation (d) Periodic block

717. Eucalyptus hybrid is worked with ———— system

(a) Simple Coppice Systems

(b) Coppice of Two rotations

(c) Shelter wood coppice

(d) Coppice with standard systems

718. Deodar forest is commonly worked with ———— system in Himachal Pradesh

(a) Eberhard's wedge

(b) Shelter wood strip

(c) Irregular shelter wood systems

(d) Indian irregular shelter wood systems

719. Shelter wood/selection system is suitable for ———— bearer species

(a) Shade (b) Light

(c) Temperature (d) All the above

720. Forest growing on rocky and precipitous slopes are worked with ———— system

(a) Shelter wood (b) Selection

(c) Clear felling (d) Uniform

721. *Dendrocalamus strictus* forest in Uttar Pradesh and Madhya Pradesh are worked with ———— system

(a) Selection cutting combine with cleaning and cultural operations

(b) Selection

(c) Shelter wood

(d) Coppice

722. In two storied high forest system, the crop in each storey is approximately ———— aged

(a) Even (b) Uneven

(c) a and b (d) none

723. Group Selection System is categorized under system of ———— regeneration

(a) Diffused (b) Concentrated

(c) a and b (d) None

724. The interval between seeding felling and the final felling is called ———— interval

(a) Seeding felling

(b) Secondary felling

(c) Final felling

(d) Regeneration interval

725. The time which elapses between successive main fellings on the same area is called ————

(a) Felling cycle (b) Final felling

(c) Seeding felling (d) Secondary felling

726. In Culm Selection System, the felling cycle will be ————.

(a) 4 years (b) 6 years

(c) 8 years (d) 10 years

727. *Acacia catechu* forests of Himachal Pradesh in worked with ———— system

(a) Simple Coppice

(b) Shelter wood coppice

(c) Coppice of two rotation

(d) Coppice selection

728. Chipko Movement is an example of forest conservation through

(a) Tehri-Garhwal district

(b) Uttaranchal

(c) Public awareness and participation

(d) Political issue

729. Chipko movement is related to

(a) Forest conservation

(b) Soil conservation

(c) Water conservation

(d) Wetland conservation

730. Chipko movement was launched for protection of

(a) Forests (b) Grass lands

(c) Wet lands (d) Live stocks

731. Conservation of species in its natural habitat is

(a) In-situ (b) Ex-situ

(c) In-vitro (d) Both(b) and(c)

732. Deforestation brings about

(a) Soil erosion

(b) Weed control

(c) Decreases drought

(d) Increased sunlight

733. Deforestation is caused due to

(a) Silviculture

(b) Construction of roads

(c) Rainfall

(d) Plantation of trees

734. Deforestation is the major causal agent of

(a) Depletion of natural resources

(b) Environmental pollution

(c) Desertification of habitat

(d) Genetic erosion

735. Deforestation will decrease

(a) Soil erosion (b) Land slides

(c) Soil fertility (d) Rainfall

736. Development of botanical garden is what type of conservation

(a) Ex-situ (b) A common

(c) An easy (d) In-situ

737. Extensive planting of trees to increase forest cover is called

(a) Deforestation (b) Agroforestry

(c) Social forestry (d) Afforestation

738. For mapping, Remote sensing equipment is mounted on

(a) Aircraft and satellite

(b) Ship and spacecraft

(c) Satellite and spacecraft

(d) All the above

739. Forests control drought by

(a) Preventing soil erosion

(b) Increasing oxygen

(c) Increasing humidity and rainfall

(d) Preventing floods

740. Fresh water present on earth is
(a) 97.5% (b) 0.01%
(c) 2.5 % (d) 1.97%

741. Fresh water problem arises due to
(a) Globaly available stocks are insufficient
(b) Uneven distribution on earth
(c) No regeneration of required quantity through natural hydrological cycle
(d) Any one of above

742. In India, common type of forest is
(a) Tropical thon forests
(b) Sal and Teak forests
(c) Tropical most deciduous forest
(d) Tropical dry deciduous forest

743. In silvicultural system, group system was developed by
(a) Gayer (b) Van rich cotta
(c) Emberhard (d) Thevor

744. The silvicultural system in which regeneration fellings instead of being done uniformly all over the compartment are carried out in scattered group is defined as
(a) Uniform systems
(b) Selection systems
(c) Shelter wood strip systems
(d) Group Systems

745. Which system is called as Andaman's Canopy Lifting Shelter wood system?
(a) Selection systems
(b) Indian Irregular shelter wood systems
(c) Group systems
(d) Uniform systems

746. Silvicultural systems in which clear felling are done in the form of strips which progress successively in one direction is known as
(a) Alternate strip system
(b) Clear strip system
(c) Shelterwood strip system
(d) Clear-felling with standards

747. Deodar forests of Uttar Pradesh and Himachal Pradesh are generally worked under
(a) Clear strip system
(b) Indian Irregular Shelterwood system
(c) Coppice system
(d) Selection system

748. Center for International Forestry Research (CIFOR) is in
(a) Nairobi, Kenya
(b) Vienna, Austria
(c) Bogor, Indonesia
(d) Ibadan, Nigeria

749. Which of the silvicultural system is used for the regeneration and management of high forests
(a) Clear felling
(b) Shelter wood
(c) Selection felling
(d) All of the above

750. The silvicultural system followed for black wattle *(Acacia mearnsii)* in the Nilgiris is
(a) Clear felling followed by natural regeneration
(b) Clear felling followed by artificial regeneration
(c) Uniform system
(d) Simple coppice system

751. Blue colour in the regeneration map represents
(a) Excellent regeneration
(b) Good regeneration
(c) Moderate regeneration
(d) Fair regeneration

752. Red colour in the regeneration map represents
(a) Excellent regeneration
(b) Good regeneration
(c) Moderate regeneration
(d) Fair regeneration

753. Sandal wood forest is worked under
(a) Group Selection
(b) Natural selective systems
(c) Group Systems
(d) Indian irregular shelter wood systems

754. --------- is a silvicultural system in which equal or equiproductive areas of mature crops are successively clear felled in single operation
(a) Shelter wood systems
(b) Clear felling systems
(c) Selection systems
(d) Coppice systems

755. Clear felling or clear strip system is suitable for
(a) Shade demanders
(b) Shade bearers
(c) Light demanders
(d) Light bearers

756. Rose wood tree includes
(a) Blackwood rosewood
(b) Indian rosewood
(c) Malabar black wood
(d) All the above

ANSWERSHEET
3. Silviculture

1. **(a)**	2. **(a)**	3. **(a)**	4. **(a)**	5. **(b)**	6. **(c)**	7. **(a)**	8. **(c)**	9. **(a)**	10. **(b)**
11. **(b)**	12. **(b)**	13. **(c)**	14. **(b)**	15. **(a)**	16. **(d)**	17. **(b)**	18. **(c)**	19. **(d)**	20. **(d)**
21. **(a)**	22. **(c)**	23. **(c)**	24. **(d)**	25. **(d)**	26. **(d)**	27. **(c)**	28. **(b)**	29. **(a)**	30. **(a)**
31. **(d)**	32. **(a)**	33. **(c)**	34. **(c)**	35. **(c)**	36. **(a)**	37. **(c)**	38. **(d)**	39. **(c)**	40. **(a)**
41. **(a)**	42. **(b)**	43. **(c)**	44. **(c)**	45. **(c)**	46. **(b)**	47. **(b)**	48. **(a)**	49. **(a)**	50. **(c)**
51. **(d)**	52. **(b)**	53. **(a)**	54. **(c)**	55. **(d)**	56. **(b)**	57. **(c)**	58. **(d)**	59. **(a)**	60. **(a)**
61. **(c)**	62. **(b)**	63. **(a)**	64. **(a)**	65. **(d)**	66. **(d)**	67. **(c)**	68. **(b)**	69. **(a)**	70. **(a)**
71. **(d)**	72. **(b)**	73. **(b)**	74. **(a)**	75. **(a)**	76. **(c)**	77. **(d)**	78. **(b)**	79. **(c)**	80. **(c)**
81. **(c)**	82. **(d)**	83. **(b)**	84. **(d)**	85. **(a)**	86. **(c)**	87. **(d)**	88. **(d)**	89. **(a)**	90. **(b)**
91. **(c)**	92. **(c)**	93. **(d)**	94. **(d)**	95. **(a)**	96. **(b)**	97. **(c)**	98. **(b)**	99. **(b)**	100. **(b)**
101. **(d)**	102. **(b)**	103. **(b)**	104. **(c)**	105. **(a)**	106. **(d)**	107. **(c)**	108. **(b)**	109. **(b)**	110. **(b)**
111. **(b)**	112. **(d)**	113. **(d)**	114. **(a)**	115. **(d)**	116. **(d)**	117. **(d)**	118. **(a)**	119. **(b)**	120. **(a)**
121. **(a)**	122. **(d)**	123. **(c)**	124. **(d)**	125. **(d)**	126. **(c)**	127. **(d)**	128. **(b)**	129. **(b)**	130. **(c)**
131. **(b)**	132. **(a)**	133. **(b)**	134. **(c)**	135. **(a)**	136. **(b)**	137. **(d)**	138. **(b)**	139. **(a)**	140. **(a)**
141. **(a)**	142. **(a)**	143. **(b)**	144. **(d)**	145. **(c)**	146. **(d)**	147. **(d)**	148. **(c)**	149. **(b)**	150. **(a)**
151. **(b)**	152. **(c)**	153. **(c)**	154. **(b)**	155. **(c)**	156. **(a)**	157. **(d)**	158. **(a)**	159. **(b)**	160. **(a)**
161. **(d)**	162. **(a)**	163. **(b)**	164. **(c)**	165. **(a)**	166. **(d)**	167. **(b)**	168. **(d)**	169. **(c)**	170. **(d)**
171. **(b)**	172. **(c)**	173. **(b)**	174. **(a)**	175. **(c)**	176. **(d)**	177. **(c)**	178. **(d)**	179. **(d)**	180. **(c)**

181. (d)	182. (b)	183. (d)	184. (c)	185. (a)	186. (b)	187. (a)	188. (c)	189. (c)	190. (c)
191. (b)	192. (a)	193. (a)	194. (d)	195. (b)	196. (b)	197. (d)	198. (b)	199. (b)	200. (c)
201. (c)	202. (b)	203. (c)	204. (c)	205. (a)	206. (c)	207. (c)	208. (c)	209. (a)	210. (b)
211. (b)	212. (b)	213. (a)	214. (c)	215. (d)	216. (b)	217. (a)	218. (d)	219. (a)	220. (d)
221. (a)	222. (c)	223. (d)	224. (b)	225. (c)	226. (a)	227. (a)	228. (c)	229. (b)	230. (c)
231. (d)	232. (d)	233. (a)	234. (d)	235. (b)	236. (d)	237. (b)	238. (a)	239. (a)	240. (a)
241. (d)	242. (a)	243. (d)	244. (c)	245. (c)	246. (a)	247. (d)	248. (b)	249. (b)	250. (a)
251. (c)	252. (b)	253. (b)	254. (a)	255. (a)	256. (c)	257. (a)	258. (a)	259. (b)	260. (c)
261. (a)	262. (c)	263. (a)	264. (a)	265. (b)	266. (b)	267. (a)	268. (a)	269. (a)	270. (d)
271. (c)	272. (b)	273. (b)	274. (c)	275. (a)	276. (a)	277. (c)	278. (d)	279. (a)	280. (b)
281. (a)	282. (b)	283. (b)	284. (a)	285. (d)	286. (b)	287. (d)	288. (b)	289. (b)	290. (d)
291. (b)	292. (c)	293. (b)	294. (d)	295. (b)	296. (c)	297. (a)	298. (c)	299. (c)	300. (c)
301. (b)	302. (a)	303. (c)	304. (d)	305. (b)	306. (a)	307. (d)	308. (c)	309. (a)	310. (b)
311. (c)	312. (b)	313. (a)	314. (a)	315. (a)	316. (b)	317. (b)	318. (b)	319. (a)	320. (b)
321. (c)	322. (c)	323. (a)	324. (a)	325. (c)	326. (a)	327. (a)	328. (b)	329. (a)	330. (a)
331. (a)	332. (b)	333. (b)	334. (c)	335. (c)	336. (a)	337. (a)	338. (b)	339. (c)	340. (a)
341. (a)	342. (b)	343. (a)	344. (a)	345. (c)	346. (a)	347. (d)	348. (b)	349. (c)	350. (c)
351. (a)	352. (b)	353. (c)	354. (a)	355. (a)	356. (a)	357. (a)	358. (a)	359. (a)	360. (c)
361. (a)	362. (a)	363. (a)	364. (c)	365. (b)	366. (a)	367. (a)	368. (a)	369. (a)	370. (a)
371 (a)	372. (b)	373. (b)	374. (d)	375. (b)	376. (a)	377. (a)	378. (a)	379. (a)	380. (c)
381. (a)	382. (a)	383. (a)	384. (c)	385. (b)	386. (a)	387. (a)	388. (a)	389. (a)	390. (d)
391. (c)	392. (a)	393. (a)	394 (a)	395. (b)	396. (a)	397. (a)	398. (a)	399. (c)	400. (a)
401. (a)	402. (a)	403. (a)	404. (a)	405. (a)	406. (d)	407. (b)	408. (a)	409. (d)	410. (d)
411. (a)	412. (a)	413. (b)	414. (d)	415. (a)	416. (a)	417. (b)	418. (a)	419. (b)	420. (d)
421. (a)	422. (d)	423. (c)	424. (a)	425. (b)	426. (c)	427. (d)	428. (d)	429. (d)	430. (b)
431. (b)	432. (a)	433. (a)	434. (a)	435. (b)	436. (c)	437. (d)	438. (c)	439. (d)	440. (a)
441. (b)	442. (b)	443. (a)	444. (d)	445. (a)	446. (d)	447. (d)	448. (d)	449. (a)	450. (b)
451. (a)	452. (c)	453. (a)	454. (a)	455. (a)	456. (d)	457. (b)	458. (a)	459. (c)	460. (a)
461. (d)	462. (c)	463. (a)	464. (a)	465. (a)	466. (a)	467. (a)	468. (b)	469. (a)	470. (b)
471. (a)	472. (a)	473. (a)	474. (c)	475. (a)	476. (d)	477. (d)	478. (c)	479. (d)	480. (a)
481. (a)	482. (b)	483. (b)	484. (a)	485. (d)	486. (a)	487. (b)	488. (c)	489. (c)	490. (a)
491. (a)	492. (d)	493. (a)	494. (b)	495. (a)	496. (a)	497. (b)	498. (a)	499. (b)	500. (d)
501. (c)	502. (a)	503. (a)	504. (a)	505. (d)	506. (a)	507. (b)	508 (a)	509. (d)	510. (d)
511. (a)	512. (a)	513. (a)	514. (d)	515. (a)	516. (a)	517. (a)	518. (d)	519. (d)	520. (d)

521. **(d)** 522. **(d)** 523. **(a)** 524. **(d)** 525. **(a)** 526. **(a)** 527. **(a)** 528. **(a)** 529. **(a)** 530. **(b)**
531. **(a)** 532. **(c)** 533 **(a)** 534. **(a)** 535. **(c)** 536. **(a)** 537. **(a)** 538. **(a)** 539. **(b)** 540. **(c)**
541. **(a)** 542. **(d)** 543. **(a)** 544. **(a)** 545. **(a)** 546. **(a)** 547. **(b)** 548. **(a)** 549. **(a)** 550. **(a)**
551. **(c)** 552. **(b)** 553. **(a)** 554. **(a)** 555. **(a)** 556. **(b)** 557. **(c)** 558. **(c)** 559. **(b)** 560. **(b)**
561. **(a)** 562. **(c)** 563. **(a)** 564. **(a)** 565. **(b)** 566. **(a)** 567. **(b)** 568. **(c)** 569. **(a)** 570. **(c)**
571. **(a)** 572. **(c)** 573. **(a)** 574. **(a)** 575. **(b)** 576. **(c)** 577. **(b)** 578. **(b)** 579. **(a)** 580. **(d)**
581. **(d)** 582. **(a)** 583. **(a)** 584. **(a)** 585. **(c)** 586. **(a)** 587. **(a)** 588. **(a)** 589. **(a)** 590. **(b)**
591. **(c)** 592. **(a)** 593. **(b)** 594. **(a)** 595. **(a)** 596. **(b)** 597. **(c)** 598. **(a)** 599. **(a)** 600. **(a)**
601. **(a)** 602. **(d)** 603. **(c)** 604. **(a)** 605. **(a)** 606. **(b)** 607. **(d)** 608. **(a)** 609. **(b)** 610. **(d)**
611. **(a)** 612. **(d)** 613. **(a)** 614. **(a)** 615. **(b)** 616. **(b)** 617. **(a)** 618. **(b)** 619. **(a)** 620. **(b)**
621. **(c)** 622. **(d)** 623. **(b)** 624. **(a)** 625. **(c)** 626. **(a)** 627. **(a)** 628. **(a).** 629. **(d).** 630. **(d)**
631. **(d)** 632. **(d)** 633. **(d)** 634. **(a)** 635. **(b)** 636. **(a)** 637. **(d)** 638. **(a)** 639. **(a)** 640. **(b)**
641. **(c)** 642. **(c)** 643. **(a)** 644. **(a)** 645. **(c)** 646. **(d)** 674. **(b)** 648. **(a)** 649. **(d)** 650. **(a)**
651. **(a)** 652. **(d)** 653. **(a)** 654. **(d)** 655. **(d)** 656. **(b)** 657. **(d)** 658. **(a)** 659. **(b)** 660. **(b)**
661. **(a)** 662. **(b)** 663. **(d)** 664. **(b)** 665. **(c)** 666. **(c)** 667. **(d)** 668. **(d)** 669. **(c)** 670. **(d)**
671. **(c)** 672. **(d)** 673. **(c)** 674. **(b)** 675. **(a)** 676. **(b)** 677. **(b)** 678. **(c)** 679. **(d)** 680. **(b)**
681. **(a)** 682. **(a)** 683. **(a)** 684. **(a)** 685. **(d)** 686. **(b)** 687. **(d)** 688. **(d)** 689. **(d)** 690. **(d)**
691. **(d)** 692. **(d)** 693. **(d)** 694. **(d)** 695. **(c)** 696. **(a)** 697. **(a)** 698. **(a)** 699. **(b)** 700. **(a)**
701. **(c)** 702. **(a)** 703. **(b)** 704. **(b)** 705. **(a)** 706. **(a)** 707. **(b)** 708. **(a)** 709. **(a)** 710. **(c)**
711. **(a)** 712. **(d)** 713. **(a)** 714. **(a)** 715. **(a)** 716. **(a)** 717. **(a)** 718. **(d)** 719. **(a)** 720. **(b)**
721. **(a)** 722. **(a)** 723. **(a)** 724. **(b)** 725. **(a)** 726. **(a)** 727. **(d)** 728. **(c)** 729. **(a)** 730. **(a)**
731. **(a)** 732. **(a)** 733. **(b)** 734. **(c)** 735. **(d)** 736. **(a)** 737. **(d)** 738. **(a)** 739. **(c)** 740. **(c)**
741. **(b)** 742. **(d)** 743. **(a)** 744. **(d)** 745. **(b)** 746. **(b)** 747. **(b)** 748. **(c)** 749. **(d)** 750. **(a)**
751. **(b)** 752. **(d)** 753. **(b)** 754. **(b)** 755. **(c)** 756. **(d)**

4

Agroforestry

KEY POINTS

- The word Social forestry was coined by Westoby and used in the Ninth Commonwealth Forestry Congress in 1968
- Farm forestry was defined by NCA (1976) as the practice of forestry in all its aspects in and around the farms or village lands integrated with other farm operations
- Social forestry project phase I was launched during 1981 with the assistance from Swedish International Development Authority (SIDA) from 1981 to 1986
- Social forestry project phase II was implemented during 1988-89 to 1992-93
- TCIP - Tree Cultivation Incentive Programme
- TCEP - Tree Cultivation Extension Programme
- NAP - National Afforestation Programme
- Agroforestry is a collective name for land-use systems involving trees combined with crops and/or animals on the same unit of land.
- Nair (1987) has classified the agroforestry systems based on the four criteria viz., Structural Basis, Functional basis, Socio economic Basis and Ecological basis
- The taungya (taung = hill, ya = cultivation) is a Burmese word coined in Burma in 1850s.
- The taungya system was introduced into India by Brandis in 1890
- First taungya plantations were raised in 1896 in North Bengal
- In southern India, the taungya system is called 'kumri'
- Alley cropping is also known as hedgerow intercropping
- Shelter-belt is the belts/blocks consisting of several rows of trees established at right angles to the prevailing wind
- Wind-breaks are strips of trees and/or shrubs planted to protect fields, homes, canals or other areas from wind and blowing soil or sand
- Silvipasture system is classified in to three categories viz., Protein bank, Livefence of fodder trees and hedges and Trees and shrubs on pasture
- Agrisilvopastural system is grouped into two categories viz., Home gardens and Woody hedgerows
- Home gardens also called as Multitier system or Multitier cropping
- Home gardens are characterized by high species diversity and usually 3-4 vertical canopy strata

- Agroecological zone in india – 8
- Agroclimatic zones in india – 15
- Macro design and diagnosis covers the entire ecological zone with in a country
- Micro design and diagnosis covers the within or part of the chosen ecological zone

Suitability of Agroforestry systems

Conditions/Area	Agroforestry systems
Humid tropical/ subtropical lands	Home gardens, Trees on range lands and pastures, Improved fallow in shifting cultivation and multi purpose wood lots
Semiarid and arid lands	Silvipastoral systems, Wind breaks and shelter belts
Tropical High lands	Woody perennials for soil conservation, Improved fallow
Tropical soil productivity	Permanent inter cropping with N fixing trees, Alley cropping, Trees for soil conservation
Increasing farm income	Multi storey cropping, Agri horti silvicultural, Agri silvicultural systems, Alley cropping
Arid lands	Agrisilvicultural, Agrisilvipastoral system, Block plantation of fuel wood, Shelter belts, Wind break
Hilly region	Silvihorticultural, silvipastoral, Agrisilvicultural, Agrihorticultural systems.
Wetlands	Agrisilvicultural, Agrisilvipastoral system, Agrisilvi aquaculture, Agri silvi aquapastoral systems, Aquaculture

MULTIPLE CHOICE QUESTIONS

1. The sustainable land use systems involving trees combined with crops and / or animals on the same unit of land is termed as

(a) Agroforestry

(b) Sustainable forest management

(c) Social Forestry (d) Farm Forestry

2. When the use of land is directed towards satisfying basic needs and is managed by the owner is said to be

(a) Commercial Agroforestry systems

(b) Intermediate Agroforestry systems

(c) Subsistence Agroforestry systems

(d) All the above

3. The practice of forestry in and around the farms lands is called as

(a) Social Forestry

(b) Community forestry

(c) Farm Forestry (d) All the above

4. The practice of forestry in areas devoid of tree growth and other vegetation outside the conventional forest areas is

(a) Social Forestry

(b) Extension Forestry

(c) Farm Forestry (d) Afforestation

5. The practice of forestry with the object of raising flowering trees is called as

(a) Recreation forestry

(b) Urban forestry

(c) Social forestry (d) Farm Forestry

6. Plantations grow as liner strips of lands on the sides of public roads, canals and railway lines are called as

(a) Linear strip plantations

(b) Commercial forestry

(c) Avenue Plantations

(d) Urban forestry

7. Home Garden Agroforestry system is practiced in
(a) Humid tropical climates
(b) Semi arid lands
(c) High lands
(d) Sub humid low lands

8. Shading effect of trees can be overcome by
(a) Canopy management
(b) Root management
(c) Water
(d) Fertilizer

9. Alley cropping is also known as
(a) Hedge row intercropping
(b) Woody Hedges
(c) Wood lots
(d) Soil conservation hedges

10. Multistorey arrangement reduces
(a) Yield
(b) Nutrition
(c) Competition
(d) All the above

11. The oldest known agroforestry systems is
(a) Alley cropping
(b) Taungya systems
(c) Home gardens
(d) Shifting cultivation

12. Social forestry programme was launched in the
(a) Fifth five year plan
(b) Sixth five year plan
(c) Seventh five year plan
(d) Eighth five year plan

13. The collection period for neem (*Azadiracta indica)* seed is
(a) January - March
(b) March – May
(c) May – June
(d) June – July

14. In the manufacture of Braille paper ----------------------- species is used
(a) *Eucalyptus globulus*
(b) *Eucalyptus tereticornis*
(c) *Eucalyptus camandulensis*
(d) *Eucalyptus urophylla*

15. *Indigofera tinctoria* is used in
(a) Tannin industry
(b) Dye industry
(c) Resin
(d) Gum

16. Which classification of agroforestry is more important and commonly adopted?
(a) Socioeconomic classification
(b) Structural classification
(c) Floristic classification
(d) Ecological classification

17. In which agroforestry system, agricultural crops is of secondary importance?
(a) Silvipasture
(b) Agrisilviculture
(c) Silviagriculture
(d) Hortisilviculture

18. The most prevalent agroforestry systems in high rainfall areas of Kerala is
(a) Silvipasture
(b) Silviagriculture
(c) Horti-silvi-pasture
(d) Home gardens

19. Which among these tree species are best suited for alley cropping?
(a) *Eucalyptus tereticornis*
(b) *Casuarina equisetifolia*
(c) *Cassia siamea*
(d) *Ailanthus excelsa*

20. Which among these factors have degraded the Indian grasslands the most?
(a) Adverse site factors
(b) Poor management
(c) Poor productivity
(d) Over grazing

21. Carrying capacity is more related to
(a) Agrisilviculture
(b) Improvement of grasslands
(c) Improvement of nutritive value of grasslands
(d) None of these

22. Trade/common name of *Dalbergia latifolia* is
(a) Rose Wood (b) Beef Wood
(c) Soft wood (d) Hard wood

23. Animals feed in fenced pastures with control access is terms as
(a) Paddoching (b) Pasture reseeding
(c) Grazing (d) Browsing

24. Indian Grassland and Fodder Research Institute is located at
(a) Dehradun (b) Jhansi
(c) Jodhpur (d) Nigeria

25. Rotation of aerable crops with two or more years of sown pasture is called as
(a) Mixed farming
(b) Mixed cropping
(c) Land use systems
(d) Lay farming

26. Forest products contribution to gross domestic product (GDP) is
(a) 1% (b) 1.7%
(c) 2% (d) 2.7%

27. Tree planted in silvi-agriculture should be oriented in the
(a) East west direction
(b) North south direction
(c) East south direction
(d) West south direction

28. Most common agroforestry system practiced in arid regions are
(a) Agrisilvicultural systems
(b) Silvopastural systems
(c) Agrosilvopastoral systems
(d) Other systems

29. Social forestry was first coined by
(a) Westoby (b) Shah
(c) Brandis (d) Nair

30. Social forestry was first adopted successfully in
(a) Gujarat (b) Tamil Nadu
(c) Kerala (d) Rajasthan

31. Van Mohotsav is a festival of
(a) Wild Life
(b) Tree Planting
(c) Timbers
(d) All the above

32. Van Mohotsav was started in
(a) 1847 (b) 1947
(c) 1850 (d) 1950

33. National Wasteland Development Board came into existence during the period
(a) 1970 (b) 1975
(c) 1980 (d) 1985

34. The Plantation forestry in India was started during
(a) 1942 (b) 1842
(c) 1947 (d) 1847

35. The major species used for planting on tank bed is
(a) *Acacia leucophloea*
(b) *Prosopis juliflora*
(c) *Acacia nilotica*
(d) *Eucalyptus* spp.

36. Which type of tree is better for noise abatement?
(a) Evergreen trees
(b) Deciduous trees
(c) Semi evergreen trees
(d) All the above

37. Heliotropic type of branching should be preferred for
(a) Roadside Planting
(b) Canal Planting
(c) Railway Planting
(d) All the above

38. Neem and Mango are preferred for
(a) Railway Line Planting
(b) Canal Bank Plantation
(c) Tank bed Plantation
(d) Avenue Planting

39. Social forestry project in Tamil Nadu was first launched with the assistance of
(a) CIDA (b) SIDA
(c) JBIC (d) All the above

40. SIDA Project phase–I was implemented during
(a) 1981-86 (b) 1987-88
(c) 1988-89 (d) 1992-93

41. Protection and reforestation of degraded forests with the involvement of local people is
(a) Sustainable forest management
(b) Joint Forest Management
(c) Social Forestry
(d) Community forestry

42. The unit of management under JFM is
(a) Hamlet (b) Division
(c) Beat (d) Village

43. The people's representation body on each of the identified management unit of JFM is
(a) Sustainable forest management
(b) Joint Forest Management
(c) Social Forestry
(d) Community forestry

44. TCIP stands
(a) Tree cultivation intensive programme
(b) Tree cultivation incentive program
(c) Tree cultivation inducement programme
(d) Tree cultivation income programme

45. SIDA project phase II was implemented during
(a) 1981 – 86 (b) 1987-88
(c) 1988-89 (d) 1992-93

46. Tamil Nadu afforestation project was implemented with financial assistance from
(a) Denmark (b) Japan
(c) Netherland (d) Sweden

47. When an immediate supply of a single necessary factor falls below the combined demand of the plant is called
(a) Allelopathy (b) Allelospoly
(c) a and b (d) None

48. The direct or indirect effect of one plant upon another through the production of chemical inhibitors is called
(a) Allelopathy (b) Allelospoly
(c) a and b (d) None

49. Allelopathy denotes
(a) Postive interaction
(b) Negative interaction
(c) a and b
(d) None

50. Allelochemicals produced by plants are
(a) Primary plant metabolites
(b) Secondary plant metabolites
(c) a and b
(d) None

51. Allelochemicals are released by means of
(a) Volatilization (b) Leaching
(c) Exudation (d) All the above

52. *Prosopis cineraria* and *Cenchrus ciliaris* is a good fodder tree plus grass combination for
(a) Semi-arid climate with red soil
(b) Temperate climate
(c) Arid climate with sandy soil
(d) Semi-arid block cotton soil

53. Tree competes with the neighboring crops mainly for
(a) Light (b) Moisture
(c) Nutrients (d) All the above

54. Allelopathy is the best example for
(a) Association (b) Ammensalism
(c) Symbiosis (d) Photo corporation

55. The broad and well known classification of agroforestry systems was given by
(a) Chundawat (b) Gautam
(c) Nair (d) Dwivedi

56. A common example of the zonal pattern is
(a) Hedgerow inter cropping
(b) Home gardens
(c) Silvopastural systems
(d) Agrosilvopastural systems

57. The fundamental attributes of all agroforestry systems are
(a) Recover depleted soil Nutrients
(b) Increase crop yields
(c) Production and Protection
(d) Stabilize the soil structure

58. Integration of trees with grasses is known as
(a) Apiculture
(b) Silvipasture
(c) Aquaforestry
(d) Multipurpose wood lots

59. Taungya was first introduced by
(a) Dr. Brandis (b) Nair
(c) Clements (d) L.S. Khanna

60. Commercial agroforestry systems is used
(a) Production of perennial cash crops
(b) Satisfying basic needs
(c) Sale of the production of output
(d) All the above

61. The deliberated use of land for the production of crops including trees and crops is called as
(a) Agri horticulture
(b) Agripasture
(c) Agrisilvipasture
(d) Agrisilviculture

62. The main aim of raising crops in departmental taungya is to
(a) Reduce the weed growth
(b) Provide additional income
(c) Maximum utilization of site
(d) All the above

63. The practice of managing rows of closely planted woody plants with annual crops planted in alleys in between hedges is called as
(a) Hedge row intercropping
(b) Woody Hedgerows
(c) Soil Conservation hedges
(d) Multipurpose tree gardens

64. Shelter belts should possess a typical
(a) Triangular shape
(b) Linear Belt
(c) Square shape
(d) Rectangular shape

65. Strips of trees planted to protect fields, homes, canal from wind is referred to
(a) Shelter belts
(b) Wind breaks
(c) Alley Cropping
(d) All the above

66. Planting nectar producing trees on the boundary of the agricultural field is called as
(a) Apiculture with trees
(b) Aquaforestry
(c) Multipurpose wood lots
(d) Woody Hedgerows

67. The hydrological unit that drains at a common point is called
(a) Watershed (b) Catchment
(c) Drainage Basin (d) All the above

68. A unit of landscape that contains all the drainage areas and channels contributing to a single stream / river system
(a) Watershed (b) Dam
(c) Seashore (d) River beds

69. The drainage area of tertiary stream or river is
(a) 400 to 40000 ha
(b) 4000 to 40000 ha
(c) 40000 to 400000 ha
(d) None

70. The drainage area of catchments is
(a) 400 – 200 ha
(b) 2000 – 4000 ha
(c) > 40000 ha
(d) More than one lakh hectare

71. The drainage area of a sub-catchment is
(a) < 40000 ha (b) > 40000 ha
(c) > 100000 ha (d) 4000- 40000 ha

72. The continuous circulation of water among the hydrosphere, atmosphere and lithosphere is a called as
(a) Nutrient cycle
(b) Hydrological cycle
(c) a and b (d) None

73. The drainage area of sub watershed is
(a) 200 – 400 ha
(b) 200 – 4000 ha
(c) 200 – 2000 ha
(d) 2000 – 4000 ha

74. The drainage area of a mini watershed is
(a) 200 – 400 ha (b) 200 – 4000 ha
(c) 200 – 2000 ha (d) 400 – 2000 ha

75. The drainage area of a micro water shed is
(a) <400 ha (b) > 400 ha
(c) 400 – 2000 ha (d) 400 – 4000 ha

76. The practice of rotational cropping helps in
(a) Improving Soil Structure
(b) Maintaining soil fertility
(c) Soil & Water conservation
(d) All the above

77. Erosion Permitting and erosion resisting crops grows alternatively at right angle to the slope of the land is called as
(a) Linear Planting
(b) Strip planting
(c) Cover Planting
(d) All the above

78. The process of taking up nutrients from deeper soil profiles and deposting them on the surface layer is referred to
(a) Nutrient uptake
(b) Nutrient absorption
(c) Nutrient Pumping
(d) Nutrient cycling

79. The process of cycling of nutrients from soil to the plant and back to the soil is called as
(a) Nutrient uptake
(b) Nutrient absorption
(c) Nutrient Pumping
(d) Nutrient cycling

80. The process by which the complex organic substances are broken down to simple inorganic elements is called as
(a) Mineralization
(b) Nitrification
(c) Ammonification
(d) All the above

81. NWDB classified wastelands into
(a) Two types
(b) Three types
(c) Four types
(d) Ten Types

82. Extend of wastelands in India
(a) 63.82 m ha
(b) 36.52 m ha
(c) 83.52 m ha
(d) 38.52 m. ha

83. The land comprising deep gorges, gullies in deep alluvium and draining into a major river is known as
(a) Sheet s
(b) Drill
(c) Eroded land
(d) Gullied / Ravinous land

84. The last stage of water erosion is the formation of
(a) Sheet (b) Drill
(c) Gullies (d) Ravines

85. The wastelands formed due to canal seepage and improper drainage is
(a) Gullies and Ravines lands
(b) Water logged and marshy lands
(c) Mangroves
(d) Sand dunes

86. The lands which are rich in soluble salts and exchangeable sodium
(a) Saline Soil
(b) Alkaline Soil
(c) Saline – Alkaline
(d) All the above

87. The suitable species for reclamation of saline and sodic soil is
(a) *Prosopis juliflora*
(b) *Albizia lebbeck*
(c) *Azadirachta indica*
(d) *Casuarina equisetifolia*

88. The most suitable species for reclamation of sandy soils is
(a) *Casuarina equisetifolia*
(b) *Acacia* spp.
(c) *Melia dubia*
(d) *Azadirachta indica*

89. A family of procedure for the diagnosis of land management problems and potentials and the design of agroforestry solution is
(a) Sustainability (b) Productivity
(c) Adoptability (d) Agroforestry D&D

90. Agroforestry D&D covers entire ecological zone within a country is
(a) Macro D &D (b) Micro D & D
(c) a and b (d) None

91. Agroforestry D & D covers within the chosen ecological zone is
(a) Macro D &D (b) Micro D & D
(c) a and b (d) None

92. The main criteria for good agroforestry design are
(a) Productivity (b) Sustainability
(c) Adaptability (d) All the above

93. The major species used for planting on tank bed is
(a) *Acacia leucophloea*
(b) *Acacia nilotica*
(c) *Prosopis* spp.
(d) *Eucalyptus* spp.

94. Tree means water, water is bread and bread is life was first time quoted by
(a) Westoby
(b) Brandis
(c) Sunderlal Baguguna
(d) Nair

95. The international donor agencies for social forestry in India
(a) CIDA
(b) USAID
(c) SIDA
(d) All the above

96. The cross sectional shape of any shelter belt is
(a) Parallel (b) Neloid
(c) Cylinder (d) Pyramid

97. The practice of forestry with the object of obtaning maximum gain per unit area called as
(a) Mixed Forestry
(b) Multiple use forestry
(c) Intensive forestry
(d) All the above

98. The practice of forestry for the simultaneous use of a forest area for two or more purpose called as
(a) Multiple use Forestry
(b) Mixed forestry
(c) Multipurpose trees and shrubs
(d) Multispecies tree garden

99. Shelter belt have (Pick the odd man out)
(a) Single tree species
(b) Multiple tree species
(c) Used to arrest wind
(d) Prevent sand dune formation

100. Jhuming is the other name for
(a) Shifting cultivation
(b) Taungya
(c) a and b
(d) None

101. Components of social forestry
(a) Farm forestry
(b) Extension Forestry
(c) Recreational forestry
(d) All the above

102. Hedge row intercropping is
(a) Agrisilvicultural systems
(b) Silvopastural systems
(c) Agrosilvopastural systems
(d) other systems

103. Example for MPTs
(a) Casuarina (b) Acacia
(c) Gliricidia (d) All the above

104. The ratio of height and width roughly in shelter -belt
(a) 1:10 (b) 1:15
(c) 1:20 (d) 1:25

105. The home garden is characterized by
(a) 2-3 Vertical strata
(b) 3-4 vertical strata
(c) 3-5 vertical strata
(d) 2-5 vertical strata

106. A well designed windbreak can offer protection up to --------------- times of its height towards leeward side
(a) 10 (b) 20
(c) 25 (d) 15

107. Essential attributes of agroforestry system are
(a) Productivity (b) Sustainability
(c) Adoptability (d) All the above

108. Based on the nature of components, Nair (1985) classified the agroforestry system into ---------- groups
(a) Four (b) Five
(c) Six (d) Sixteen

109. Home gardens found in tropics are one of the classic example of
(a) Agrisilvicultural systems
(b) Silvopastoral systems
(c) Agrosilvopastoral systems
(d) Apiculture with trees

110. Taungya system was originated in
(a) North Bengal
(b) Kerala
(c) Tamil Nadu
(d) Myanmar

111. Head quarters of International Center for Research in Agroforestry is located at
(a) Ibadan (b) Nairobi
(c) Canada (d) Denmark

112. An example for NFTs
(a) Teak (b) Leuceana
(c) Neem (d) All the above

113. The first state in India, created a separate social Forestry wing in the forest department was
(a) Kerala (b) Gujarat
(c) Tamil Nadu (d) Karnataka

114. In Southern India, the taungya system is called as
(a) Kumari (b) Jhum
(c) Podu (d) Dao or axe and fire

115. Taungya is a --------------------- word
(a) Gujarat (b) Assamese
(c) Nepali (d) Burmese

116. Social forestry was first recommended by
(a) National Commission on Agriculture
(b) Van mahotsav
(c) Nineth Commonwealth Forestry Congress
(d) All the above

117. Joint forest management evolved in early
(a) 1950 (b) 1960
(c) 1970 (d) 1980

118. The litteral meaning of agroforestry is
(a) Agri + Forest (b) Agri + Horti
(c) Agri + Fish (d) Agri + Honeybees

119. Ninth commonwealth congress was held in
(a) 1960 (b) 1968
(c) 1988 (d) 1952

120. Which one of the following states is having more arid areas?
(a) Gujarat
(b) Haryana
(c) Andhra Pradesh
(d) Rajasthan

121. The scientific name of pons cypress is
(a) *Taxodium distichum*
(b) *Taxodium scandens*
(c) *Salix alba*
(d) *Celtis australis*

122. Characteristics of multipurpose tree suitable for agroforestry is
(a) High adaptability
(b) Nitrogen fixing
(c) Coppicing ability
(d) All the above

123. Social forestry project phase II was implemented during 1988-89 to 1992-93 with a financial outlay of
(a) Rs. 591 million
(b) Rs. 584 million
(c) Rs. 854 million
(d) Rs. 951 million

124. The National Afforestation programme is being implemented through a …….. tier structure.
(a) 2 (b) 3
(c) 4 (d) 5

125. Institute of Tropical forestry is situated at
(a) Pune (b) Puerto Rica
(c) Jhansi (d) Kenya

126. Basic set of elements or components of Agroforestry are
(a) Tree (b) Herbs
(c) Animals (d) All the above

127. Social forestry aids were received from
(a) SIDA (b) CIDA
(c) USAID (d) All the above

128. "The Journal of Agroforestry Today" is published from
(a) Kenya (b) Denmark
(c) Japan (d) Canada

129. National Agroforestry Policy implemented in the year …….
(a) 2014 (b) 2012
(c) 2010 (d) 2011

130. An example for live fence is
(a) Euphorbia (b) Gliricidia
(c) Acacia (d) Sesbania

131. Nitrogen fixing ability of Casuarina is attributable to
(a) Azotobacter (b) Frankia
(c) Rhizobium (d) Azospirillum

132. The system in which woody component is combined with agricultural crop is known as
(a) Agri-Silvi (b) Agri horti
(c) Silvi pastoral (d) Apiculture

133. Shifting cultivation is most prevalent in
(a) North East India
(b) Eastern part of India
(c) Northern part of India
(d) All the above

134. Tropical Forest Research Institute is located in
(a) Jhansi (b) Jabalpur
(c) Agra (d) Jodhpur

135. In north eastern hill region, shifting cultivation is called as
(a) Jhum (b) Podu
(c) Kumari (d) Slash and burn

136. Most common agroforestry system practiced in arid regions are
(a) Agrisilvicultural systems
(b) Silvipastoral systems
(c) Agrosilvipastoral systels
(d) Aquaforestry

137. Which forestry practice is not included in the scope of social forestry
(a) Farm forestry
(b) Extension forestry
(c) Recreational forestry
(d) Protection forestry

138. Which social forestry programmes is purely managed by private efforts
(a) Farm Forestry
(b) Community forestry
(c) Rehabilitation of degraded forest area
(d) Urban forestry

139. Based on temporal arrangement of AF system, 'homegarden' is the example of
(a) Coincident (b) Concomitant
(c) Interpoleted (d) Relay cropping

140. Central Arid Zone Research Institute is located at
(a) Jabalpur (b) Jhansi
(c) Jodhpur (d) Dehradun

141. Protein bank is a sub system of
(a) Agri silviculture (b) Silvi pastoral
(c) Agri horti
(d) Agri silvi pastoral

142. The first Taungya plantation was established in
(a) Maharastra (b) Gujarat
(c) North Bengal (d) Rajasthan

143. ICFRE is located in
(a) New Delhi (b) Dehradun
(c) Jhansi (d) Ibadan

144. Green gold is
(a) Sal (b) Sandal
(c) Bamboo (d) Teak

145. Home garden is highly suitable for
(a) Humid/sub humid region
(b) Arid / Semi arid region
(c) Hilly Region
(d) High lands

146. Generally Shelter belt assumes the shape of
(a) Quadrangle (b) Rectangle
(c) Triangle (d) Tropizoidal

147. Energy plantation is established for
(a) Pulp wood production
(b) Fuel wood production
(c) Soil Conservation
(d) Wind break

148. TAP is funded by
(a) SIDA (b) CIDA
(c) DANIDA (d) JBIC

149. When was JFM concept initiated in India
(a) 1980 (b) 1970
(c) 1973 (d) 1972

150. Who initiated JFM concept?
(a) Sundar Lal (b) K.M. Munshi
(c) A.K. Banerjee (d) B.K. Nair

151. Who initiated Van Mahostava?
(a) Brandis
(b) K.M. Munshi
(c) A.K. Banerjee
(d) Westoby

152. When was TAP programme initiated in Tamil Nadu?
(a) 1981 – 1986 (b) 1987 – 88
(c) 1988 - 89 (d) 1997-2001

153. Social forestry is a forestry which aims at producing flow of protection and recreation benefits for the community. This definition was give by
(a) Westoby
(b) Brandis
(c) Sunderlal Baguguna
(d) SIDA

154. Social Forestry project Phase–I launched in
(a) 1981 (b) 1985
(c) 1988 (d) 1992

155. Social Forestry project Phase–II launched in
(a) 1981 (b) 1985
(c) 1988 (d) 1992

156. The average size of tree patta land could be
(a) 2.0 (b) 1.2
(c) 4.0 (d) 5.0

157. NRCAF is located in
(a) Agra (b) Jhansi
(c) Jodhupur (d) Jabalpur

158. Benefit sharing concept is a core objective in
(a) Joint Forest Management
(b) Sustainable forest management
(c) Community forestry
(d) Social Forestry

159. Joint forest management concept was first implemented at
(a) Jabalpur (b) Ayyalur
(c) Arabari (d) Jhansi

160. Wind breaks are
(a) Strips of trees (b) Line of trees
(c) a and b (d) None

161. Example of Allelo chemicals
(a) Gallic acid (b) Vanilic acid
(c) Syringic acid (d) All the above

162. The Diagnosis and Design (D & D) survey in agroforestry was given by
(a) P.K.R. Nair (b) Dwivedi
(c) Chundawat (d) Gautam

163. In North Indian Agroforestry the combination of ---------------- is common
(a) Rice with Poplar
(b) Rice with Eucalyptus
(c) Rice with Melia
(d) All the above

164. The combination wheat with eucalyptus was common in
(a) Punjab (b) Rajasthan
(c) Karnataka (d) Orissa

165. In Kerala the most commonly found agroforestry systems is
(a) Alley cropping
(b) Multispecies tree garden
(c) Protein bank
(d) Home garden

166. The silvipastoral system in Kangeyam region of Tamil Nadu consists of
(a) *Cenchrus ciliaris* & *Acacia nilotica*
(b) *Cenchrus ciliaris* & *Acacia leucophloea*
(c) *Eulaliopsis binata* & *Acacia nilotica*
(d) *Eulaliopsis binata* & *Acacia leucophloea*

167. Most commonly used tree species for shade in tea estates is
(a) *Grevillea robusta*
(b) *Eucalyptus* spp.
(c) *Melia dubia*
(d) All the above

168. In Joint Forest Management programme, the unit of management is
(a) Village
(b) Division
(c) Range
(d) Beat

169. In Joint Forest Management programme, the objectives are decided at
(a) Division
(b) Village level
(c) Range
(d) Beat

170. Waste land
(a) Land with <12% productivity
(b) Land with >12% productivity
(c) Land with <20% productivity
(d) Land with >20% productivity

171. ICRAF stands for
(a) International Council for Research in Agro Forestry
(b) Indian Council for Research in Agro Forestry
(c) Indian Centre for Rainfall Assessment in Forest
(d) International Centre for Rainfall Assessment in Forest

172. National watershed development programme for rural areas (NWDPRA) has been launched in the year
(a) 1997 (b) 1995
(c) 1972 (d) 1982

173. Loyal timber of the poor is
(a) Prosopis (b) Teak
(c) Acacia (d) Bamboo

174. Indian lilac is the common name for
(a) *Melia azadirach*
(b) *Tectona grandis*
(c) *Casuarina equisetifolia*
(d) *Acacia nilotica*

175. The queen of timbers is
(a) Rose wood (b) Vagai
(c) Teak (d) Eucalyptus

176. King of timber is
(a) Coconut (b) Teak
(c) Neem (d) Casuarina sps

177. In *Populus deltoides* based Agrisilvicultural systems, the suitable intercrop is
(a) Ginger (b) Paddy
(c) Wheat (d) Sugarcane

178. Calorific value (k cal/kg) of *Acacia catechu* is
(a) 3124 (b) 4400
(c) 4700 – 4900 (d) 5142 -5244

179. Calorific value (k cal/kg) of *Prosopis juliflora* is
(a) 3800 (b) 5000 – 5500
(c) 5200 – 5400 (d) 5400 - 5600

180. Which one of the following is a specialized agroforestry systems?
(a) Agrisilviculture
(b) Silvipasture
(c) Aquaforestry
(d) Hortisilviculture

181. ICAR had launched All India Co-ordinated Research Project on Agroforestry in India during
(a) 1974 (b) 1979
(c) 1981 (d) 1983

182. The characteristics of arid areas are
(a) Salinity (b) Saline water
(c) Alkalinity (d) All of the above

183. The tree species having xerophytic characteristics are suitable for
(a) Humid Bengal Assam Basin
(b) Arid Zones
(c) Sub Humid areas
(d) Humid to semi arid areas

184. Which one of the following species is preferred for water logged areas
(a) Coconut (b) Eucalyptus
(c) Neem (d) Casuarina spp.

185. Which one of the following crop has high salt tolerant capacity
(a) Wheat (b) Rice
(c) Sunhemp (d) Mustard

186. Important silvicultural operation in agroforestry is
(a) Pruning (b) Thinning
(c) Weeding (d) Climber cutting

187. Which is not a factor in which the agroforestry systems are classified?
(a) Floristic
(b) Function
(c) Agro climatic zones
(d) None

188. Classification of agroforestry based on arrangement of components in respect of time (Pick the odd out)
(a) Coincident (b) Mixed dense
(c) Concomitant (d) Interploted

189. Agroforestry practice of growing *Albizia lebbeck* with Sorghum is called

(a) Hydromorphic system

(b) Xexomorphic system

(c) Mesophytic system

(d) None

190. Swidden is a synonym for one of the following agroforestry system. Identify the same

(a) Taungya

(b) Home garden

(c) Parkland system

(d) Shifting cultivation

191. An agroforestry system suitable for small orchards prone to wind damage is

(a) Shelterbelt

(b) Swidden

(c) Windbreak

(d) Multistory cropping

192. A major thrust to "social forestry concept" was given by

(a) NCA report

(b) Forest Conservation Report

(c) National Forest Policy

(d) Indian Forest Act

193. How may agroclimatic zones are there in Tamil Nadu?

(a) 15 (b) 7

(c) 9 (d) 8

194. Silvipasture means

(a) Growing trees with grass

(b) Growing trees with annuals

(c) Growing trees with fruit crops

(d) Growing trees with medicinal plants

195. HDSR is applied for which plantation?

(a) Timber

(b) Pulpwood

(c) Fuel wood

(d) Agroforest

196. Which of this is not an advantage of agroforestry system?

(a) Maximum Production

(b) Improve soil fertility

(c) Maximize adverse effect of climatic factors

(d) Aid industrial growth

197. In order to get maximum net return from agroforestry we have to

(a) Choose suitable tree species

(b) Choose high yielding agricultural crops

(c) Plant clonally propagated tree seedlings

(d) Adopt suitable tree crop combination

198. Planting of forest trees in agroforestry systems (Pick the odd out)

(a) Reduces temperature

(b) Reduces evaporation

(c) Increase irrigation requirement

(d) Increase humidity

199. TAP project is implemented with the assistance of........

(a) Germany

(b) France

(c) Japan

(d) Sweedish

200. Shelterbelt have (Pick the odd out)

(a) Single species

(b) Multiple tree species

(c) Used to arrest wind

(d) Prevent sand dune formation

ANSWERSHEET
4. Agroforestry

1. (a)	2. (c)	3. (c)	4. (b)	5. (a)	6. (a)	7. (a)	8. (a)	9. (a)	10. (c)
11. (d)	12. (a)	13. (d)	14. (a)	15. (b)	16. (b)	17. (c)	18. (d)	19. (c)	20. (d)
21. (b)	22. (a)	23. (a)	24. (b)	25. (d)	26. (b)	27. (a)	28. (a)	29. (a)	30. (a)
31. (b)	32. (d)	33. (d)	34. (b)	35. (c)	36. (a)	37. (a)	38. (d)	39. (b)	40. (a)
41. (b)	42. (d)	43. (b)	44. (b)	45. (c)	46. (b)	47. (c)	48. (a)	49. (c)	50. (b)
51. (d)	52. (c)	53. (d)	54. (b)	55. (c)	56. (a)	57. (c)	58. (b)	59. (a)	60. (c)
61. (d)	62. (d)	63. (a)	64. (a)	65. (b)	66. (a)	67. (a)	68. (a)	69. (b)	70. (d)
71. (b)	72. (b)	73. (d)	74. (d)	75. (a)	76. (d)	77. (b)	78. (c)	79. (d)	80. (b)
81. (a)	82. (a)	83. (d)	84. (d)	85. (b)	86. (b)	87. (a)	88. (a)	89. (d)	90. (a)
91. (b)	92. (d)	93. (b)	94. (a)	95. (c)	96. (b)	97. (c)	98. (a)	99. (b)	100. (a)
101. (d)	102. (a)	103. (d)	104. (a)	105. (b)	106. (c)	107. (d)	108. (a)	109. (c)	110. (d)
111. (b)	112. (b)	113. (b)	114. (a)	115. (d)	116. (a)	117. (c)	118. (a)	119. (b)	120. (d)
121. (b)	122. (d)	123. (c)	124. (a)	125. (b)	126. (d)	127. (d)	128. (a)	129. (a)	130. (a)
131. (b)	132. (a)	133. (a)	134. (b)	135. (a)	136. (a)	137. (d)	138. (a)	139. (c)	140. (a)
141. (b)	142. (c)	143. (b)	144. (c)	145. (a)	146. (c)	147. (b)	148. (d)	149. (b)	150. (c)
151. (b)	152. (d)	153. (a)	154. (a)	155. (c)	156. (a)	157. (b)	158. (a)	159. (c)	160. (a)
161. (d)	162. (a)	163. (a)	164. (a)	165. (d)	166. (b)	167. (a)	168. (a)	169. (b)	170. (c)
171. (b)	172. (b)	173. (a)	174. (a)	175. (c)	176. (b)	177. (c)	178. (d)	179. (b)	180. (c)
181. (d)	182. (d)	183. (b)	184. (b)	185. (b)	186. (a)	187. (c)	188. (b)	189. (b)	190. (d)
191. (c)	192. (a)	193. (d)	194. (a)	195. (c)	196. (c)	197. (d)	198. (c)	199. (c)	200. (a)

5

Forest Mensuration

KEY POINTS

- Mensuration is derived from the Latin word Mensura
- Mana meaning measure of capacity or volume
- Tula mana meaning measure of weight
- Avamana – linear measure
- Kala mana – measure of time
- Forest mensuration is that branch of forestry which deals with the determination of dimensions, form, volume, age and increment of single trees, stands or whole woods, either standing or after felling
- Prior to the passage of "The standards of weights and measures Act" is 1956.
- The British or FPS system enacted during the year 1956
- Metric systems or CGS system which was introduced in 1962
- Important aim of forest mensuration is reasonable or relative accuracy
- Volume is the important measurement for forest management
- 640 acres = 1 square mile

Parameter	Unit of measurement	
	British Systems	Metric Systems
Height	Feet without decimal	Meter with first decimal
Diameter	Inches with first decimal	Centimeter with odd decimal
Girth	Feet and inches	Meter and Centimeter
Area	Square feet	Square meter or centimeter
Volume	Cubic feet	Cubic meter

- In CGS to FGS systems 1 km = 0.62137 mile
- In CGS to FGS systems 1 ha = 2.47105 acres
- Girth class- 5 cm, 15 cm, 30 cm
- Diameter class – 2cm, 5cm, 10 cm
- Height class – 1, 3, 5m
- LASER – Light Amplification by Stimulated Emission of Radiation

- The circumference of a circle = $2\pi r$ or πd
- Area of circle is = πr^2
- Based on two diameter area of circle is = $\pi((a+b)/2)^2$
- Based on girth the area of circle is = $g^2/4\pi$
- The breast height adopted by most of the European countries is 1.3 m
- The adopted breast height for measuring height and Diameter in India is 1.37 on above the ground level
- True area of the ellipse = πab (Basal area)
- Ration between diameter and girth of circle is $1/\pi$ or 0.3182
- Conversion factor of teak is (d/g) – 0.3160
- Conversion factor of sal is (d/g) – 0.3119
- Conversion of GOB into GUB is $g = g' - 2\pi t$
- Seth and Lohani have prepared bark thickness table and Bark percentage tables
- The maximum spread of the crown along its widest diameter is called crown width
- Height – generally considered as an index of fertility
- Total height – tip of leading shoot to the ground level
- Bole height - distance between ground level and crown point
- Commercial bole height – it is the height of bole that is usually fit for utilization as timber
- Height of standard timber bole – it is the height of the bole from the ground level up to the point where diameter over bark is 20cm
- Stump height – height of stump above ground
- Crown Length: It is the vertical measurement of the crown of a tree from the tip to the point half way between the lowest green branches forming green crown all round and the lowest green branch on the bole.
- Crown Height: It is the height of the crown as measured vertically from the ground level to the point half way between the lowest green branch and the green branches forming green crown all round.
- Horizontal projection on the ground of the tree crown called crown cover
- Hypsometers, altimeters and clinometers are used to measure height
- Instruments based on similar triangle principle is used in Christen's hypsometer, Smythesis hypsometer, calipers
- Instruments based on similar triangle principle is used in Brandis hypsometer, Abney's level, Haga altimeter, Topographic Abney's level and Relaskop

Instruments		Uses
Wooden Scale	:	Measuring diameter of stump and also used in stump and stem analysis
Callipers	:	Measuring diameter of standing trees and logs
Tape	:	Used to measure girth of trees and logs
Wheeler pentaprism	:	Diameter at any height measurement
Swedish bark gauge	:	Bark thickness

Hypsometers	:	Used for determining the height of standing tree
Altimeters	:	Determine heights of trees
Clinometers	:	It measures angle of slope. Any instrument which measures angles of slope can be used for height of tree by trigonometrical methods
Brandis hypsometer	:	Measuring the height of trees
Topographical Abney's level	:	Measuring the height of trees
Haga Altimeter	:	Measuring heights of trees
Spiegel Relaskop	:	Developed by Bitterly and used for measurement of basal area of crops
Spiegel Relaskop Tele relaskop, Simple angle gauge	:	Instruments are used in horizontal point sampling
Conimeter		Used in vertical sampling

- Relaskop instrument developed by Bitterlich
- Form is defined as the rate of taper of a log or stem
- Taper is the decrease in diameter of a stem of a tree or of a log from base upwards
- Metzger's theory or girder theory to explain variations in taper from tree to tree
- Form factor - the ratio of the volume of a tree or its parts to the volume of a cylinder having the same length and cross section as the tree
- Normal Form Factor (F) = V /Sh
- Artificial Form Factor is a breast height form factor
- Basal area measured at breast height and volume refers to the whole tree (above and below) the point of measurement is called Artificial form factor
- Artificial form factor is not reliable but most of the time used
- Basal area measured any convenient height and volume refers to part above the point is called Absolute form factor
- Form height is defined as the product of form factor and total height of tree
- Form quotient is defined as the ratio between the mid-diameter and the d.b.h.

  $$\text{Form quotient} = \frac{\text{Mid diameter}}{\text{d.b.h.}}$$

- Form point is the point in the crown as which wind pressure is estimated to be center
- In India and Britain, volumes of logs are often calculated by the quarter girth formula $= (g/4)^2 xl$
- The quarter girth formula gives only 78.5% of the true volume
- $\text{Density} = \dfrac{\text{Wood Density}}{\text{Water Density}}$
- $\text{Specific gravity (sp. Gr.) of wood} = \dfrac{\text{Weight of wood}}{\text{Weight of the same volume of water}}$

- $\text{Volume} = \frac{\text{Weight (in gms)}}{\text{Specific gravity}}$
- In F.P.S. System, $\text{Volume} = \frac{\text{Weight (in lbs)}}{\text{Specific gravity} \times 62.5}$ cft
- Volume table is a table showing for a given species the average contents of trees, logs or sawn timber for one or more given dimensions.
- Volume table showing the volume of a given species for single or multiple dimensions.
- The volume table dimensions may be only dbh, dbh and height or dbh, height and some measure of form or taper of the tree.
- According to the number of variables the volume table classified as three viz., Single Variable (dbh); Two variables (dbh and height) and Three variables (dbh, height and tree form)
- According to the scope of their application the volume table classified as three
  - General Volume Table - usually based on two variables, dbh and total tree height
  - Regional Volume Table
  - Local Volume Table - Generally based on one independent variable, i.e. dbh
- According to the outturn given by them volume table classified as five viz., Standard Volume Table, Commercial Volume Table, Sawn Outturn Tables, Assortment Tables, Sawn Outturn Assortment Table
- The volume tables are prepared by the following methods viz., Graphical method, Regression equations method, alignment chart method
- The tariff table gives volume based on breast height girth alone and its applicable to a particular stand is determined from a limited number of sample trees
- The portion of the tree stem or log which is unmatchable is called cull
- Biomass refers to the weight of the above ground vegetative matter produced per unit area
- Regression equation for single stemmed Weight = $a + b\,D^2\,H$
- Regression equation for multi stemmed Weight $W = a + b\,NS\,D^2\,H$
- General volume table regression equation is $V = a+bD^2H$
- $\text{Age} = \frac{1}{p1s}$ where s is a constant equal to $\frac{\log p_1 - \log p_2}{\log d_2 - \log d_1}$
- Crop diameter - diameter corresponding to the mean basal area of a uniform, generally pure crop
- Mean diameter - diameter corresponding to the mean basal area of a group of trees or a stand
- $\text{Mean basal area} = \frac{n_1 s_1 + n_2 s_2 + n_3 s_3}{n_1 + n_2 + n_3 + \text{-----}}$

  Where n_1, n_2, n_3 .... are the number of trees in each diameter class and s_1, s_2, s_3 ... are the basal areas of mean trees of different diameters classes
- $\text{Mean diameter } D = \frac{p_1 d_1 + n_2 d_2 + n_3 d_3 +}{n_1 + n_2 + n_3 + \text{--------}}$

where n_1, n_2, n_3 .... are the number of trees in each diameter class and d_1, d_2, d_3 ....are the average diameters of these diameter classes

- Crop height = $-\dfrac{s_1h_1 + s_2h_2 + s_3h_3 + \text{-----}}{s_1 + s_2 + s_3 + \text{-----}}$ (Lorey's formula)

  where s1, s2, s3, are the total basal areas of each of the diameter classes and h1, h2, h3, are the mean heights of trees of these diameter classes

- Even aged Crop age = $\dfrac{s_1a_1 + s_2a_2 + s_3a_3 + \text{------}}{s_1 + s_2 + s_3 + \text{------}}$

  where s_1, s_2, s_3, are the basal areas of the even-aged groups and a_1, a_2, a_3 are the ages of the even-aged groups

- Uneven aged mean crop age =- $\dfrac{V_1 + V_2 + V_3 + \text{-----}}{V_1/a_1 + V_2/a_2 + V_3/a_3 + \text{-----}}$ (Smalian & Heyer's formula)

  Where V_1, V_2, V_3, are the total volumes of even-aged groups and a_1, a_2, a_3 are the average ages of the groups

- Uneven aged mean crop age = $\dfrac{s_1a_1 + s_2a_2 + s_3a_3 + \text{-----}}{S}$

  Where, s_1, s_2, s_3 are the total basal areas of even-aged groups and a_1, a_2, a_3 are the average ages of these groups; S is the total basal area of the whole forest

- Uneven aged mean crop age = $\dfrac{n_1a_1 + n_2a_2 + n_3a_3 + \text{------}}{n_1 + n_2 + n_3 + \text{------}}$

  Where, n_1, n_2, n_3 are the number of trees in each age groups and a_1, a_2, a_3 are the average ages of even aged groups

- Uneven aged mean crop age = $\dfrac{m_1a_1 + m_2a_2 + m_3a_3 + \text{------}}{m_1 + m_2 + m_3 + \text{------}}$

  Where, m_1, m_2, m_3----- are the areas occupied by different age classes and a_1, a_2, a_3------ are the average ages of even aged groups

- Uneven aged mean crop age = $\dfrac{a_1 + a_2 + a_3 + \text{-----}}{n}$

  Where, a_1, a_2, a_3------- are the ages of sample trees representing different age classes
- n is the total number of sample trees
- Determination of volume of pulp wood (tones) = Tree volume in m^3/ 1.4158
- The growth of trees with annual rings can be determined by Stump Analysis, Stem Analysis and Increment Boring methods
- Stump analysis is the analysis of a stump cross-section by measuring annual rings in order to estimate the age of the tree and its past rate of diameter and basal area growth
- Forest Inventory is the tabulated, reliable and satisfactory tree information, related to the required unit, respectively units, of assessment in hierarchic order
- Enumeration of the desired species above the specified diameter limit is carried out over the entire area of the forest unit is called total enumeration

- Enumeration is to be done only in a representative portion of the whole forest is called partial or sample enumeration
- Two kinds of sampling are used in forest inventories viz., Random and Non random sampling method
- Random sampling methods are classified in to six. Simple random sampling, Stratified random sampling, Multi-stage sampling, Multiphase sampling, Sampling with varying probability and List sampling
- Non random sampling methods are classified in to three viz., selective sampling, systematic sampling and sequential sampling
- Sampling intensity of tropical wet evergreen forest is – 10%, Tropical moist deciduous forest is - 2½% and Sub tropical pine forest is – 5%
- Point sampling can be either, horizontal or vertical depending on whether diameter (or basal area) or height has to be estimated
- Horizontal sampling has been widely used and has revolutionized forest inventory practices
- Diopter is a measure of the value of an angle expressed as its sine. One diopter is approximately equivalent to an angle of 0.57° or 34.36 minutes
- Yield table is not applicable to uneven-aged forest
- Maximum crown area expressed as a percentage of one hectare is called maximum crown area
- Volume yield table is an yield table which expresses outturn in terms of volumes
- Money yield table is an yield table constructed from volume yield table in which outturn is expressed in terms of money instead of volume
- Basal area per ha = Number of trees /ha X Average basal area
- Crop volume per ha = Basal area /ha X Crop height X Form Factor
- Stand table is a table showing the distribution of stems by diameter classes for each of the series of crop diameters.
- Remote sensing (RS), also as earth observation
- Remote Sensing is the science of acquiring, processing and interpreting images that record the interaction between electromagnetic energy and matter
- The Global Positioning System (GPS) is a satellite based navigation system that can be used to locate positions anywhere on earth.
- A Geographic Information System (GIS) is a computer system for capturing, storing, querying, analyzing, and displaying geographic data

MULTIPLE CHOICE QUESTIONS

1. Mensuration is derived from--------------
   (a) Russia (b) Latin
   (c) Greek (d) a & b
2. One hectare--------------------.
   (a) 2.47 acres (b) 2.54 acres
   (c) 2.49 acres. (d) 2.59 acres
3. In metric system, girth of a tree is always taken in
   (a) Metres & to the nearest centimeter.
   (b) Foot and to the nearest inches.
   (c) Centimeters to the nearest millimeter.
   (d) Only in foot.

4. The FPS system was enacted during the year

(a) 1956 (b) 1966
(c) 1962 (d) 1967

5. The CGS or metric system enacted during the year

(a) 1956 (b) 1966
(c) 1962 (d) 1967

6. The Diameter class (cm) followed in India

(a) 1,2,5 (b) 2,5,10
(c) 1,3,5 (d) 3,5,10

7. The girth (cm) class followed in India

(a) 1,2,5 (b) 2,5,10
(c) 5,15,30 (d) 3,5,10

8. The height class (m) followed in India are

(a) 1, 3, 5 (b) 1, 2, 5
(c) 1, 3, 10 (d) 1, 2, 10

9. Height class followed for tree attaining a height of 25 m

(a) 5 (b) 10
(c) 2 (d) 1

10. Which of the following is not correctly matched

(a) Mana – Measure of volume
(b) Tulamana – Measure of weight
(c) Avamana – Measure of diameter
(d) Kalapramana – Measure of time

11. The breast height measurement of 1.3 adopted in

(a) Burma (b) America
(c) Europe (d) India

12. Basal Area of a tree refers to

(a) Cross sectional area at stem height
(b) Cross sectional area at base
(c) Cross sectional area at top height
(d) Cross sectional area at breast height.

13. In metric system, diameter of a tree is always taken in

(a) Meter and to the nearest centimeters
(b) Centimeters and to the nearest 1/5 of a centimeter
(c) Centimeter and to the nearest millimeter
(d) In inches to the nearest tenth of an inch

14. The maximum spread of the crown along its widest diameter is

(a) Crown length
(b) Crown height
(c) Crown width
(d) Crown diameter

15. Relationship between the diameter and girth of trees is

(a) $1/\pi$ (b) 0.3182
(c) 7/22 (d) All the above

16. The formula for calculating volume of log

(a) g / 4 X l (b) $(g^2/4)$ x l
(c) $(g/4)^2$ xl (d) None

17. The instrument designed based on the principal of similar triangle

(a) Christen Hypsometer
(b) Brandis Hypsometer
(c) Altimeter
(d) Abney level

18. Instrument used for measuring the height of the tree based on trigonometrical principles.

(a) Christen's hypsometer
(b) Smythies hypsometer
(c) Improvised callipers
(d) Brandis hypsometer.

19. Abneys level is used for__

(a) Measurements of height
(b) Measurement of volume
(c) Measurements of height and contour survey
(d) Past rate of growth of a tree

20. ------------ insturement used for measure the basal area of the crop
(a) Reloskop
(b) Abneys level
(c) Brandis hypsometer.
(d) Improvised calliper

21. Pick the Newton's formula
(a) $(S_1 + 4\ Sm + S_2 / 6) \times l$
(b) Sm x l
(c) S_l
(d) $S_1 + S_2 / 2 \times l$.

22. Factors affecting the overall weight of a quantity of wood are.
(a) Density and moisture content
(b) Bank and foreign material
(c) Moisture content, Bank and foreign material.
(d) All of these

23. Pressler's increment borer is used to determine
(a) Increment of a tree.
(b) Height of a tree
(c) Volume of a tree
(d) Age of a tree

24. Metzer's theory is related to
(a) Stem form (b) Form height
(c) Sampling (d) None of these

25. In slopping ground, diameter of tree is measured on
(a) Down hill side (b) Uphill side
(c) Leaning side (d) None of these

26. Stump analysis is done to determine
(a) Age and diameter relation
(b) Age and Height relation
(c) Age and Volume relation
(d) All of these

27. The method of sampling adopted in Bamboo enumeration is
(a) Multistage sampling
(b) Multiphase sampling
(c) Random sampling
(d) List sampling.

28. The breast height adopted by most of the European countries is
(a) 1.3 m (b) 1.37m
(c) 1.27m (d) 3.0m

29. The normal size of the sample plot is
(a) 0.1 to 0.2 ha. (b) 0.1 to 0.5 ha.
(c) 0.1 to 1.0 ha (d) 0.1 to 2.0 ha.

30. The Instrument, which is used to measure angle of slope, is
(a) Altimeter (b) Clinometer
(c) Hypsometer (d) Abney's level

31. Artificial Form Factor is also known as
(a) Absolute form factor
(b) Breast height form factor
(c) Normal form factor
(d) Tree form factor

32. True Form factor is also known as
(a) Absolute form factor
(b) Breast height form factor
(c) Normal form factor
(d) Tree form factor

33. Crop diameter refers to
(a) Mean diameter of tree in a stand
(b) Mean diameter of Dominant trees
(c) Mean diameter of suppressed trees
(d) Diameter corresponding to the mean basal area

34. The Standards of Weights and Measures Act enacted during the year
(a) 1956 (b) 1966
(c) 1957 (d) 1967

35. The product of form factor and total height of tree
(a) Form -quotient
(b) Form- Class
(c) Tree- Height (d) Form- Height

36. Crop diameter is
(a) Mean diameter of tree in a stand
(b) Diameter corresponding to the mean basal area of uniform crop
(c) Mean diameter of dominant trees
(d) None of these

37. Basal Area of tree implies
(a) Cross sectional area at stump height
(b) Cross sectional area at breast height
(c) Cross sectional area at base.
(d) None of these.

38. Age of trees with annual rings is determined by
(a) Swedish bark gauge
(b) Wheeler penta prism
(c) Pressler's increment borer
(d) Blume-leiss hypsometer

39. Average annual increment for any short period is called as
(a) CAI (b) PAI
(c) MAI (d) IP

40. The relationship between CAI/MAI (pick the odd out)
(a) CAI is greater than MAI, the MAI is raising
(b) To begin MAI keeps below CAI
(c) CAI attains its maximum before MAI
(d) MAI is fully under CAI

41. Which instrument is used to determine the age of a tree
(a) Abney's level (b) Pressler's borer
(c) Relascope (d) Altimeter

42. Effective variables for volume table preparation
(a) Single variable
(b) Double variable
(c) Trible variable
(d) Multi variable

43. The volume tables are combined from DBH alone is called
(a) General volume tables
(b) Regional volume tables
(c) Local volume tables
(d) Standard volume tables.

44. Crop height is determined by
(a) Andre's formula
(b) Heyer's formula
(c) Gumpel's formula
(d) Lorey's formula

45. Proportionate distribution of trees in selection forests is attributable to
(a) Mayers law
(b) Faustman law
(c) De Liocourts law
(d) None of these

46. Optimal number of sample plots for each thinning regime in yield table preparation is
(a) 500 (b) 600
(c) 700 (d) 800

47. The formula for converting d.o.b. into d.u.b. is
(a) $g = g' - 2\pi t$ (b) $g' = g - 2\pi h$
(c) $g = d' - 2\pi t$ (d) $g = g' - 2\pi h$

48. Smalian's formula for calculating volume of log is
(a) Sl (b) Sm x l
(c) $\frac{S_1 + S_2}{2} \times 1$ (d) $\frac{S_1 + S_2 + \sqrt{S_1 S_2}}{3} \times 1$

49. Calculate the basal area per ha if the number of tallied trees was 14 and the BAF of the wedge prism used is three
(a) 42 m^2 (b) 14 m^2
(c) 21 m^2 (d) None

50. Normal form quotient is defined as
(a) Mid diameter / d.b.h.
(b) $\frac{\text{Diameter at one half its height above breast height}}{\text{d.b.h.}}$
(c) $\frac{\text{Diameter at any point}}{\text{d.b.h.}}$
(d) Mid diameter x d.b.h.

51. Who coined the term form-Quotient?
(a) Schiffel (b) Tor-Jonson
(c) Schneider (d) Pressler.

52. In Normal Form factor, basal area is measured at
(a) Breast height
(b) Any Convenient height
(c) A constant proportion of the total height
(d) None of these.

53. Who proposed the theory to explain variations in taper from tree to tree.
(a) Metzger (b) Huber
(c) Newton (d) Smalian.

54. The instrument 'wedge prism' is mainly used in
(a) Plot sampling
(b) Horizontal point sampling
(c) Vertical point sampling
(d) All of the above

55. One diopter is approximately equivalent to an angle of
(a) 0.57° (b) 0.47°
(c) 0.75° (d) 0.74°

56. Linear growth model is preferred only for.
(a) Nursery stage (b) Sapling stage
(c) Pole stage (d) Tree stage

57. The standard length of a log expect the last section is
(a) 5 m (b) 4 m
(c) 3 m (d) 6 m

58. If the canopy density is between 0.75 and 1.0, then the forest is referred as
(a) Closed (b) Dense
(c) Thin (d) Open

59. Yield tables in which data are given for different grades of thinning
(a) Volume yield table
(b) Money yield table
(c) Single yield table
(d) Multiple yield tables

60. Yield table is generally prepared for
(a) Regular forest
(b) Irregular forest
(c) Reserved forest
(d) Mixed forest

61. The quarter girth formula to calculate volume of log is
(a) $g^2l / 4$ (b) $4/g^2 \times l$
(c) $g^2l / 16$ (d) $16/g^2 \times l$

62. What are the variables require enough for general volume table preparation?
(a) Diameter only
(b) Height only
(c) Diameter and height
(d) Diameter, height and form

63. Sampling units are taken at two or more stages and sampling units gets smaller at each successive stage of sampling is called as
(a) Unrestricted sampling
(b) Stratified sampling
(c) Multistage sampling
(d) Multiphase sampling

64. The height corresponding to the mean diameter of 250 biggest diameters per hectare as read from Diameter-height curve is
(a) Top height
(b) Crop height
(c) Mean height
(d) Mean basal area

65. The age-diameter, age-height, age-volume relations throughout the life of the trees are analyzed by
(a) Stump analysis
(b) Stem analysis
(c) Pressler's increment borer
(d) Mathematical formula

66. The instrument, which is used to measure diameter at any convenient height is
(a) Tree calliper
(b) Tailor's tape
(c) Wheeler's Penta Prism
(d) Relaskop

67. Volume of stem from ground level down to 20 cm diameter over bark is
(a) Commercial volume
(b) Standard stem small wood
(c) Standard stem timber
(d) Sawn out turn

68. In point sampling, P.P.S. is referred to as
(a) Poly areal plot sampling
(b) Probability point sampling
(c) Probability plot sampling
(d) Probability proportional to size

69. Normally the figures of fractional quality running between the quality classes
(a) 0.0 to 1.0 (b) 0.0 to 10
(c) 0.0 to 0.2 (d) 0.0 to 2.0

70. Which one is not considered as yield table
(a) Volume yield table
(b) Volume table
(c) Money yield table
(d) Stand table

71. It is a measure of the relative productive capacity of a site for a particular species is called as
(a) Site quality
(b) Site index
(c) Canopy density
(d) Crown density

72. An instrument, which is particularly used for vertical point sampling is
(a) Wedge prism (b) Spiegal relaskop
(c) Telerelskop (d) Conimeter

73. Generally temporary sample plots are used for
(a) Enumeration survey
(b) Preparation of yield table
(c) Measurement of crop volume and increment
(d) Comparison of growth in different qualities of locality

74. Schneider's formula is used for determining
(a) Height of the tree
(b) Growth of the tree
(c) Volume of the tree
(d) Increment percentage

75. Which sampling method is random sampling method
(a) Selective sampling
(b) Systematic sampling
(c) Stratified sampling
(d) Sequential sampling

76. When the tree is forked above the breast height, it is counted as,
(a) One tree
(b) Each fork treated as separate tree
(c) Tree counted as one or two depend upon place measurement
(d) Cannot be determined

77. Site index is the relation between
(a) Age and site
(b) Age and diameter
(c) Age and height
(d) Age and volume

78. The DBH measured at United Kingdom is at
(a) 1.37m (b) 1.30m
(c) 1.39m (d) 1.35m

79. On sloping ground the DBH is measured
(a) Down hill side
(b) Uphill side
(c) Both up and down hill side
(d) None of the above

80. The figures obtained by Remote sensing visualizes facts by
(a) Orographic view
(b) Orthographic view
(c) Perspective view
(d)Perpedicular projection

81. In CGS systems to FGS system one kilometer is equal to ------------------ mile
(a) 0.6213 (b) 1.6093
(c) 0.4040 (d) 0.3861

82. --------------- stage sampling is commonly used in forest surveys
(a) One (b) Two
(c) Three (d) Four

83. Solid volume of fire wood is measured by
(a) Smalians formula
(b) Hubers formula
(c) Hunden hagen formula
(d) None of the above

84. Bole surface area calculated with the following parameters
(a) Girth and length of log
(b) Length and Breadth of log
(c) Girth & length with form substitutes
(d) a and b

85. Horizontal projection on the ground of the tree crown is called
(a) Crown cover (b) Crown width
(c) Crown length (d) Tree width

86. Vertical point sampling is attributable to
(a) Bitterlich (b) Hirata
(c) Avery (d) Chaturvedi

87. Who introduced the horizontal point sampling technique?
(a) Hirata (b) Stroced
(c) Bitterlich (d) Smalin.

88. The instrument 'wedge prism' mainly used in
(a) Plot sampling
(b) Horizontal point sampling
(c) Vertical point sampling
(d) All of the above

89. If the canopy density is between 0. 50 and 0.75, then the forest is referred as
(a) Closed (b) Dense
(c) Thin (d) Open

90. The branch of forestry which deals with the determination of dimension, form and age and increment of single trees, stands or whole woods either standing or after felling is called
(a) Dendrology
(b) Forest Management
(c) Dendrometry
(d) Forest Protection

91. What are the variables require enough for local volume table preparation?
(a) Diameter only
(b) Height only
(c) Diameter and height
(d) Diameter, height and form

92. Sampling units are used at the different phase to collect different information or same information by different methods
(a) Unrestricted sampling
(b) Stratified sampling
(c) Multistage sampling
(d) Multiphase sampling

93. If the focal length of the camera is 20 cm and the flying height above the ground level is 5000 m, calculate the photo scale
(a) 1 : 10000 (b) 1 : 15000
(c) 1 : 20000 (d) 1 : 25000

94. The following are some clinometers used in levelling. Pick the odd man out.
(a) Ghat tracer
(b) Topograpic Abney level
(c) De Lisle clinometer
(d) Tangent clinometer

95. Hyposometric levelling is a method to find the
(a) Height of mountains
(b) Depth of water bodies
(c) Altitudes
(d) Height of trees

96. Importance of Value Index (IVI) is used for
(a) Estimation of biomass
(b) Mapping of forests
(c) Phyto-sociology of forests
(d) Survey of vegetation

97. Definition of 'Productivity' is
(a) Biomass produced in one ha
(b) Biomass produced in one acre
(c) Biomass produced in one ha per year
(d) Biomass produced after one rotation

98. Volume of stem wood in round between 20 cm and 5 cm diameter over bark is called

(a) Commercial volume

(b) Standard stem small wood

(c) Standard stem timber

(d) Sawn out turn

99. Volume of stem from ground level down to thin end of diameter over bark is

(a) Commercial volume

(b) Standard stem small wood

(c) Standard stem timber

(d) Sawn out turn

100. The Latin name for Andaman Padauk tree is

(a) *Veteria indica*

(b) *Pterocarpus dalbergioides*

(c) *Shorea robusta*

(d) *Dalbergia latifolia*

101. The butt end of a tree resembles a frustum of

(a) Trapezium (b) Paraboloid

(c) Cone (d) Neiloid

102. Top portion of tree resembles a frustum of

(a) Neiloid (b) Paraboloid

(c) Cone (d) Trapezium

103. Middle portion of tree resembles a frustum of

(a) Neiloid (b) Paraboloid

(c) Cone (d) Trapezium

104. Measure of the value of the angle expressed as its Sine is termed as

(a) Diopter

(b) BAF

(c) Calibration distance factor

(d) Plot radius factor

105. The sample plot with minimum perimeter for a given area is

(a) Square (b) Rectangular

(c) Triangular (d) Circular

106. Difficulty in applying yield tables to natural forests is

(a) Lack of a representative age

(b) Uniform site quality

(c) Over stocking

(d) Over matured crop

107. Spectral signature of green vegetation in a satellite captured image is

(a) White (b) Green

(c) Blue (d) Red

108. Vertical point sampling is used for estimating

(a) Basal area

(b) No of stems / Ha

(c) Mean stand height

(d) All of the above

109. Panchromatic camera of IKNOS statellite has the highest ground resolution of

(a) 1 m (b) 1 .5 m

(c) 5.8 m (d) 8.5 m

110. Stump analysis is used to assess the relation of

(a) Age and growth

(b) Diameter and volume

(c) Height and diameter

(d) Height, diameter and volume

111. Tree caliper will give

(a) Accurate measurement

(b) Consistence measurement

(c) Absolute measurement

(d) None of these

112. Christen's hypsometer is used to measure the tree

(a) Volume (b) Taper

(c) Height (d) None of these

113. Which of the following is correctly matched

(a) Clinometer – density

(b) Altimeter – atmospheric pressure

(c) Hypsometer – height

(d) Relaskop – light intensity

114. Indian Institute of Remote Sensing is located at

(a) Dehra Dun (b) Bangalore

(c) New Delhi (d) Chandigarh

115. The ideal time for taking Aerial photograph is

(a) 9 to 11 A.M (b) 11 to 12 P.M

(c) 2 to 3 P.M (d) 3 to 4 P.M

116. Metzger's Theory is attributable to

(a) Tree stem form

(b) Tree volume

(c) Tree height

(d) Species density

117. Horizontal Point Sampling is used for determination of

(a) Basal area for hectare

(b) Number of stems per hectare

(c) Volume per hectare

(d) All of the above

118. Stem analysis is done to determine

(a) Age of a tree

(b) Height of a tree

(c) Volume of a tree

(d) Past rate of growth of a tree

119. The rotation at which MAI is maximum

(a) Rotation of height income

(b) Financial rotation

(c) Rotation of maximum volume production.

(d) All of these

120. In remote sensing, the scale for small scale photography ranges between

(a) 1 : 40000 and 1 : 70000

(b) 1 : 20000 and 1 : 40000

(c) 1 : 5000 and 1 : 20000

(d) None of these

121. The Forest Survey of India (FSI) conducts Forest Cover Survey once in every

(a) 2 years (b) 5 years

(c) 10 years (d) 4 years

122. In remote sensing, medium scale photograph has scale of

(a) 1:40000 to 1:70000

(b) 1:20000 to 1:40000

(c) 1:5000 to 1:20000

(d) 1:50000 to 1:100000

123. The productive capacity of Indian Forest is

(a) 2.1 m^3/ha (b) 0.7 m^3/ha

(c) 1 m^3/ha (d) 0.5 m^3/ha

124. The instrument not used in horizontal point sampling is

(a) Simple angle gauge

(b) Dumpy level

(c) Spiegel releskope

(d) Wedge prism

125. Form factor is worked out using the formula

(a) F = V / Sh (b) F = Sh / V

(c) F = V / S (d) F = V / h

126. The diameter corresponding to 250 biggest diameters per hectare is called

(a) Crop diameter

(b) Mean diameter

(c) Top diameter

(d) Average diameter

127. The term normal form quotient is attributable to

(a) Tor Jonson (b) Schiffel

(c) Hojer (d) Behre

128. Instrument used measuring the height of the tree based on trigonometical principles

(a) Christen's hypsometer

(b) Smythies hypsometer

(c) Improvised calipers

(d) Brandis hypsometer

129. Wooden scale is used to measure

(a) Diameter of standing tree

(b) Diameter of felled tree

(c) Girth of standing tree

(d) Girth of felled tree

130. Which instrument is used to measure diameter at any height of a tree?
(a) Abney's level
(b) Christen's hypsometer
(c) Tree calliper
(d) Wheeler's pentaprism

131. Solid volume of the stack can be determined by
(a) Timber calculators
(b) Xylometric method
(c) Specific gravity method
(d) none of these

132. Local volume table is applicable to
(a) Working circle
(b) Block
(c) Compartment
(d) Range.

133. Girth is equal to
(a) πd2 (b) πd
(c) 2π (d) πd/4

134. The error involved in using caliper is
(a) Positive (b) Negative
(c) Both a and b (d) None of these

135. The formula for basal area
(a) ah/2 (b) $\pi d^2/_2$
(c) $\pi\, d^2/_4$ (d) πab

136. Horizontal point sampling is used for determination of
(a) Volume / ha
(b) Basal area / ha
(c) Height
(d) None of these

137. Random sampling (tick the odd man)
(a) Simple
(b) Stratified
(c) Multistage
(d) Systematic

138. Non random sampling (Pick the odd man out)
(a) Systematic (b) Selective
(c) Sequential (d) List

139. A table showing for a given species average weight of trees or logs for one or more dimensions like girth, height or DBH is termed as
(a) Stand table (b) Money table
(c) Biomass table (d) Yield table

140. The limit of standard small wood
(a) Down to 20 cm diameter over bark
(b) Down to 20 inch diameter under bark
(c) Down to 5 cm diameter under bark
(d) Down to 5 cm diameter over bark

141. A table showing the distribution of stems by diameter classes is referred as
(a) Stand table (b) Local volume table
(c) Yield table (d) None of these

142. The abney level is used to measure ___.___
(a) Tree height (b) Dbh
(c) Board feet (d) All of the above

143. Calipers are used to measure _______.
(a) Tree height (b) Diameter
(c) Board feet (d) Land area

144. A tool used to help to read aerial photographs is the _______.
(a) Haga altimeter (b) clinometer
(c) stereoscope (d) Abney level

145. Which tool is used for timber cruising?
(a) Wedge prism (b) Increment borer
(c) Cant hook (d) Chain saw

146. The haga altimeter will measure _____.
(a) Tree height (b) Dbh
(c) Basal area (d) Board feet

147. A log of marketable size which is made unmarketable because of rot, defect or poor form is called a _______.
(a) Slash (b) Cull
(c) Scratch (d) Defects timber

148. A log rule is a table or formula used to determine tree:
(a) Age (b) Diameter
(c) Height (d) Volume

149. A bark gauge measures _____
(a) Diameter of the tree trunk
(b) Ratio of bark to wood
(c) The decibels generated when a felled hits the ground
(d) Tree bark thickness

150. A wedge prism is used in a type of inventory called:
(a) Cluster sampling
(b) Double sampling
(c) Plot sampling (d) Point sampling

151. On a topographic map which of the following is a dark, heavy line with an elevation figure?
(a) Contour index (b) Contour interval
(c) Contour line
(d) Horizontal control station

152. The Biltmore stick is used to measure _____
(a) Tree height (b) Diameter
(c) Board feet (d) Land area

153. The area of the end of a log is called the _____ area.
(a) Basal (b) End
(c) Stump (d) Butt

154. To perform a basal area sample, a forester would use which of the following tools:
(a) Diameter tape (b) wedge Prism
(c) Relkascope (d) All of the above

155. The diameter of trees is usually measured at:
(a) Breast height
(b) 4 and 1/2 feet above the ground
(c) 1.37m
(d) All of the above

156. Which of the following is/are an advantage of prisms?
(a) Fast
(b) Simple to use properly
(c) Reasonably accurate
(d) All of the above

157. DBH can be measured by all of the following tools except _____.
(a) Diameter tape
(b) Biltmore stick
(c) Merit hypsometer
(d) Calipers

158. Which of the following will NOT measure the height of a tree?
(a) Merit hypsometer
(b) Abney level
(c) Clinometer (d) Log scale stick

159. Diameter measurements for sawlogs are taken at:
(a) The large end of the log
(b) The log center
(c) The small end of the log
(d) The stump height mark

160. When using the prism, a tree to be counted should be:
(a) All trees in the plot regardless of offset
(b) Offset completely
(c) Partially offset
(d) All of the above

161. The general term used to describe instruments used to measure the diameter of a tree is:
(a) Altimeter (b) Clinometer
(c) Dendrometer (d) Hypsometer

162. A hand-held instrument used for measuring tree heights, slopes and vertical angles is _____.
(a) Abney level (b) Clinometer
(c) Stereoscope
(d) Merritt hypsometer

163. A wedge prism is used to measure the _____ of trees in a stand.
(a) Basal area
(b) Board foot volume
(c) Diameter
(d) Cubic foot volume

164. When performing a forest inventory with a wedge prism you are selecting trees by:
(a) Species
(b) Their dbh
(c) Their distance form you
(d) Using a table of random numbers to identify the trees

165. An increment borer is NOT used to do following?
(a) Estimate the age of a tree
(b) Determine the growth rate of a tree
(c) Eliminate knots in lumber
(d) Count the annual rings

166. Tree diameter growth is related to______.
(a) Age of the tree
(b) Availability of moisture & nutrients
(c) Stand density
(d) All of the above

167. _________ is used to estimate basal area.
(a) Clinometer (b) Hypsometer
(c) Planimeter (d) Wedge prism

168. A Jacob staff is used with a _____ to increase accuracy of tree measurement readings.
(a) Diameter tape (b) Planimeter
(c) Relaskop (d) Stereoscope

169. How many square feet are there in an acre?
(a) 40,000 (b) 43,522
(c) 43,560 (d) 10,000

170. Which of the following is the most limiting factor in the growth and development of an individual tree?
(a) Nutrients (b) Sunlight
(c) Rainfall (d) Temperature

171. Foresters recognize which of the following as the basic management unit in a forest?
(a) Stand (b) Lot
(c) Section (d) None of the above

172. DBH is:
(a) Diameter at breast height
(b) 4.5 feet above the ground
(c) A standard accepted in forestry worldwide
(d) All of the above

173. Crop volume per ha is calculated using ______ formula
(a) No. of trees / ha x average basal area
(b) Girth / ha x crop height
(c) Basal area / ha x crop height x farm factor
(d) Basal diameter /ha x crop height

174. Which of the following tools would NOT be used to determine tree height?
(a) Clinometer (b) Abney level
(c) Hypsometer (d) None of these

175. Which is the closest to a hectare?
(a) 1 acre (b) 2.5 acres
(c) 5.2 acres (d) 7 acres

176. The radius of a one-fifth acre plot is:
(a) 52.7 feet (b) 37.2 feet
(c) 27.5 feet (d) 73.2 feet

177. Of the tools for measuring DBH, which is the most precise?
(a) Caliper
(b) Biltmore stick
(c) Diameter tape
(d) Carpenter's rule

178. The scaling diameter for saw logs is normally measured at:
(a) The big end inside bark
(b) The big end outside bark
(c) The small end inside bark
(d) The small end outside bark

179. The term stumpage refers to:
(a) Stumps left after harvesting
(b) The diameter of trees measured 1 foot off the ground
(c) The value of trees standing in a woodlot
(d) A measure of tree volume

180. Which of the following instrument measures angle as well as direction?
(a) Clinometer
(b) Abney level
(c) Relkascope
(d) None of the above

181. Pick out the Non random sampling
(a) Selective sampling
(b) List sampling
(c) Multistage sampling
(d) All of the above

182. Management of forests for multiple uses involves which of the following uses:
(a) Timber
(b) Recreation
(c) Water
(d) All of the above

183. Which one is correct?
(a) CAI - It is the increment that a tree puts on from origin upto that age at which the tree is cut
(b) PAI- It is the average annual increment for any short period.
(c) MAI- It is the increment which a tree puts on a single year
(d) TAI - It is the total increment upto a given age divided by that age.

184. Basal area of a woodlot is given in what units?
(a) Square feet/acre
(b) Cubic feet/acre
(c) Board feet/acre
(d) Number of trees/acre

185. Merchantable height is:
(a) Commonly measured with a log scale
(b) Measured at the first fork in the tree depending on the species
(c) Measured to the minimum diameter based on products to be taken
(d) Not normally considered in the volume table
(e) Both B and C

186. The number of stems falls off from one diameter class to the next in geometrical progression, which means that the percentage reduction in the stem number from one diameter class to the next is constant. this is referred to as
(a) Brasnett's Law
(b) Knuchel's Law
(c) De Liocourt's Law
(d) Osmaston's Law

187. Tick the odd man out among the following technical terms connected with increment
(a) Mean Annual increment
(b) Increment percent
(c) Current annual percent
(d) Terminal annual increment

188. If girth over bark (g) and bark thickness (t) are respectively 30 cm and 2cm. Find out the girth under bark (g'). Assume ? to be 3.0.
(a) 6 cm (b) 12 cm
(c) 18 cm (d) 36 cm

189. One of the following instruments used for tree height measurement is not based on trigonometric principle. Identify the same.
(a) Brandis hypsometer
(b) Smythie's hypsometer
(c) Blume Leiss hypsometer
(d) Abney's level

190. The following are some methods developed for volume measurement in the sample plot. Identify the odd man out.
(a) Huber's method
(b) Urich's method
(c) Gumple's method
(d) Hartig's method

191. In remote sensing, ideal time for photography will be
(a) 9 to 11 A.M and 1 to 2 P.M
(b) 10 to 12 A.M and 2 to 4 P.M
(c) 9 to 10 A.M and 1 to 2 P.M
(d) 10 to 12 A.M and 1 to 2 P.M

192. Which one of the following is not a formula for tree form ?
(a) Behre's formula
(b) Hojer's formula
(c) Metzger's formula
(d) Smalian's formula

193. Volume table based on diameter, height and form quotient is called as
(a) Form class volume table
(b) General volume table
(c) Standard volume table
(d) Commercial volume table

194. The following are some of the terms for point sampling. Identify the odd man out
(a) Angle count cruising
(b) Pointless cruising
(c) Variable plot cruising
(d) Polyplot cruising

195. The following are some kinds of volume tables. Identify the odd man out.
(a) Multiple yield table
(b) Variable yield table
(c) Single yield table
(d) Money yield table

196. Crown length is vertical measurement of the crown of a tree from the tip of the crown of a tree to the
(a) Lowest green branch on the bole
(b) Lowest green branch forming green crown all around
(c) Lowest branch green or dead
(d) Point halfway between the lowest green branch forming green crown all around and the lowest green branch on the bole

197. Which height is used for determine the volume of a tree crop?
(a) Mean height
(b) Mean height or crop height
(c) Top height
(d) Mean height or top height

198. Thickness of bark can be estimated by
(a) Increment borer
(b) Swedish bark gauge
(c) Relascope
(d) Callipers

199. The volume table gives the volume in round wood down to variable stated thin end diameters is
(a) Commercial volume table
(b) Standard volume table
(c) General volume table
(d) None of these

200. The instrument used for measuring diameter at unreadable height of standing tree is
(a) Tape (b) Callipers
(c) Dendrometer (d) All of the above

201. The error caused by the instrument, the user and the weather condition is called
(a) Instrumental error
(b) Personal error
(c) Error due to observation
(d) Error due to leaning of trees

202. Pick the odd one out from the following volume tables classified based on the scope of application
(a) General volume table
(b) Regional volume table
(c) Local volume table
(d) Sawn out turn table

203. When CAI is equal to MAI
(a) Maximum average volume per unit area per annum is obtained
(b) Average volume per unit area is lowest
(c) The CAI is at its peak
(d) MAI is at its minimum

204. Which of the following gives the correct volume of standing trees?
(a) DBH alone
(b) DBH, Height and form factor
(c) DBH and Height
(d) Height and form factor

205. Which of the following statement is false? Form factor is used for
(a) Volume estimate of individual trees
(b) Estimation of volume of a group of trees
(c) Study the laws of growth
(d) Study stem form of trees

206. Land cover classification using three levels of the scale of 1:100000 was developed by which of the following agency:
(a) IIRS (b) ISRO
(c) NRSA (d) RRSC

207. Remote sensing system which provide their own source of energy for illumination are known as
(a) Optical sensors
(b) Active sensors
(c) Passive sensors
(d) Thermal sensors

208. Land surface feature has specific response to radiation and it is termed as
(a) Refraction
(b) Reflection
(c) Spectral signature
(d) Radiation

209. Scattering which occurs when the atmosphere particles are about the same size s the wave length of the radiation
(a) Mie scattering
(b) Non-selective
(c) Rayleigh scattering
(d) None of these

210. IIRS is located at
(a) Hyderabad (b) New Delhi
(c) Dehradhun (d) Bangalore

211. Ability of the sensors to differentiate different wavelength is known as
(a) Spartial resolution
(b) Spectral resolution
(c) Radiometric resolution
(d) Tempral resolution

212. Scanning system which collects the data over a range of wavelengths is called as
(a) TM (b) MSS
(c) WIFS (d) None of these

213. Radiation which emitted from the Sun is in accordance with
(a) Kirchoff's Law
(b) Stefan-Boltzman law
(c) Wien's displacement law
(d) None of these

214. Which is the first Indian satellite
(a) IRS-IC
(b) IRS-IA
(c) IRS-1
(d) IRS-p1

215. Which of the following wavelength band is used for vegetational studies
(a) Visible only (b) Visible & IR
(c) UV & IR (d) IR, UV

216. Relative brightness of objects is an image is referred to
(a) Texture (b) Tone
(c) Pattern (d) None of these

217. Which of the following is the recently launched satellite with 1m spatial resolution
(a) IRS-IE (b) LANDSAT-7
(c) IKNOS (d) SPOT-III

218. Which of the following is very useful in detecting and mapping forest fires
(a) Visible sensors
(b) Thermal sensors
(c) Microwave sensors
(d) None of these

219. Aerial photograph which has been taken with optical axis of approximately perpendicular to the horizontal plane & has been deviation upto 4:0 is
(a) Oblique photograph
(b) High oblique photograph
(c) Vertical
(d) Low oblique photograph

220. RS & GIS has application in
(a) Biodivesity assessment
(b) Vegetation Classification
(c) Wild life management
(d) All of the above

221. Large scale photograph has scale of
(a) 1:40000 to 1:70000
(b) 1:20000 to 1:40000
(c) 1:5000 to 1:20000
(d) 1:50000 to 1:100000

222. SOI Topo sheet is a
(a) Perspective projection
(b) Orthogonal projection
(c) 3D-Projection
(d) None of these

223. NNRMS head quarters in India is located at
(a) Hyderabad (b) Bangalore
(c) Dehradhun (d) Ahemedabad

224. Agency for aerial photography in India is
(a) NRSA (b) IIRS
(c) RRSC (d) ISRO

225. High oblique photographs have tilt of
(a) >4 (b) >60
(c) >30 (d) <30

226. Enhanced Thematic Mapper (ETM) has
(a) 7-bands (b) 4-bandsw
(c) 3-bands (d) 8-bands

227. The oldest application of RS technology is in
(a) Agriculture (b) Military
(c) Forestry (d) All of these

ANSWERSHEET
5. Forest Mensuration

1. **(b)**	2. **(a)**	3. **(a)**	4. **(a)**	5. **(c)**	6. **(b)**	7. **(c)**	8. **(a)**	9. **(a)**	10. **(c)**
11. **(c)**	12. **(d)**	13. **(b)**	14. **(c)**	15. **(d)**	16. **(c)**	17. **(a)**	18. **(d)**	19. **(c)**	20. **(a)**
21. **(a)**	22. **(d)**	23. **(d)**	24. **(a)**	25. **(b)**	26. **(a)**	27. **(b)**	28. **(a)**	29. **(a)**	30. **(b)**
31. **(b)**	32. **(c)**	33. **(d)**	34. **(a)**	35. **(d)**	36. **(b)**	37. **(b)**	38. **(c)**	39. **(b)**	40. **(d)**
41. **(b)**	42. **(b)**	43. **(c)**	44. **(d)**	45. **(c)**	46. **(b)**	47. **(a)**	48. **(c)**	49. **(a)**	50. **(a)**
51. **(a)**	52. **(c)**	53. **(a)**	54. **(b)**	55. **(a)**	56. **(a)**	57. **(c)**	58. **(b)**	59. **(d)**	60. **(a)**
61. **(c)**	62. **(c)**	63. **(c)**	64. **(a)**	65. **(b)**	66. **(c)**	67. **(c)**	68. **(d)**	69. **(c)**	70. **(b)**
71. **(a)**	72. **(d)**	73. **(a)**	74. **(d)**	75. **(c)**	76. **(a)**	77. **(c)**	78. **(b)**	79. **(b)**	80. **(c)**
81. **(a)**	82. **(b)**	83. **(d)**	84. **(a)**	85. **(a)**	86. **(b)**	87. **(c)**	88. **(b)**	89. **(c)**	90. **(c)**
91. **(a)**	92. **(d)**	93. **(a)**	94. **(b)**	95. **(c)**	96. **(d)**	97. **(c)**	98. **(b)**	99. **(a)**	100. **(b)**
101. **(d)**	102. **(c)**	103. **(b)**	104. **(a)**	105. **(d)**	106. **(a)**	107. **(d)**	108. **(c)**	109. **(a)**	110. **(a)**
111. **(a)**	112. **(c)**	113. **(c)**	114. **(a)**	115. **(a)**	116. **(a)**	117. **(a)**	118. **(d)**	119. **(c)**	120. **(a)**
121. **(a)**	122. **(b)**	123. **(b)**	124. **(b)**	125. **(a)**	126. **(c)**	127. **(b)**	128. **(d)**	129. **(b)**	130. **(d)**
131. **(b)**	132. **(c)**	133. **(b)**	134. **(c)**	135. **(c)**	136. **(b)**	137. **(d)**	138. **(d)**	139. **(d)**	140. **(a)**

141. (a)	142. (a)	143. (b)	144. (c)	145. (a)	146. (a)	147. (b)	148. (d)	149. (d)	150. (d)
151. (a)	152. (b)	153. (a)	154. (b)	155. (d)	156. (d)	157. (c)	158. (d)	159. (a)	160. (c)
161. (d)	162. (b)	163. (a)	164. (b)	165. (c)	166. (d)	167. (d)	168. (b)	169. (c)	170. (c)
171. (a)	172. (d)	173. (c)	174. (d)	175. (b)	176. (a)	177. (c)	178. (a)	179. (c)	180. (d)
181. (a)	182. (d)	183. (b)	184. (a)	185. (e)	186. (c)	187. (d)	188. (c)	189. (b)	190. (c)
191. (a)	192. (d)	193. (a)	194. (d)	195. (b)	196. (d)	197. (a)	198. (b)	199. (a)	200. (c)
201. (a)	202. (d)	203. (a)	204. (b)	205. (a)	206. (c)	207. (b)	208. (b)	209. (a)	210. (c)
211. (b)	212. (b)	213. (b)	214. (b)	215. (b)	216. (b)	217. (c)	218. (b)	219. (c)	220. (d)
221. (c)	222. (d)	223. (a)	224. (a)	225. (c)	226. (a)	227. (b)			

6

Forest Management

KEY POINTS

- Forest Management is the application of business methods and technical forestry principles to the operation of a forest property
- Management of forests broadly involves three tasks viz. 1. Control of growing stock, its structure and composition 2. Distribution and marketing of produce 3. Administration of forest property
- India's First forest policy was enunciated in 1894
- In 1894 forest policy public benefits as a sole objective
- Indian republic formulated its first national forest policy in 1952
- State-owned forests in India is 95.8%
- National Commission on Agriculture was constituted on 1970
- NCA, 1976 suggested a modified functional classification as production forest, protection forest and social forest.
- Based on legal the forests are broadly classified as Reserved, Protected, Village and Un-classes forests
- Territorial classification of forests are classified into Blocks, Compartments and Sub-Compartments
- Block is a main territorial division of the forest
- Compartment is the smallest permanent working plan unit of management
- Sub compartment is the silvicultural unit of management
- Beat is the smallest functional territorial unit
- Based on silvicultural management, forests are classified into Working Circles, Felling Series, Cutting Section, Coupes, Periodic Blocks
- The periodic block is divided into two viz., single or floating and Fixed or Permanent
- Periodic block = Rotation/ Regeneration period
- The time that elapses between successive main fellings on the same area is called felling cycle
- DFO is the kingpin of forest administration in India
- Working plan is an "A written scheme of management aiming at continuity of policy, controlling the treatment of a forest"

- National Forest Policy of 1952 suggested a functional classification of Indian Forests into Protection Forests, National Forests, Village Forests and Tree-lands
- The units of forest management is a Working Plan
- The number of annual coupes (annual felling areas), equal to the number of years in the rotation
- Rotation is called production period
- Rotation is the period which elapses between the formation of a wood and the time when it is finally cut over
- Physical rotation is the rotation which coincides with the natural lease of life of a species on a given site.
- Silvicultural rotation is the rotation through which a species retains satisfactory vigour of growth and reproduction on a given site
- Main objective of technical rotation is the yield maximum material of specified size or specific use
- The total increment upto a given age divided by that age is Mean Annual Increment
- The rotation of maximum volume production coincides where the MAI is maximum.
- The material that a forest can yield annually (or periodically) in perpetuity is called sustained yield
- Progressive yield concept is developed by german forester Hartig
- Utilization Percent is the percent ratio of normal yield to the N.G.S.
- $S_e = \frac{X}{0.0p}$; where p - rate of interest and S_e is the soil expectation value
- $Se = \frac{Y_r + T_a\, 1.0\, p^{r.a} + ... T_q\, 1.0\, p^{r.q} - C.\, 1.0p^r}{1.0p^r - 1} - E$ (Faustmann's soil expectation value formula)
- Rotations controlling the supply of certain services i.e., the Silvicultural and Physical Rotation.
- Rotations controlling the output of material forest products in form or quantity i.e.. the Technical and maximum volume rotations.
- Rotations controlling the financial returns, i.e., the Rotations of maximum Gross or net Income and the Financial Rotation.
- Conversion is a change from one silvicultural system or one species to another
- Conversion period as the period during which a change from one silvicultural system to another is effected
- Normal series of age gradations, Normal growing stock and normal increment is called trinity of norms
- Increment is the increase in growth of a tree or a crop with age
- The Mean or average of all CAI is known as Mean Annual Increment
- The periodic annual increment over a short preceding period is called as CAI
- The MAI at rotation upto a given age divided by that age
- Increment Percent = $\frac{\text{Increment}}{\text{Volume}}$ X 100 = $\frac{V - v}{V + v}$ X $\frac{2}{n}$ (Pressler's Formula)

- Increment Percent = p = $\frac{400}{nD}$ (Schneider's Formula)
- World average Mean annual increment is 2.1 m^3/ha
- Mean annual increment of Indian forest is 0.7 – 0.9 m^3/ha
- Value of a tree increase is t= v X p
- Quality increment per cent p = $\frac{(K-k)}{(K+k)} \times \frac{200}{n}$
- Price increment per cent $\frac{P_2 - P_1}{P_2 + P_1} \times \frac{200}{n}$
- Mean annual increment for fast growing species is 10m^3/ha/annum
- In fast growing younger plantation, the height increment must not be less than 60 cm/ annum
- Growing stock (G.S) is the sum of all the trees growing in the forest or a specified part of it
- Growing stock determined by using total or complete enumeration, partial or sample enumeration, sample plot methods and with the help of volume tables.
- Determination of normal growing stock
- Middle of the growing season NGS = (I X r/2)
- Beginning of the next growing season NGS = (I X r/2) – (I/2)
- End of the growing season will be NGS = (I X r/2) + (I/2)
- De Liocourt's Law give the proportionate distribution in an Ideal Selection forest or its balanced composition
- Yield regulation is the determination of the yield and the prescribed means of realizing it
- Von Mantel's Formula Annual Yield = $\frac{2NGS}{r}$
- Von Mantel's Formula called as formula of glorious simplicity
- Masson's Ratio or Exploitation Percent Y = $\cdot\frac{200}{r} = \frac{2}{r}\%$
- Howard's Modification annual yield Y = $\frac{3V}{8r}$
- Cotta's Formula Annual Yield Y_a= $\frac{V}{P} + \frac{i}{2}$
- Smythie's Modification Annual yield Y = $\frac{2v}{r-x}$
- Biolley's Check Method Increment Z = V_2 — V_1 + N - P, where, P is the new recruits.
- Annual Increment $\frac{Z}{n} = \frac{V_2 + N - (V_1 + P)}{n}$

  Where, n = interval period of enumeration
- Burma Modification Annual Yield Y = $\frac{9y}{4r}$

- The Quotient Y_n/G_n is called utilization factor by Hundeshagan
- Y_n X $100/G_n$ is called utilization percent
- JFM - Joint Forest Management
- The policies and objectives of Joint Forest Management are detailed in the Indian comprehensive National Forest Policy of 1988 and the Joint Forest Management Guidelines of 1990 of the Government of India
- Joint Forest Management originated in West Bengal at the Arabari Forest Range in 1971
- 50% members of the JFM general body should be women
- 33% of the membership in the JFM Executive Committee/Management Committee should be filled from amongst the women members
- Working Plan is a written scheme of management aiming at continuity of policy and action and controlling the treatment of a forest
- Working Plan is an instrument of forest management
- Working plan officer is a special duty officer at the rank of deputy conservator of forest.
- Working plan period varied from 10 to 15 years

Crop composition — **Classification**

- > 75 % by one species - Pure Crop
- 50-75% by one species - Main Crop
- 25-50 % by one species - Mixed Crop
- < 25 % by one species - Miscellaneous Crop

Forest Type — **Colour**

- Sal - Blue
- Teak - Burnt sienna
- Barren land - Yellow
- Mixed forest - Green
- Cultivation area - Uncloured

Density — **Canopy Class**

- 1.0 - Closed
- 0.8 – 0.99 - Dense
- 0.6 – 0.79 - Fairly dense
- 0.4 – 0.59 - Thin
- 0.2 – 0.39 - Open
- < 0.2 - Sparse

Age class Density — **Symbol**

- Seedling crop - 0
- Thicket of seedling - 00

- Pole crop - +
- Mature high forest - -
- Over Mature forest - =
- Dendrometry is the measurement of trees and their volumes without felling; by the use of dendrometers.

Cumulative Weight age Value	Regeneration status	Colour for Regeneration Map
81-100	Excellent	Green
61-80	Good	Blue
41-60	Moderate	Yellow
21-40	Fair	Red
0-20	Deficient	Blank

Maps and their scale

Management maps	-	1: 50000
Enumeration maps	-	1: 50000
Forest type maps	-	1: 50000
Working plan maps	-	1: 50000
Stock maps	-	1: 15000
Regeneration maps	-	1:5000 or 1:4000

Group Designation	Reference Map Colour
Tropical Wet-Evergreen	Violet
Tropical Semi-evergreen	Purple
Tropical Moist-deciduous	Cobalt blue
Littoral and swamp forest	Prussian blue
Tropical Dry Deciduous	Yellow
Tropical Thorn	Red
Tropical dry evergreen	Pink
Sub-tropical Broad-leaved Hill forests	Vermillion
Sub-tropical Pine	Orange
Sub-tropical Dry Evergreen	Burnt umber
Northern wet temperate	Olive green
Himalayan Moist temperate	Green
Himalayan Dry Temperate	Yellow
Sub-alpine	Burnt sienna
Alpine	Sepia

MULTIPLE CHOICE QUESTION'S

1. The practical application of the scientific, technical and economical principles of forestry is called
(a) Forest Management
(b) Forest Mensuration
(c) Forest Economics
(d) None

2. Primarily a person engaged in the profession of forestry is called ------------
(a) Ranger (b) Forester
(c) Guard (d) All the above

3. The basis for establishing a management plan for a forest is
(a) Money (b) Access
(c) Objectives (d) Time

4. Reserved forest are constituted under the
(a) Indian forest Act, 1927
(b) National Forest Policy, 1988
(c) Forest Policy, 1952
(d) None

5. British rulers reserved the forest for
(a) Industrial needs
(b) To meet the needs of local people
(c) To protect nature
(d) None

6. Forest Conservation Act was enacted during
(a) 1880 (b) 1980
(c) 1988 (d) 1998

7. The smallest permanent working plan unit of management
(a) Block (b) Compartment
(c) Annual coupe (d) Periodic block

8. ------------------ is the head of the forest administration in India
(a) PCCF (b) DFO
(c) Ranger
(d) Inspector General of Forest (IGF)

9. ------------------- is a very important unit in the management and administration of forest
(a) Range (b) Beat
(c) Circle (d) Division

10. --------------- is the smallest functional territorial unit and is the foundation of Indian forest administration
(a) Beat (b) Range
(c) Circle (d) Blocks

11. ----- is the main territorial division of forest
(a) Compartment (b) Block
(c) Periodic block (d) Coupe

12. Based on territorial classification which forest is temporary nature?
(a) Sub-Compartment
(b) Periodic block
(c) Coupe (d) None of these

13. Who is the kingpin of forest administration in India?
(a) District Forest Officer (DFO)
(b) PCCF
(c) Ranger (d) ACF

14. -------------- is the backbone of forest management
(a) Even aged Forest
(b) Maximum sustained yield
(c) Regular Forest
(d) All of the above

15. What is a tool of forest management?
(a) Working Plan
(b) Working circle
(c) Felling series (d) Coupe

16. Unit of a working plan in India is
(a) Forest division
(b) Felling series
(c) Cutting section
(d) Coupe

17. A written scheme of management aiming at continuity of policy, controlling the treatement of a forest is called

(a) Working circle

(b) Working plan

(c) Felling series

(d) Coupe

18. Working circles divided into -----------

(a) Felling series (b) Division

(c) Cutting section (d) Coupe

19. Working plans are invariably based on ------------ principle

(a) Sustained yield

(b) Progressive yield

(c) All the above

(d) None

20. Felling series is divided in to ---------------

(a) Cutting section (b) Division

(c) Periodic block (d) None

21. Felling series is divided into number of ----------------

(a) Cutting Section

(b) Periodic block

(c) Working circle

(d) Annual coupes

22. The material that a forest can yield annually or periodically in perpetuity is called

(a) Sustained yield

(b) Progressive yield

(c) All the above (d) None

23. The concept of progressive yield originally advocated by

(a) Hartig (b) Smith

(c) Von montle (d) Howards

24. The concept of progressive yield envisages raising

(a) Productivity of soil and crop

(b) Only productivity of the soil

(c) Only productivity of the crop

(d) None

25. An age class with one year as the interval is called

(a) Age class (b) Age Gradation

(c) All the above (d) None

26. A forest area forming a part or whole of a working plan area is called

(a) Division

(b) Periodic block

(c) None

(d) Working circle

27. The period which a forest crop takes between its formation and final felling is known as

(a) Increment

(b) Rotation or production period

(c) Conversion period

(d) None

28. The term rotation strictly speaking is correctly applicable to

(a) Regular crops only

(b) Irregular crop

(c) Both

(d) None of these

29. Physical rotation is applicable to

(a) Protection

(b) Amenity forests

(c) Road side avenues

(d) All the above

30. Silvicultural rotation may be useful in forests managed primarily for

(a) Aesthetic

(b) Recreational purpose

(c) a and b

(d) None

31. Production of maximum material of specific dimensions or quality is the aim of

(a) Technical rotation

(b) Financial rotation

(c) Silvicultural rotation

(d) None

32. Pulp and paper industries favours adaptation for which type of rotation
(a) Technical rotation
(b) Financial rotation
(c) Silvicultural
(d) Rotation of Maximum Volume Production

33. Which type of rotation yields largest volume per unit are per annum
(a) Technical rotation
(b) Financial rotation
(c) Silvicultural
(d) Rotation of Maximum Volume Production

34. For production of firewood best rotation is
(a) Rotation of Maximum Volume Production
(b) Technical rotation
(c) Physical rotation
(d) Financial rotation

35. Common practice in forestry is to adopt a combination of
(a) Financial rotation and Technical rotation
(b) Technical and Rotation of highest income
(c) Maximum Volume Production and Financial rotation
(d) None

36. From the overall national point of view which type of rotation is important
(a) Technical rotation
(b) Rotation of maximum
(c) Silvicultural
(d) Rotation of Highest Income / Revenue

37. The rotation which is most profitable
(a) Physical rotation
(b) Technical rotation
(c) Rotation of maximum volume production
(d) Financial Rotation

38. Which type of rotation is useful for aesthetic and recreational purpose?
(a) Silvicultural rotation
(b) Physical rotation
(c) Technical rotation
(d) Rotation of maximum volume production

39. Normal forest under a given set of condition does not depend of this factor
(a) A normal increment
(b) A normal distribution of age classes
(c) A normal distribution of tree species
(d) A normal growing stock

40. Maximum average volume per unit area per annum in Eucalyptus can be obtained at
(a) MAI = CAI (b) CAI > MAI
(c) CAI = PAI (d) MAI = PAI

41. Who should draw up the working plan?
(a) Local DFO
(b) Working Plan Officer
(c) Conservator of Forests
(d) PCCF

42. Faustamann's formula is used for calculating
(a) Soil Expectation value
(b) Sustained yield
(c) Progressive yield
(d) None

43. Length of rotation is affected by
(a) Rate of growth
(b) Silvicultural characteristics of species
(c) Economic and social conditions
(d) All of the above

44. Which rotation controlling the output of material/forest products in form or quality
(a) Silvicultural
(b) Physical Rotation
(c) a and b
(d) None

45. Soil expectation value (Se) is---------------
(a) X/0.0p
(b) Faustmann's Formula
(c) a and b
(d) None of these

46. Which rotations controlling the financial returns
(a) Silvicultural and Physical Rotation
(b) Rotation of Maximum gross or net income and Financial rotation
(c) Technical and Maximum volume rotation
(d) All the above

47. The concept of normal forest was first introduced by
(a) German forester's
(b) American forester
(c) French forester
(d) None

48. Which is the forest management is territorial division
(a) Block (b) Beat
(c) Compartment (d) Coupe

49. The CVP index developed by Paterson indicates
(a) Stand Productivity
(b) Site Productivity
(c) Commercial Productivity
(d) None

50. What is the percentage of forest owned by Indian government?
(a) 60 (b) 75
(c) 85 (d) 95

51. A favorable balance between timber growth and harvest (growth equals or exceeds harvest) is referred to as:
(a) Sustained yield timber growth
(b) Increment
(c) Growing stock
(d) None

52. The volume or number of stems that can be removed annually or periodically is called
(a) Yield (b) Rotation
(c) Growing stock (d) Increment

53. All material from thinning is called
(a) Intermediate Yield
(b) Final yield
(c) Yield capacity
(d) None

54. The yield from a Normal forest is called
(a) Normal yield
(b) Growing stock
(c) Abnormal forest
(d) None

55. In Howard's modification crop is measured down to diameter corresponding to the rotation of
(a) 3/8 (b) 2/3
(c) 1/3 (d) 1/2

56. Which of the forms basis for yield regulation in French method of 1883
(a) Volume (b) Diameter
(c) Increment (d) All the above

57. Von Montal's formula is best applicable to
(a) Regular even aged normal forest
(b) Irregular uneven aged forest
(c) Abnormal forest
(d) None

58. Masson's formula is also known as
(a) Exploitation Percent
(b) Masson's Ratio
(c) a and b (d) None

59. In which method of yield regulation, enumeration is carried out down to the diameter corresponding to half the rotation
(a) Howard's Formula
(b) Masson's formula
(c) Von mantle formula
(d) All the above

60. French method of 1883 is also called as
(a) Quartier bleu method
(b) Exploitation %
(c) U.P. Safeguarding formula
(d) All the above

61. Which is taken as unit of W.P.M in India
(a) Working circle (b) Coupe
(c) Beat (d) Felling series

62. The scale of the stock maps is
(a) 1:15000 (b) 1:10000
(c) 1:1500 (d) None

63. The regeneration survey maps have scale of
(a) 1:5000 (b) 1:4000
(c) a and b (d) None

64. The scale of working plan map is
(a) 1:50000 (b) 1:40000
(c) 1:10000 (d) None

65. The enumeration maps have scale of
(a) 1:50000 (b) 1:20000
(c) 1:10000 (d) 1:5000

66. The scale of forest type maps is
(a) 1:5000 (b) 1:20000
(c) 1:10000 (d) 1:50000

67. The term capital in forest management usually refers to
(a) Growing Stock (b) Sustained yield
(c) Progressive yield (d) All the above

68. Normal growing stock in clear felling systems is
(a) (I X r) / 2
(b) [(I X r) / 2] + (I/2)
(c) [(I X r) / 2] - (I/2)
(d) All the above

69. Basic pre-requisites to achieve the objectives of sustained yield is
(a) Normal increment
(b) Age gradations
(c) Growing stock
(d) All the above

70. Commonly used reducing factor in India is
(a) Volume (b) Diameter
(c) Area (d) Area and volume

71. Term rotation is strictly applicable to
(a) Regular crop only
(b) Irregular crop
(c) Both
(d) None

72. The main objective of yield regulation is to achieve ---------------- from the forests
(a) Sustained yield
(b) Growing stock
(c) Increment
(d) None

73. Basis of yield regulation in Von Mantels formula is
(a) Area (b) Increament
(c) Volume only (d) Area and volume

74. Von Mantels formula of yield regulation is
(a) 2GS/r (b) 200/r
(c) 8V/3r (d) 9V/4r

75. Stock maps are prepared by working plan rangers with scale of
(a) 1:15000 (b) 1:10000
(c) 1:20000 (d) 1:5000

76. The production period for any species is slightly longer on
(a) Richer sites (b) Marginal sites
(c) Poorer sites (d) Both b & c

77. Normally felling cycle in selection forests is
(a) 5 years (b) 10 years
(c) Both a and b (d) 5-10 years

78. Technological development have greater implications on
(a) Sustained yield
(b) Progressive yield
(c) Normal yield
(d) None

79. Standard length of a log except the last section is

(a) 4.5 m (b) 3 m

(c) 1.5 m (d) 1 m

80. One cubic meter of timber in metric system is equal to

(a) 39.000 cft in FPS system

(b) 30.000 cft in FPS system

(c) 35.3147 cft in FPS system

(d) None

81. Recently formulated national forest policy of 1988 laid emphasis on

(a) Maximum timber production

(b) Sustainable management

(c) Joint forest management

(d) None

82. Crop diameter is

(a) Mean diameter of trees in a stand

(b) Diameter corresponding to the mean basal area of a uniform crop

(c) Mean diameter of the dominant trees

(d) None

83. JFM first implemented at

(a) Jabalpur (b) Ayyalur

(c) Arabari (d) Jhansi

84. Lowest administrative unit

(a) Beat (b) Range

(c) Compartment (d) Section

85. Which type of rotation is most favourable for natural reproduction

(a) Silvicultural rotation

(b) Physical rotation

(c) Technical rotation

(d) Financial Rotation

86. A term generally applied to the determination of the yield and the prescribed means realizing it is called

(a) Financial yield

(b) Yield regulation

(c) Yield maximization

(d) None

87. A material that a forest can yield annually in perpetuity

(a) Total yield

(b) Sustained yield

(c) Periodic yield

(d) Intermittent yield

88. The concept of progressive yield envisage

(a) Raising the productivity of soil

(b) Judicious tending

(c) Appropriate silvicultural treatments

(d) All of the above

89. Which social forestry programmes is purely managed by private efforts

(a) Farm forestry

(b) Community forestry

(c) Rehabilitation of degraded forest area

(d) Urban forestry

90. Which forestry practice is not included in the scope of social forestry

(a) Farm forestry

(b) Extension forestry

(c) Recreational forestry

(d) Protection forestry

91. The rotation followed for blue gum in the Nilgiris forest division was

(a) 8 (b) 10

(c) 12 (d) 14

92. The smallest permanent working plan unit in India is

(a) Coupe (b) Felling series

(c) Compartment (d) Circle

93. Type of rotation (pick the odd man out)

(a) Regular (b) Silvicultural

(c) Technical (d) Financial

94. The rotation followed for sandalwood in South India is

(a) Physical rotation

(b) Technical rotation

(c) Financial rotation

(d) Silvicultural rotation

95. The smallest functional territorial unit in forest management is
(a) Circle (b) Division
(c) Section (d) Beat

96. The period required to regenerate the whole of Periodic block is called as
(a) Regeneration interval
(b) Regeneration period
(c) Rotation period
(d) Periodic block

97. TAP programme in Tamil Nadu was funded by
(a) Sweden (b) Canada
(c) Japan (d) USA

98. Enumeration maps are prepared in the scale of
(a) 1:10,000 (b) 1:15,000
(c) 1:20000 (d) 1:50000

99. Howard's formula for yield regulation is
(a) 8v/3r (b) 9v/4r
(c) 6v/3r (d) 2gs/r

100. The rotation in which a species retains satisfactory vigour of growth and reproduction on a given site is
(a) Silvicultural rotation
(b) Financial rotation
(c) Technical rotation
(d) Rotation for maximum volume production

101. The concept of progressive yield was attributable to
(a) Osmaston (b) Hartig
(c) Nautiyal (d) None

102. An uneven aged crop containing theoretically all age classes is said to be
(a) Selection forest
(b) Shelter wood forest
(c) Even aged forest
(d) None

103. The oldest forest act of Tamil Nadu state is enacted during
(a) 1894 (b) 1864
(c) 1882 (d) 1927

104. The total increment up to a given age divided by that age is called
(a) Current annual increment
(b) Periodic annual increment
(c) Mean annual increment
(d) Final mean annual increment

105. Which of the following is correct?
(a) To begin with the CAI keeps below MAI
(b) The CAI attains the maximum before the MAI
(c) The MAI is falling when the CAI is still raising
(d) While the MAI is more than CAI, MAI is rising

106. The MAI at rotation age is called
(a) Current annual increment
(b) Mean annual increment
(c) Final mean annual increment
(d) Periodic annual increment

107. Presslers formula to find out increment percent is
(a) $\frac{V-v}{V+v}\times\frac{200}{n}$ (b) $\frac{V+v}{V-v}\times\frac{200}{n}$
(c) $\frac{V-v}{V+v}\times\frac{n}{200}$ (d) $\frac{V-v}{V+v}\times\frac{n}{200}$

108. Schneiders formula to find out increment percent is
(a) P = 200/nD (b) P = 400/nD
(c) P = D/200n (d) P = 400/D

109. The total volume of trees in a fully stocked forest with normal distribution of age classes for a given rotation is called
(a) Normal growing stock
(b) Growing stock
(c) Regular growing stock
(d) Even growing stock

110. The yield from normal forest is called
(a) Normal yield
(b) Sustained yield
(c) Yield capacity
(d) Final yield

111. A material that a forest can yield annually in perpetuity is called
(a) Normal yield
(b) Sustained yield
(c) Progressive yield
(d) Fixed yield

112. Which among the following statements is incorrect regarding sustained yield concept?
(a) Local labour is fully employed
(b) Market can be developed by sustained supplies
(c) There is no wide fluctuation in the yield of wood
(d) Wood used industries have assured supply of raw material

113. How many types of rotation are recognized in forestry?
(a) 2 (b) 4
(c) 6 (d) 8

114. Howard's modification is based on
(a) Volume only
(b) Area only
(c) Area and volume
(d) Increment

115. Increment method of yield regulation is based on
(a) Area only
(b) Volume only
(c) Area and volume only
(d) Increment only

116. Brandis diameter class method of yield regulation also called as
(a) Indian method
(b) US method
(c) United kingdom method
(d) Swiss method

117. GS=Ix r x c. Here c refers to
(a) Adjustment factors
(b) Flurry constant
(c) Swiss constant
(d) Bavarian constant

118. Working plan map is generally prepared on the scale
(a) 1: 50000 (b) 1: 60000
(c) 1: 40000 (d) 1: 25000

119. The first Inspector general of forest was
(a) Sir.Dietrich Brandis
(b) Howard
(c) Chaturvedi
(d) Alfred moller

120. The National Commission on Agriculture submitted its report during
(a) 1970 (b) 1976
(c) 1967 (d) 1980

121. The crop composition 50 - 75% of teak may be classified as
(a) Pure teak
(b) Mainly teak
(c) Mixed teak
(d) Misc. with some teak

122. Pick the odd man out
(a) Normal growing stock
(b) Normal series of age gradations
(c) Normal increment
(d) Normal distribution of quality classes

123. The time that elapses between the successive main fellings of the same area is called
(a) Felling cycle
(b) Felling series
(c) Felling interval
(d) Regeneration fellings

124. The Office Incharge of Administrative Unit Section
(a) Ranger (b) Forester
(c) Guard (d) Watcher

125. Legally the forest in the country are classified as
(a) National Forest and Village Forest
(b) Reserve Forest and Protected Forest
(c) Farm Forest and Social Forest
(d) Moisture deciduous & Dry deciduous

126. The productivity of Indian Forest is
(a) 2.1 m3/ha (b) 0.5 m3/ha
(c) 0.7 m3 ha (d) 1.5 m3/ha

127. It is a measure of the relative productive capacity of a site for a particular species is called as
(a) Site quality (b) Site index
(c) Canopy density
(d) Crown density

128. Site index is the relation between
(a) Age and site
(b) Age and diameter
(c) Age and height
(d) Age and volume

129. Joint Forest Management was first introduced in
(a) West Bengal (b) Tamil Nadu
(c) Orissa (d) Uttaranchal

130. The Working Plan Officer is
(a) Deputy Conservator of Forests
(b) Conservator of Forests
(c) Assistant Conservator of Forests
(d) Ranger

131. In working plan, tropical wet evergreen are coloured with
(a) Violet (b) Purple
(c) Yellow (d) Green

132. In working plan, tropical thorn are coloured with
(a) Violet (b) Red
(c) Yellow (d) Green

133. The first Independent National Forest Policy was enunciated during
(a) 1950 (b) 1952
(c) 1954 (d) 1956

134. Forest and wildlife were brought under concurrent list as per
(a) 40th amendment
(b) 42nd amendment
(c) 44th amendment
(d) 48th amendment

135. The Head Quarters of Inspector General of Forest
(a) Dehra Dun (b) Missouri
(c) Delhi (d) Mumbai

136. The sum of all the trees growing in a forest is called
(a) Selection forest
(b) Growing stock
(c) Normal forest
(d) None of these

137. The MAI at rotation age is called
(a) Current annual increment
(b) Mean annual increment
(c) Final mean annual increment
(d) Periodic annual increment

138. When CAI is equal to MAI
(a) Maximum average volume per unit area per annum is obtained
(b) Minimum average volume per unit area per annum is obtained
(c) All of the above
(d) None of these

139. CAI/MAI relationship (pick the odd out)
(a) CAI is greater than MAI, the MAI is raising
(b) To begin MAI keeps below CAI
(c) CAI attains its maximum before MAI
(d) MAI is fully under CAI

140. Average annual increment for any short period
(a) CAI (b) PAI
(c) MAI (d) IP

141. Fast growing species is one which yields an increment of
(a) 7 m^3/ha/yr (b) 10 m^3/ha/yr
(c) 15 m^3/ha/yr (d) 20 m^3/ha/yr

142. The rotation followed for teak in moist deciduous forest
(a) 40 to 60 years
(b) 20 to 40 years
(c) 60 to 80 years
(d) 80 to 100 years

143. Simmons formulae is a based on the following method of yield regulation

(a) Area (b) Increment
(c) Volume (d) All of these

144. Schneider formula is to determine

(a) Volume increment per cent
(b) Height increment per cent
(c) Diameter increment per cent
(d) Age increment

145. The first Forest Policy was enunciated during

(a) 1864 (b) 1874
(c) 1884 (d) 1894

146. Functionally forests are classified into

(a) Two types (b) Three types
(c) Four types (d) None of these

147. Working circle is a classification based on

(a) Administrative
(b) Territorial
(c) Management
(d) Functional

148. Third World Forestry Congress was held in

(a) Dehra Dun (b) Helsinki
(c) Delhi (d) Kyoto

149. Masson's formula is based on

(a) Volume
(b) Increment
(c) Volume and increment
(d) Diameter

150. Symthies safe guarding formula is based on

(a) Volume
(b) No of trees
(c) Volume and increment
(d) Diameter

151. Indian Institute of Forest Management is located at

(a) Bhopal (b) Dehradun
(c) Bangalore (d) Ludhiana

152. Which one of the following is correctly matched?

(a) CAI -It is the increment that a tree puts on from origin upto that age at which the tree is cut
(b)PAI-It is the average annual increment for any short period.
(c) MAI-It is the increment which a tree puts on a single year
(d) TAI- It is the total increment upto a given age divided by that age.

153. Legally the forest in the country are classified as

(a) Reserve and protected Forest
(b) Protection and social forest
(c) Production and village forest
(d) None of these

154. Joint forest management was first introduced in

(a) West Bengal, 1988
(b) Tamil Nadu, 1988
(c) Karnataka, 1988
(d) Gujarat, 1988

155. The working plan officer is

(a) Deputy Conservator of Forest
(b) Conservator of forest
(c) District forest officer
(d) Chief conservator of forest

156. In working plan, tropical wet evergreen are coloured with

(a) Violet (b) Red
(c) Blue (d) Black

157. In working plan, tropical thorn forest are coloured with

(a) Purple (b) Yellow
(c) Blue (d) Red

158. In working plan, Himalayan dry temperate forest are coloured with

(a) Purple (b) Yellow
(c) Blue (d) Red

159. Average annual increment for any short period
(a) MAI (b) PAI
(c) MAI and PAI (d) None of these

160. The rotation at which MAI is maximum
(a) Rotation of highest income/Revenue
(b) Rotation of maximum volume production
(c) Financial rotation
(d) All the above

161. A Change of one Silvicultural system or one species to another is called ------------
(a) Conversion (b) Conversion period
(c) Rotation (d) All the above

162. The period during which a change from one Silvicultural system to another is called ------------
(a) Conversion Period
(b) Rotation
(c) a and b (d) None of these

163. Who is called the Trinity of Norms for normal forest
(a) Osmaston
(b) Sir.Dietrich Brandis
(c) Hartig (d) Nautiyal

164. What is the Trinity of Norms in forestry
(a) Normal series of Age gradations
(b) Normal growing stock
(c) Normal increment
(d) All the above

165. A forest which corresponds in every way to the objects of management is called ---
(a) Normal Forest
(b) Increment
(c) Abnormal forest
(d) None of these

166. The crop composition > 75% of teak may be classified as
(a) Pure teak (b) Mainly teak
(c) Mixed teak
(d) Misc. with some teak

167. The sum of all trees growing in the forest or a specified part is called
(a) Increment
(b) Rotation
(c) Growing Stock
(d) None of these

168. Tangible benefits are --------
(a) Wood
(b) Industrial raw material
(c) a and b
(d) None of these

169. Intangible Benefits are
(a) Protective (b) Regulative
(c) Socio cultural (d) All the above

170. Forests are classified into various categories for purposes of ----------
(a) Description
(b) Administration
(c) Management and Record
(d) All the above

171. The NCA (1976) has suggested a modified functional classification --------
(a) Production Forest
(b) Protection Forest
(c) Social Forest
(d) All the above

172. Based on legal forests are classified -----
(a) Reserved
(b) Protected
(c) Un-classed forest
(d) All the above

173. Based on Territorial forests are classified ----------
(a) Block
(b) Compartment
(c) Sub-compartment
(d) All the above

174. India formulated its first National Forest Policy in the year
(a) 1949 (b)1952
(c) 1964 (d)1971

175. The percent ratio of normal yield to the normal growing stock is called ------------

(a) Utilization percent

(b) Exploitation Percent

(c) Masson's Ratio

(d) All the above

176. National Forest Policy (N.F.P), 1988 envisages that _________ per cent of the total geographical area of India should be under forest.

(a) 33 (b) 40

(c) 42 (d) 66

177. CAI &MAI curves are used to determine -- ---------

(a) Rotation age

(b) Regeneration period

(c) Increment

(d) Growing stock

178. The rotation followed for sandal wood in south India is

(a) Physical rotation

(b) Financial rotation

(c) Technical rotation

(d) Rotation for maximum volume production

179. Rotation of roadside and avenue trees is

(a) Technical rotation

(b) Financial rotation

(c) Rotation for maximum volume production

(d) Physical rotation

180. The rotation followed for blue gum in the Nilgiris forest division was

(a) 10 (b) 2

(c) 5 (d) 8

181. The forest survey of India (FSI) conducts Forest cover survey once in every

(a) Two Years

(b) Ten Years

(c) Five Years

(d) None of these

182. Schneider's formula is used for determining

(a) Increment Percentage

(b) Quality increment

(c) Price increment

(d) All the above

183. Who was the first person in India to be named a "Conservator of Forests"

(a) Brandis (b) Cleghorn

(c) Watson (d) None

184. Social Forestry programme was launched in the

(a) Fifth five year plan

(b) Fourth five year plan

(c) Sixth five year plan

(d) Seventh five year plan

185. Quality increment percent is ----------

(a) $p = \frac{V-v}{V+v} \times \frac{200}{n}$

(b) $\frac{V+v}{V-v} \times \frac{200}{n}$

(c) $\frac{V-v}{V+v} \times \frac{n}{200}$

(d) $\frac{V-v}{V+v} \times \frac{n}{200}$

186. Masson's formula for yield is ---------

(a) 200/r (b) 2/r

(c) a and b (d) None of these

187. Smythies Modification formula for yield is-------------

(a) 2v/(r-x) (b) 2GS/r

(c) 200/r (d) None of these

188. Burma formula for yield is -----------------

(a) 9v/4r (b) 2v/(r-x)

(c) 8v/3r (d) None of these

189. Floating periodic block is also known as

(a) Single allotment method

(b) Revocable allotment method

(c) Quartier Bleu Method

(d) All the above

190. FPB was coloured blue on the map, the rest of the area was uncoloured called as

(a) Quartier Bleu

(b) Quartier Blanc

(c) Uncoloured quartier bleu

(d) All the above

191. Which yield regulation method is widely adopted in India

(a) Annual coupes by gross area

(b) Annual coupes by reduced area

(c) a and b

(d) None of these

192. Brandis diameter class method is otherwise called as

(a) Indian method

(b) French Method

(c) German method

(d) None of these

193. A favorable balance between timber growth and harvest (growth equals or exceeds harvest) is referred to as:

(a) sustained yield timber growth

(b) silviculture

(c) timber stand improvement

(d) all of the above

194. The discount rate at which NPV is zero is called

(a) BC ratio (b) IRR

(c) GNP (d) None

195. In forest, the standard colour given for the regeneration status Excellent (81-100%) is

(a) Green (b) Blue

(c) Yellow (d) Red

196. In forest, the standard colour given for the regeneration status Fair (21-40%) is

(a) Green (b) Blue

(c) Yellow (d) Red

197. In forest, the standard colour given for the regeneration status Deficient (0-20%) is

(a) Green (b) Blue

(c) Black (d) Red

198. Joint Forest Management (JFM) is a concept of developing partnerships between fringe forest user groups

(a) Forest Department (Fd)

(b) Forest based Industry

(c) Local Panchayat

(d) None of these

199. Social Forestry Include

(a) Form Forestry

(b) Extension Forestry

(c) Recreation Forestry

(d) All of these

200. Joint forest management evolved in early

(a) 1950 (b) 1960

(c) 1970 (d) 1980

ANSWERSHEET
6. Forest Management

---

1. (a)	2. (b)	3. (c)	4. (a)	5. (a)	6. (b)	7. (b)	8. (d)	9. (a)	10. (a)
11. (b)	12. (a)	13. (a)	14. (b)	15. (a)	16. (a)	17. (b)	18. (a)	19. (a)	20. (a)
21. (d)	22. (a)	23. (a)	24. (a)	25. (b)	26. (d)	27. (b)	28. (a)	29. (d)	30. (c)
31. (a)	32. (d)	33. (d)	34. (a)	35. (c)	36. (d)	37. (d)	38. (a)	39. (c)	40. (a)

41. **(b)**	42. **(a)**	43. **(d)**	44. **(c)**	45. **(c)**	46. **(b)**	47. **(a)**	48. **(a)**	49. **(b)**	50. **(d)**
51. **(a)**	52. **(a)**	53. **(a)**	54. **(a)**	55. **(d)**	56. **(a)**	57. **(a)**	58. **(c)**	59. **(a)**	60. **(a)**
61. **(a)**	62. **(a)**	63. **(c)**	64. **(a)**	65. **(a)**	66. **(a)**	67. **(a)**	68. **(d)**	69. **(d)**	70. **(c)**
71. **(a)**	72. **(a)**	73. **(c)**	74. **(a)**	75. **(a)**	76. **(d)**	77. **(d)**	78. **(b)**	79. **(b)**	80. **(c)**
81. **(b)**	82. **(b)**	83. **(c)**	84. **(a)**	85. **(a)**	86. **(b)**	87. **(b)**	88. **(d)**	89. **(a)**	90. **(d)**
91. **(b)**	92. **(c)**	93. **(a)**	94. **(a)**	95. **(d)**	96. **(b)**	97. **(c)**	98. **(d)**	99. **(a)**	100. **(a)**
101. **(b)**	102. **(a)**	103. **(c)**	104. **(c)**	105. **(b)**	106. **(c)**	107. **(a)**	108. **(d)**	109. **(a)**	110. **(a)**
111. **(b)**	112. **(c)**	113. **(c)**	114. **(a)**	115. **(d)**	116. **(a)**	117. **(b)**	118. **(a)**	119. **(a)**	120. **(b)**
121. **(b)**	122. **(d)**	123. **(a)**	124. **(b)**	125. **(b)**	126. **(c)**	127. **(a)**	128. **(c)**	129. **(a)**	130. **(a)**
131. **(a)**	132. **(b)**	133. **(b)**	134. **(b)**	135. **(c)**	136. **(b)**	137. **(c)**	138. **(a)**	139. **(d)**	140. **(b)**
141. **(b)**	142. **(a)**	143. **(c)**	144. **(c)**	145. **(d)**	146. **(c)**	147. **(c)**	148. **(b)**	149. **(a)**	150. **(b)**
151. **(a)**	152. **(b)**	153. **(a)**	154. **(a)**	155. **(a)**	156. **(a)**	157. **(d)**	158. **(b)**	159. **(b)**	160. **(b)**
161. **(a)**	162. **(a)**	163. **(a)**	164. **(d)**	165. **(a)**	166. **(a)**	167. **(c)**	168. **(c)**	169. **(d)**	170. **(d)**
171. **(d)**	172. **(d)**	173. **(d)**	174. **(b)**	175. **(a)**	176. **(a)**	177. **(a)**	178. **(a)**	179. **(d)**	180. **(a)**
181. **(a)**	182. **(a)**	183. **(c)**	184. **(a)**	185. **(a)**	186. **(c)**	187. **(a)**	188. **(a)**	189. **(c)**	190. **(b)**
191. **(a)**	192. **(a)**	193. **(a)**	194. **(b)**	195. **(a)**	196. **(d)**	197. **(c)**	198. **(a)**	199. **(d)**	200. **(c)**

7

Forest Utilization

KEY POINTS

- Forest utilization - Process of harvesting, converting and dispersing of forest produce and other resources of forest.
- Logging: Extraction, felling, conversion and transport
- Bill hook: It's a tool often used for felling bamboo small poles and for clearing, chopping of bush growth underneath.
- Axe: Primitive tool used in the logging operation.
- Saws: Hand saws are the most commonly used tool for felling and conversion.
- Face: The edges of the tooth which faces the cutting direction.
- Back: The opposite edge of a tooth.
- Space: The distance from tooth to tooth measured at points of two adjacent teeth.
- Gullet: The entire opening between two adjacent teeth.
- Pitch: The angle between the face of a tooth and the line passing through the point of the teeth.
- Set: The extent to which the teeth are bent to either side of the plane of the blade.
- Kerf: The width of the saw act.
- Gauge: The thickness of the blade.
- Wedges: Wedges are used in felling trees and to prevent jamming of the saw in cross cutting and for longitudinal splitting.
- Metal wedges are made of steel or iron and are generally upto 30 cm length, 8 cm wide and weight upto 5 kg.
- Cant-hook: This is used as lever in rolling, stopping and turning logs.
- Picaroons: Picaroons are short poles 85-100 cm long, with a recurred pike or a hook, for drawing or pulling small logs.
- Debarking spade: Debarking spade is with a bent blade which is used for debarking logs.
- Measuring stick: The measuring stick is used for measuring log lengths, which has marking ends.
- Log hook: Log hooks are used for dragging, lifting and rolling.
- Cross cut saws: The cross cut saws are generally used for felling of trees and conversion into logs.
- Peg toothing: The peg toothing consists of teeth similar to each other and of the form of isosceles triangle.

- Raker toothing: In the raker toothing, the three tasks are divided between the two different types of teeth of quite different forms.
- Pruning saws: For cutting of branches on growing trees and to obtain knot free timber.
- A skidder is a type of heavy vehicle used in a logging operation for pulling cut trees out of a forest.
- A forwarder is a vehicle that carries logs from the stump to a roadside landing.
- Squares - Completely squared, rectangular, no wanes.
- Waney baulks - Where part of rounded surface of the log is left at one or more, or even all, corners.
- Half baulks - When baulks are split into two, down the center, each piece is termed as half baulk.
- Hakries - The rounded logs split with axe.
- Poles - The rounded products of varying lengths.
- Beams = A converted wood of more than 4 m length and not less than 15 cm in cross section.
- Scantlings - Square or rectangular, generally under 15 cm in section.
- Wire ropes: A wire rope consists of numerous, steel wire combine into stands which are in turn grouped around a central core to form a flexible one fall rope.
- Blocks - Pieces of short lengths with square surfaces.
- Planks-Wooden slabs up to a maximum of 5cm thickness with varying length & width.
- Boom is a Dutch word for floating obstruction, a barricade across a river to intercept the singly floating timber coming down a river, with a view of catching, collecting or diverting them
- Ergonomics - Ergonomics is the science of work and its productivity, its production in society's total result of production and ways of measuring productivity.
- Reduced impact logging (RIL) systems are currently being developed in Brazil and other countries in response to concerns over the ecological and economic sustainability of harvesting natural tropical forest stands.
- Grading: Intelligent choice of timber for specific purpose is which is generally understood by the term 'grading'.
- Sliced veneer can also be produced in semi-rotary machine.
- Semi-rotary veneer is produced on rotary veneer lathe.
- Plywood is a term applied to glue wood construction built of veneers in such a manner that the grain of each veneer is at right angles to that of the adjacent veneer in the assembly.
- Plywood: It is a sheet material manufactured by bounding wood veneers by use of binder and pressing the veneer and assembling in such a way that the grains of adjacent veneers are right angle to the each other.
- Laminated wood: The wood layers (laminae) are laid parallel to their grain, glued and fastened together.
- Core board: It is a composite board built up of a core composed of strips of various dimensions glued together or joined together to form a slab, which is in turn glued between two or more outer veneers with the direction of the grain of the core strips running at right angles to that of the adjacent veneers.
- Batter board – strips of wood are not more than 7.5 cm wide

- Block board – size is not more than 2.5 cm
- Lamin board – thickness not more than 7.0 mm
- Sandwich boards - Built up board having a core of light material, faced on both sides with relatively thin layers of material having high strength properties.
- The term "Impregnated Wood" is a collective name for woods impregnated with all possible impregments i.e., wax, paraffins, resins, oils, etc.. with the object of improving dimensional stability and preventing water absorption.
- Compressed wood - In this case, the wood is compressed without impregnating it with any synthetic resins.
- Chemically modified wood has been developed to make it less hygroscopic, by use of chemicals, which replace the hydroxyl group in cellulose, simultaneously inducing stability and decrease in cellulose, simultaneously inducing stability and decrease in shrinkage and swelling.
- A fiberboard is a sheet of material made from fibres of wood.
- A particle board is a board or sheet constituted from fragments of wood and other lingo-cellulosic materials bonded with organic binders with the help of one or more agents like heat, pressure, humidity, catalyst, etc.
- Particles of waste wood, in the form of flakes, shavings, splinters, etc., or chips and particles obtained by converting the raw material in a chipper which are usually screened to get particles of more or less uniform size.
- Wood-softwoods: *Abies pindrow, Picea smithiana, Pinus roxburghii*
- Hardwoods: *Boswellia serrata*, Mulberry
- Single stage bleaching – calcium hypochloride is used
- Multistage bleaching – chlorine gas, calcium hypochlorite, chloric dioxide
- Rayon is meaning for artificial silk. First commercial rayon was produced by Bemiganal.
- Match wood Suitable species: *Bombax ceiba, Sterculia villosa, Trewia nudiflora, Mangifera indica, Alstonia scholaris, Spondias pinnata, Anthocephalus cadamba, Ailanthus excelsa.*
- Destructive distillation usually begins at 250°C and is finished about 350°C when practically all of the acetic acid, wood alcohol and wood tar have been evolved.

MULTIPLE CHOICE QUESTION'S

1. The process of harvesting, conversion and disposal of standing forest produce is called
(a) Forest Mensuration
(b) Forest Management
(c) Forest utilization
(d) Forest Economics

2. Aim of the forest utilization is
(a) Complete and efficient utilization of standing timber
(b) Extraction of timbers
(c) Storage of timbers
(d) Marketing of timbers

3. The tree price generally represents ---------- value of the saleable products
(a) 1-2 % (b) 10-22%
(c) 25-31 % (d) 25-50%

4. Total cost of felling, conversion and transport are approximately ---------- % of the total cost of timber in the market
(a) 20 % (b) 30%
(c) 40% (d) 50%

5. The wastages in felling and conversion will be more in ---------- region
(a) Hilly (b) Waste land
(c) Forest areas (d) Terrain

6. The total volume of usable rough timber from the average tree is only
(a) 17% to 22% (b) 27% to 45 %
(c) 37% to 52% (d) 47 % to 65 %

7. Logging branch at Forest Research Institute, Dehra Dun came into existence during
(a) April, 1947 (b) June, 1947
(c) April, 1957 (d) June, 1957

8. Logging branch at Forest Research Institute, Dehra Dun established due to the recommendations of
(a) Dr. A. Huber (b) Dr. A. Koroleff
(c) a and b (d) None of these

9. Logging Training Centre was established at
(a) Batote (b) Dehradun
(c) Delhi (d) Mumbai

10. Logging training centre at Batote was established by
(a) Dr. A. Huber
(b) Dr. A. Koroleff
(c) Dr. H.G. Winklemann
(d) All of the above

11. Logging Training Centre at Batote was created during
(a) 1948 -1949 (b) 1958 – 1959
(c) 1968-1969 (d) 1978-1979

12. Logging Training Centre project implemented throughout India mainly for
(a) Improvement in logging methods
(b) Timber harvesting
(c) a and b
(d) None of these

13. The extent of utilization of any species closely related with its
(a) Market price (b) Demand
(c) Supply price (d) All of the above

14. The felling tools and implements were characterized by
(a) Strength (b) Durability
(c) Efficiency (d) All of the above

15. Axes are used mainly for
(a) Felling
(b) Trimming
(c) Splitting and grubbing
(d) All of the above

16. The portion of axe head in front of the eye is known as
(a) Blade (b) Back
(c) Face (d) Thickness

17. The portion of axe head behind the eye is known as
(a) Blade (b) Back
(c) Face (d) Thickness

18. The sides of the blade should be
(a) Back (b) Convex
(c) Curves (d) Thickness

19. The edge of the blade should be
(a) Back (b) Convex
(c) Curves (d) Thickness

20. Generally the weight of the axe will be
(a) to 1.5 Kg (b) 1.5 to 2.0 Kg
(c) 2.5 to 3.0 Kg (d) to 2.5 Kg

21. In India, the axe handles will be in ---------- shape
(a) Round (b) Square
(c) Rectangular (d) Oval

22. European and American axes have ------------ shaped handle
(a) Round (b) Square
(c) Rectangular (d) Oval

23. Bill hook is often used or felling
(a) Teak (b) Bamboos
(c) Terminalia spp. (d) All of the above

24. Saws are widely used for
(a) Felling (b) Cross cutting
(c) Ripping of logs (d) All of the above

25. The edge of the tooth which faces the cutting direction is called
(a) Blade (b) Back
(c) Face (d) Thickness

26. The opposite edge of the tooth is called
(a) Blade (b) Back
(c) Face (d) Thickness

27. The distance between the two adjacent teeth is known as
(a) Blade (b) Back
(c) Face (d) Sapce

28. The entire opening between two adjacent teeth is called
(a) Space (b) Gullet
(c) Pitch (d) Set

29. The extent to which the teeth are bent is termed as
(a) Space (b) Gullet
(c) Pitch (d) Set

30. The width of the saw cut is
(a) Kerf (b) Gauge
(c) Set (d) Gullet

31. The thickness of the blade is called
(a) Kerf (b) Gauge
(c) Set (d) Gullet

32. The suitable stroke length of for cross cut saws is
(a) 50 to 80 cm
(b) 60 to 80 cm
(c) 80 to 100 cm
(d) 100 to 120 cm

33. ---------- different tasks of the tooth are
(a) Cut (b) Break
(c) Carry away the fibres
(d) all of the above

34. Peg toothed saw consists of all teeth in the form of
(a) Isoceles triangles (b) Oval
(c) Square (d) Circular

35. The raker toothed saw consisting of teeth in the form of
(a) Rakers (b) Cutters
(c) a and b (d) None of these

36. Bow man saw is ---------- man saw
(a) One Man (b) Two man
(c) a and b (d) None of these

37. The pruning saw will have tooth pattern of -------------- teeth cutting
(a) Reverse (b) Straight
(c) a and b (d) None of these

38. Power saws are mainly used for
(a) Felling (b) Crosscutting
(c) a and b (d) None of these

39. Wedges are used while felling to prevent
(a) Jamming (b) Cutting
(c) Splitting (d) All of the above

40. Cant hooks are used as a
(a) Lever in rolling (b) Stopping
(c) Turning (d) All of the above

41. Pickaroons are used for
(a) Drawing
(b) Pulling of small logs
(c) a and b
(d) None of these

42. Debarking spades are used for
(a) Cutting of logs
(b) Splitting the logs
(c) Felling the logs
(d) Debarking the logs

43. Log hooks are used for
(a) Dragging (b) Lifting
(c) Rolling of logs (d) All of the above

44. The use of Stem Tightener is to prevent
(a) Splitting of stem
(b) Cutting of stem
(c) Felling of stem
(d) Debarking of stem

45. The best felling season is
(a) Winter (b) Summer
(c) Rainy (d) All of the above

46. Felling should begin with ----- of the slope
(a) Top (b) Bottom
(c) Centre (d) Any part of tree

47. The waste attributed due to stumpis
(a) 1% to 2 % (b) 2% to 5%
(c) 5% to 7% (d) 7% to 10%

48. Notch or undercut is made to decide the
(a) Direction of fall
(b) Opposite direction of fall
(c) a and b
(d) None of these

49. The undercut should be ---------- diameter of the tree
(a) 1/3 (b) 2/3
(c) 1/4 (d) 2/4

50. Felling by roots is practised in
(a) *Santalum album*
(b) *Acacia catechu*
(c) *Juglens regia*
(d) All of the above

51. The length of the log depends on
(a) Market requirement
(b) Transport
(c) a and b
(d) None of these

52. The rounded logs split with axe is called
(a) Squares (b) Hakries
(c) Poles (d) Waney baulks

53. Pieces of short lengths with square surfaces is called
(a) Beams (b) Scantlings
(c) Planks (d) Blocks

54. Quarter sawing other wise termed as
(a) Radially sawing
(b) Hand sawing
(c) Machine sawing
(d) None of these

55. In quarter sawn timber ------------ will be exposed
(a) Medullary rays
(b) Soft fibrous tissues
(c) Haof theserd fibrous tissues
(d) None of these

56. In tangentially sawn timber ------- will be exposed
(a) Medullary rays
(b) Soft fibrous tissues
(c) Hard fibrous tissues
(d) None of these

57. The transport constitutes ---------- % of the total cost of delivery
(a) 20 to 30 (b) 30 to 40
(c) 40 to 60 (d) 60 to 70

58. Minor transportation is also called
(a) Off Road Transportation
(b) Transport by land
(c) Water transportation
(d) Overhead transportation

59. Floating method of water transportation is applicable mostly to
(a) Softwoods (b) Hardwoods
(c) a and b (d) None of these

60. Telescopic floating attributed to
(a) Upper streams with plenty of water
(b) Free floating
(c) a and b
(d) None of these

61. Nullas can be effectively utilized for
(a) Telescopic floating
(b) River floating
(c) Free floating
(d) All of the above

62. Construction over and around obstruction in telescopic floating is called
(a) Pathru (b) River floating
(c) Free floating (d) All of the above

63. River floating is otherwise called as
(a) Free floating (b) Sweeping
(c) a and b (d) None of these

64. Major hazard in the river floating is
(a) Jamming (b) Splitting
(c) Debarking (d) All of the above

65. The average distance coverd per day in river floating is
(a) 1 to 2 km (b) 2 to 3 Km
(c) 3 to 4 km (d) 4 to 5 km

66. A quantity of wood firmly bounded togather is called
(a) Raft section (b) Floating
(c) Booms (d) Wet slides

67. Collection of raft section is called
(a) Raft (b) Floating
(c) Booms (d) Wet slides

68. The average number of bamboo culms per raft is around
(a) 100 culms (b) 1000 culms
(c) 10,000 culms (d) 100000 culms

69. Construction of floating obstruction across the river is termed as
(a) Raft (b) Floating
(c) Booms (d) Wet slides

70. Permanent booms are otherwise called
(a) Fixed booms
(b) Two way boom
(c) V-Shaped boom
(d) All of the above

71. Two way boom is otherwise called
(a) Single arm boom
(b) V-shaped boom
(c) a and b
(d) None of these

72. The arms of two way boom should form --- degree of angle at the apex on the down stream side
(a) 20° (b) 30°
(c) 40° (d) 50°

73. To place the boom proper in right position ------------ is used
(a) Guy ropes (b) Log line
(c) Line dori (d) All of the above

74. To extract timber from nullas where quantity of water is less ------------ used
(a) Floating (b) Rafting
(c) Booms (d) Wet slides

75. --------- transportation method can be deployed where surface transportation is not possible
(a) Transport by land
(b) Water Transportation
(c) Over head transportation
(d) Minor transportation

76. A place where wood and other forest produce is stored for disposal is
(a) Forest (b) Biomass
(c) Saw mill (d) Depot

77. A place which is mainly intended for collection and checking of forest produce is called
(a) Forest Depots
(b) Transit depot
(c) Sale depot
(d) None of these

78. An intermediate depots which may inside or outside the forest is called
(a) Forest Depots
(b) Transit depot
(c) Sale depot
(d) None of these

79. A place where the timber and other forest produce stored for disposal by sale is called
(a) Forest Depots (b) Transit depot
(c) Sale depot (d) None of these

80. Situation where continuous supply of forest produce is necessary --------- felling system can be deployed

(a) Felling by government agency

(b) Felling by private agency

(c) Government & purchasers combined

(d) Purchasers alone

81. Situation which demands care and control over felling --------------- felling system is highly suited

(a) Felling and Extraction by government agency

(b) Felling and Extraction by government & purchasers combined

(c) Felling and Extraction by purchasers alone

(d) All of the above

82. Most satisfactory system of felling and extraction is

(a) Working by government agency

(b) Working by purchasers alone

(c) Working by government and purchasers combined

(d) All of the above

83. -------- is the system of sale where the exact quantity of produce is not known

(a) Lump-sum sale

(b) Payment on outturn

(c) a and b

(d) None of these

84. -------- is the system of sale where the amount fixed with the actual produce extracted or sold

(a) Lump-sum sale

(b) Payment on outturn

(c) a and b

(d) None of these

85. A price fixed which prevents unreasonable low price during the sale of produce is

(a) Upper price (b) Reserve price

(c) a and b (d) None of these

86. Sale of forest produce based on individual consultation is called

(a) Sale by private bargain

(b) Sale to highest bidder

(c) Sale by royalty

(d) None of these

87. --------- is the method of sale of forest produce where keen competition exists

(a) Sale by private bargain

(b) Sale to highest bidder

(c) Sale by royalty

(d) None of these

88. ------ is the auction method which commences with prohibitively high price

(a) Upset price (b) Reverse price

(c) Dutch auction (d) None of these

89. Earnest Money Deposit is usually ------- % of the contract value

(a) 2 % (b) 5 %

(c) 10% (d) 14%

90. Security Deposit is ------------ % of the contract value

(a) 2 % (b) 5 %

(c) 10% (d) 14%

91. Intelligent choice of timber for specific purpose is called

(a) Logging (b) Grading

(c) a and b (d) None of these

92. In India, ---------- is the organization responsible for grading of timbers

(a) Indian Standards Institution

(b) Grading

(c) Quality (d) Storage

93. Separation of fibres along the grain is called

(a) Check (b) Shake

(c) Flutes (d) Spiral grain

94. Patches of bark either partially or wholly embedded inside the wood is called

(a) Shake (b) Flutes

(c) Spiral grain (d) Inbark

95. Difference in girth at the butt and top is called
(a) Taper (b) Flutes
(c) Knot (d) Surface cracks

96. Deviation from straightness of log is called
(a) Crookedness (b) Curvature bend
(c) a and b (d) none of these

97. Tree that shows departure from its cylindrical shape is called
(a) Knot (b) Flute
(c) Spiral grain (d) Spilt

98. A portion of branch that embedded in the wood by natural growth is called
(a) Knot (b) Flute
(c) Spiral grain (d) Spilt

99. A knot in which fibres are completely inter grown with surrounding wood is called
(a) Live knot (b) Dead knot
(c) a and b (d) None of these

100. A knot in which fibres are not completely inter grown with surrounding wood is called
(a) Live knot (b) Dead knot
(c) a and b (d) None of these

101. A type of growth in which the fibres take a spiral course about the bole is called
(a) Spiral grain (b) Crack
(c) Spilt (d) Rot

102. Separation of fibres along the longitudinal directions is called
(a) Spiral grain (b) Crack
(c) Spilt (d) Rot

103. Separation of fibres along the grain which extends rom one surface to another is called
(a) Spiral grain (b) Crack
(c) Spilt (d) Rot

104. A small soft core occurring in the structural centre is called
(a) Centre – heart (b) Hollow centre
(c) Rot (d) Shakes

105. The partial or complete or complete removal of pith with or without removal of heartwood is called
(a) Centre – heart
(b) Hollow centre
(c) Rot
(d) Shakes

106. A partial or complete separation between adjoining layers of fibres is called
(a) Centre – heart
(b) Hollow centre
(c) Rot
(d) Shakes

107. A crack originating from centre and extending radially in one or opposite direction is called
(a) Heart shake (b) Star shake
(c) Radial shake (d) Ring shake

108. Several splits or cracks originate from the centre of the timber is called
(a) Heart shake (b) Star shake
(c) Radial shake (d) Ring shake

109. All cracks appearing along the radii is called
(a) Heart shake (b) Star shake
(c) Radial shake (d) Ring shake

110. Shakes that are formed rupture of tissues in a circular direction is called
(a) Heart shake
(b) Star shake
(c) Radial shake
(d) Cup and Ring shakes

111. Complex knot formed at points where dormant buds show abnormal vigour is called
(a) Burr (b) Twisted fibre
(c) Waviness (d) Checks

112. A defect caused in timber due to inclusion with the wood of local wounds that have healed up is termed as
(a) Checks (b) Split
(c) Rindgalls (d) Radial shakes

113. Wood boring marine animals are belonging to the family

(a) Crustaceans (b) Molluses

(c) a and b (d) None of these

114. Important Molluses genera that attacks timber in sea water are

(a) Teredo (b) Brankia

(c) Martesia (d) All of the above

115. Crustaceans and Molluses that attacks timber in sea water are collectively called as

(a) Shipworms (b) Earthworms

(c) a and b (d) None of these

116. Important Crustaceans genera that attacks timber in sea water are

(a) Limnoria (b) Chof theseelura

(c) Sphaeroma (d) All of the above

117. In white rot wood is subjected to -------- and --------- which imparts brown colour to wood

(a) Oxidation (b) Hydrolysis

(c) a and b (d) None of these

118. World's best aircraft timber is

(a) *Ochrama pyramidalis*

(b) *Abies pindrow*

(c) *Artocarpus hirsute*

(d) Acer spp.

119. Wooden battery seperators are commonly made from

(a) Conifers (b) Teak

(c) Casuarina (d) Sal

120. The best wood for making the wooden bearing is

(a) *Guaiacum officinale*

(b) *Guaiacum sanctum*

(c) a and b (d) None

121. The wood that is widely used for making dugouts is

(a) *Duabanga sonneratioides*

(b) *Trewia nudiflora*

(c) a and b

(d) None

122. The best suited wood for tight cooperage is

(a) Quercus spp. (b) Teak

(c) Casuarina (d) All of the above

123. Harmonium is generally made out of

(a) *Tectona grandis*

(b) *Pinus roxburghii*

(c) *Canarium eupjyllum*

(d) *Artocarpus heterophyllus*

124. -------- is the most suited wood for pencil industry

(a) *Juniperus virginiana*

(b) *Juniperus procera*

(c) a and b

(d) None

125. Widely used species for making pattern are

(a) *Tectona grandis* (b) *Cedrus deodara*

(c) a and b (d) None

126. Species suitable for timber purposes (pick the odd man out)

(a) *Pterocarpus santalinus*

(b) *Tectona grandis*

(c) *Dalbergia sissoo*

(d) *Ailanthus excelsa*

127. Pick out the tree which yields soft fibre

(a) *Sterculia villosa*

(b) *Albizia lebbeck*

(c) *Azadirachta indica*

(d) *Acacia nilotica*

128. Wavy-grain wood is attributable to

(a) *Dalbergia latifolia*

(b) *Tectona grandis*

(c) *Pterocarpus santalinus*

(d) *Santalum album*

129. Timbers which are capable of withstanding rapid seasoning are termed as

(a) Most refractory

(b) Non refractory

(c) Medium refractory

(d) Refractory

130. The tree suitable for timber as well as fodder purpose, is

(a) *Tectona grandis*
(b) *Pterocarpus marsupium*
(c) *Acacia nilotica*
(d) *Ceiba pentandra*

131. The Latin name for golden bamboo is

(a) *Bambusa bamboos*
(b) *Bambusa vulgaris*
(c) *Bambusa tulda*
(d) *Dendrocalamus strictus*

132. Indian Kino tree is

(a) *Pterocarpus santalinus*
(b) *Pterocarpus dalbergioides*
(c) *Pterocarpus marsupium*
(d) None

133. Name the proprietary preservative containing high boiling creosote as base

(a) Curpinol (b) Hardiproof
(c) Solignum (d) Celcure

134. Henna dye is obtained from

(a) *Lawsonia inermis*
(b) *Punica granatum*
(c) *Bixa orellana*
(d) *Morinda tinctona*

135. Highly refractory wood (pick out the odd man out)

(a) *Xylia xylocarpa*
(b) *Madhuca latifolia*
(c) *Shorea robusta*
(d) *Vateria indica*

136. The Latin name for satin wood

(a) *Chloroxylon swietenia*
(b) *Lagerstroemia lanceolata*
(c) *Simarouba glauca*
(d) *Pterocarpus dalbergioides*

137. Oleo resins are obtained from

(a) *Pinus roxburghii*
(b) *Carcinia Morella*
(c) *Vateria indica*
(d) *Canarium strictum*

138. Cutch is obtained from

(a) *Acacia nilotica*
(b) *Acacia catechu*
(c) *Pterocarpus marsupium*
(d) *Acacia coincinna*

139. An example for wood oil is

(a) *Cedrus deodara*
(b) *Calophyllum inophyllum*
(c) *Eucalyptus globulus*
(d) *Lannea coramendallica*

140. Indian butter tree is

(a) *Calophyllum inophyllum*
(b) *Madhuca latifolia*
(c) *Mimusops elengi*
(d) *Pongamia pinnata*

141. Fruit-tan is obtained form

(a) *Anogeissus latifolia*
(b) *Terminalia chebula*
(c) *Cassia auriculata*
(d) *Acacia mollissima*

142. 'Chilgoza' is obtained from

(a) *Pinus roxburghii*
(b) *Pinus kesiya*
(c) *Pinus gerardiana*
(d) *Pinus ellioti*

143. Which one of the following is correctly matched?

(a) *Dysoxylum malabaricum* ____ Elm
(b) *Ulmus Wallichiana* ___ White cedar
(c) *Tectona grandis* _____ Ven teak
(d) *Picea smithiana* _____ Spruce

144. *Antiaris toxicaria* (upas) yielding a strong fibre belongs to the family

(a) Urticaceae (b) Moraceae
(c) Ulmaceae (d) Asclepiadaceae

145. The Kino of Indian pharmacopoeia often termed as 'Malabar Kino' is obtained from

(a) *Bombax malabaricum*
(b) *Dysoxylum malabaricum*
(c) *Pterocarpus marsupium*
(d) *Ailanthus grandis*

146. The common host plant of lac insect is
(a) *Azadirachta indica*
(b) *Dalbergia sissoo*
(c) *Butea monosperma*
(d) *Acacia auriculiformis*

147. The scientific name of Divi-Divi is
(a) *Mimusops littooalis*
(b) *Caesalpinia coriaria*
(c) *Berberis nepalensis*
(d) *Morinda tinctoria*

148. Trade/Common name of *Dalbergia latifolia* is
(a) Satin wood (b) Rose wood
(c) Axle wood (d) Amaltas

149. Deodar (*Cedrus deodara*) oil, which is used to cure ulcer and skin diseases is obtained from its
(a) wood (b) seed
(c) leaves (d) roots

150. *Indigofera tinctoriais* used in
(a) Medicine
(b) Pulp industry
(c) Ship building industry
(d) Dye industry

151. In the manufacture of braille paper which of the following species is used?
(a) *Eucalyptus grandis*
(b) *Eucalyptus tereticornis*
(c) *Eucalyptus globulus*
(d) *Eucalyptus citriodora*

152. Veneer is utilized for the manufacture of
(a) Plywood
(b) Black board
(c) Compregnated wood
(d) All of the above

153. Palmrosa oil is
(a) Flower oil
(b) Grass oil
(c) Root oil
(d) None of these

154. Commercial cultivation of *Saussuria lappa* is found in the state of
(a) Himachal Pradesh (b) Tamil Nadu
(c) Rajasthan (d) Orissa

155. Which of the following is correctly matched?
(a) *Cinnamomum tamela* - Cardamom
(b) *Butea monosperma* - Silk
(c) *Diospyrus melanoxylum* - Bidi leaf
(d) Rauwolfia - Climber

156. The trees species suitable for charcoal production is
(a) *Tectona grandis*
(b) *Terminalia arjuna*
(c) *Prosopis juliflora*
(d) *Mangifera indica*

157. Which of the following is correctly matched?
(a) Black babul - *Acacia leucopholea*
(b) White babul - *Acacia nilotica*
(c) Green wattle - *Acacia dalbata*
(d) Black wattle - *Acacia decurrens*

158. The part of Chinchona plant which yields quinine is
(a) Root (b) Leaf
(c) Flower (d) Bark

159. The raw materials for tanning industries are from
(a) *Acacia nilotica*
(b) *Acacia leucopholea*
(c) *Acacia mearnsii*
(d) None

160. Costus oil is obtained from
(a) *Veleriana wallichi*
(b) *Saussuria lappa*
(c) *Canarium strictum*
(d) *Pandanus tectorius*

161. The average tannin content from myrobalans is
(a) 28% (b) 30%
(c) 32% (d) 36%

ANSWERSHEET
7. Forest Utilization

1. (c)	2. (a)	3. (b)	4. (d)	5. (a)	6. (c)	7. (d)	8. (c)	9. (a)	10. (c)
11. (b)	12. (c)	13. (a)	14. (d)	15. (d)	16. (a)	17. (b)	18. (b)	19. (c)	20. (b)
21. (a)	22. (d)	23. (b)	24. (d)	25. (c)	26. (b)	27. (d)	28. (b)	29. (d)	30. (a)
31. (b)	32. (c)	33. (d)	34. (a)	35. (c)	36. (a)	37. (a)	38. (c)	39. (a)	40. (d)
41. (c)	42. (d)	43. (d)	44. (a)	45. (a)	46. (a)	47. (b)	48. (a)	49. (b)	50. (d)
51. (c)	52. (b)	53. (d)	54. (a)	55. (a)	56. (b)	57. (d)	58. (a)	59. (a)	60. (a)
61. (a)	62. (a)	63. (c)	64. (a)	65. (b)	66. (a)	67. (a)	68. (c)	69. (c)	70. (a)
71. (b)	72. (c)	73. (a)	74. (d)	75. (c)	76. (d)	77. (a)	78. (b)	79. (c)	80. (a)
81. (b)	82. (b)	83. (a)	84. (b)	85. (c)	86. (a)	87. (b)	88. (c)	89. (c)	90. (c)
91. (b)	92. (a)	93. (a)	94. (d)	95. (a)	96. (c)	97. (b)	98. (a)	99. (a)	100. (b)
101. (a)	102. (b)	103. (c)	104. (a)	105. (b)	106. (d)	107. (a)	108. (b)	109. (c)	110. (d)
111. (a)	112. (c)	113. (c)	114. (d)	115. (a)	116. (d)	117. (c)	118. (a)	119. (a)	120. (c)
121. (c)	122. (a)	123. (a)	124. (c)	125. (c)	126. (d)	127. (a)	128. (c)	129. (b)	130. (b)
131. (b)	132. (c)	133. (c)	134. (a)	135. (d)	136. (a)	137. (a)	138. (b)	139. (a)	140. (b)
141. (b)	142. (c)	143. (d)	144. (b)	145. (c)	146. (b)	147. (b)	148. (b)	149. (d)	150. (d)
151. (c)	152. (a)	153. (b)	154. (a)	155. (c)	156. (c)	157. (c)	158. (d)	159. (a)	160. (b)
161. (a)									

8

Wood Science and Technology

KEY POINTS

- Main characteristics of woody plants are Vascular tissues, Persistent and aerial stem, Secondary thickening, Lignification.
- Woods, and the trees are divided into two categories: hardwoods and softwoods
- Needlelike leaves characterize softwood tree
- Hardwoods and softwoods are included in the botanical division spermatophytes, meaning they produce seeds
- Hardwoods are in the subdivision angiospermae and softwoods are in the gymnospermae subdivision
- Example of Soft wood genera viz., *Pinus*, *Picea*, *Larix*, *Abies*, *Tsuga*, *Sequoia* (redwood), *Taxus*, *Taxodium*, *Juniperus*, *Thuja*, *Chamaecyparis* and *Libocedrus*
- Example of Hardwood genera viz., *Quercus* (oak), *Fraxinus* (ash), *Ulmus* (elm), *Acer* (maple), *Betula* (birch), *Fagus* (beech) and *Populus* (cottonwood, aspen)
- softwoods are nonporous (they do not contain vessels).
- Moisture content = $\frac{\text{Wet weight} - \text{Oven dry weight}}{\text{Oven dry weight}} \times 100$
- FSC - Forest Stewardship Council
- wood is composed of cellulose, lignin, hemicelluloses and minor amounts (usually <10%) of extraneous materials contained in a cellular structure
- Main Characteristics of wood are Cellular in structure, Hygroscopic, Biodegradable, Chemically stable, Insulatable and Combustible nature
- Wood is a collection of small cells
- Wood (xylem) is found inside a covering of bark, which is composed of an inner layer (phloem) and an outer protective layer (outer bark)
- Discontinuous rings – Growth rings sometimes fail to form around the complete cross section.
- False rings – Occasionally, normal seasonal growth is interrupted by events such as drought, late frosts, or defoliation by insects or hail
- The dark center area is known as heartwood, the lighter tissue as sapwood. It is in the sapwood that the living cells are found.
- Sapwood tissue also serves to conduct water upward in a living tree.

- Heartwood no longer functions physiologically but provides mechanical support to the tree.
- The direction parallel to the long axis of most of the long tapered fibers of wood is called the grain direction.
- Trees in which fibers are spirally arranged about the stem axis are said to have spiral grain.
- Wood is a carbohydrate composed principally of carbon, hydrogen and oxygen.
- Silica contents exceeding 0.5 per cent are relatively common in tropical hardwoods and in some species may exceed 2.0 per cent by weight.
- The elemental constituents of wood are combined into a number of organic compounds: cellulose, hemicellulose and lignin.
- Cellulose is manufactured directly from units of glucose
- Lignin is a complex and high molecular weight polymer built upon phenylpropane units.
- Lignin occurs between individual cells and within the cell walls.
- Wood cells that function primarily in the storage and conduction of food materials are known as parenchyma.
- All types of cells are characterized by secondary wall layers that are not continuous.
- The type of pit typifying nonparenchyma cell is called a borderea pit,
- A small portion of the volume of some softwood is composed of longitudinally oriented parenchyma cells.
- Vessel elements are generally much larger in diameter than other types of longitudinal cells.
- Juvenile wood has been defined as secondary xylem produced by cambial regions that are influenced by activity in the apical meristem
- Compression wood is highly undersirable in lumen or other solid wood products.
- Tension Wood is Reaction Wood of the hardwood species.
- Bark is low in strength compared to wood.
- Shrinking and swelling is an exactly reversible process in small pieces of stress-free wood.
- Plywood consists of thin sheets or veneers of wood which are glued together in a way that the grain of each veneer is at right angles to that of the adjacent veneer.
- Faces: Faces of face and back is the term used for denoting the outer piles or members in a plywood panel.
- Core: The centre ply or member is known as the core. It may comprise of veneer, timber or various combinations of veneer and timber.
- Cross bands: In its simplest form, plywood consists of an odd number of piles.
- Clipping - This process involves jointing and splicing of the veneers.
- Laminated wood is made up of layers of wood that are laid in a way that the grains of successive layers are parallel.
- Core boards are made up of a core formed by strips of wood of various dimensions that are bonded together by glues between veneers.
- Fibre boards are made from the fibres of wood or other lingo-cellulosic materials.

MULTIPLE CHOICE QUESTION'S

1. Identify the hardwood timber species
(a) Melia (b) Pinus
(c) Abies (d) Sequoia

2. The type of cells present in soft wood is
(a) Tracheids (b) Vessels
(c) Fibre (d) None of the above

3. A storage tissue in wood is
(a) Parenchyma (b) Collenchyma
(c) Schlerenchyma (d) None of above

4. Pick out the tree which yields soft fibre
(a) *Sterculia villosa* (b) *Albizia lebbeck*
(c) *Azadirachta indica*
(d) *Acacia nilotica*

5. The tree that serves as raw material for the manufacture of pencil is
(a) Adina (b) Juniperus
(c) Terminalia (d) Mesua

6. An example for wood oil is
(a) *Cedrus deodara*
(b) *Calophyllum inophyllum*
(c) *Eucalyptus globulus*
(d) *Lannea coramendallica*

7. Indian butter tree is
(a) *Calophyllum inophyllum*
(b) *Madhuca latifolia*
(c) *Mimusops elengi*
(d) *Pongamia pinnata*

8. Fruit-tan is obtained from
(a) *Anogeissus latifolia*
(b) *Terminalia chebula*
(c) *Cassia auriculata*
(d) *Acacia mollissima*

9. 'Chilgoza' is obtained from
(a) *Pinus roxburghii*
(b) *Pinus kesiya*
(c) *Pinus gerardiana*
(d) *Pinus ellioti*

10. Pick the odd man out: Defects due to rupture of tissues
(a) Twist (b) Split
(c) Check (d) Shake

11. Except one, the following refer to natural defects in a log. identify the exception
(a) Shake (b) Flute
(c) Spring (d) Check

12. Wavy-grain wood is attributable to
(a) *Dalbergia latifolia*
(b) *Tectona grandis*
(c) *Pterocarpus santalinus*
(d) *Santalum album*

13. Timbers which are capable of withstanding rapid seasoning are termed as
(a) Most refractory (b) Non refractory
(c) Medium refractory
(d) Refractory

14. One of the following species comes under highly refractory wood for seasoning
(a) *Tectona grandis* (b) *Shorea robusta*
(c) *Dalbergia sissoo* (d) *Bombax ceiba*

15. In softwood trees the conduction and mechanical functions are performed by
(a) Vessels (b) Fibres
(c) Tracheids (d) Pores

16. The Latin name for Andaman Padauk tree is
(a) *Veteria indica*
(b) *Pterocarpus dalbergioides*
(c) *Shorea robusta*
(d) *Dalbergia latifolia*

17. The Latin name for golden bamboo is
(a) *Bambusa bamboos*
(b) *Bambusa vulgaris*
(c) *Bambusa tulda*
(d) *Dendrocalamus strictus*

18. Indian Kino tree is
(a) *Pterocarpus santalinus*
(b) *Pterocarpus dalbergioides*
(c) *Pterocarpus marsupium*
(d) None

19. Name the proprietary preservative containing high boiling creosote as base
(a) Curpinol (b) Hardiproof
(c) Solignum (d) Celcure

20. The chief characteristics of an ideal wood preservative are
(a) *High toxicity*
(b) *High permanency and stability*
(c) *High penetrability*
(d) All of the above

21. Henna dye is obtained from
(a) *Lawsonia inermis*
(b) *Punica granatum*
(c) *Bixa oranalla*
(d) *Morinda tinctona*

22. Oleo resins are obtained from
(a) *Pinus roxburghii*
(b) *Carcinia Morella*
(c) *Vateria indica*
(d) *Canarium strictum*

23. Cutch is obtained from
(a) *Acacia nilotica*
(b) *Acacia catechu*
(c) *Pterocarpus marsupium*
(d) *Acacia coincinna*

24. The tree that serves as raw material for the manufacture of pencil is
(a) Adina (b) Juniperus
(c) Terminalia (d) Mesua

25. Density of trees in a forest stand is normally expressed as:
(a) Cubic feet per acre
(b) Square feet per acre
(c) Tons per acre
(d) Basal area of the stand in square feet

26. What is a choker?
(a) A wire cable or chain used to move logs
(b) An instrument for measuring tree diameters
(c) An obstruction in a stream
(d) A tree that does not fall after being cut

27. Nutrients and water absorbed from the soil move up through a tree in the:
(a) Xylem (b) Root hairs
(c) Phloem (d) All of the above

28. The practice of measuring the volume of logs or bolts after they are harvested is called:
(a) Cruising (b) Scaling
(c) Tallying (d) All of the above

29. Kerf refers to:
(a) A type of log scale
(b) Cutting width of a saw blade
(c) Log volume
(d) The distance between the chain and the bar

30. Treating wood posts and lumber with penta, creosote or CCA improves its __
(a) Attractiveness
(b) Resistance to decay and insects
(c) Strength
(d) Susceptibility to decay and insects

31. A table showing the estimated or calculated amount of lumber which may be sawed from logs of given length and diameter is called ______.
(a) Biltmore stick
(b) log rule
(c) Merritt Hypsometer
(d) Scale stick

32. Frilling involves______.
(a) Cutting the tree down
(b) Encircling a tree stem with downward, overlapping axe cuts
(c) Spraying herbicide on the tree leaves
(d) Using tubular injection

33. Which tree part listed below is considered non-living?

(a) Cambium (b) Heartwood
(c) Inner bark (d) Sapwood

34. An articulated log skidder is __________.

(a) A modified farm tractor
(b) Less maneuverable than a non-articulated skidder
(c) One with a hinge in the midsection
(d) Suitable only for gentle terrain

35. A source of revenue for a logging operation is __________.

(a) Depreciation (b) Income tax
(c) Log sales (d) Vehicle insurance

36. A false ring is __________.

(a) An extremely narrow ring
(b) An extremely wide ring
(c) A second ring for a given year
(d) A black ring

37. A Pulaski is used in firefighting for __.__

(a) Felling and clearing
(b) Grubbing and chopping
(c) Raking and sweeping
(d) Scraping and raking

38. Reducing a felled tree into marketable size logs is called __________.

(a) Bucking (b) Felling
(c) Limbing (d) Skidding

39. The function of the cambium of the tree is __________.

(a) Development of leaves
(b) Growth in height of the tree
(c) Growth of wood and bark cells
(d) Production of carbohydrates

40. Chromated zinc chloride is superior to zinc chloride as a preservative because of its ________.

(a) Lower cost
(b) Lower fire hazard
(c) Resistance to flaking
(d) Resistance to leaching

41. Chromated zinc chloride is superior to zinc chloride due to its ________.

(a) Lower cost
(b) Lower fire hazard
(c) Resistance to flaking
(d) Resistance to leaching

42. The part of the logging operation in which logs are dragged from where they were cut to a central location for loading is called ________.

(a) Bucking (b) Felling
(c) Snagging (d) Skidding

43. Water contained within the cell wall of wood is called ________.

(a) Bound water (b) Fiber water
(c) Free water (d) Suppressed water

44. Which tree part carries food made in the leaves down to branches, trunk and roots:

(a) Phloem (b) Sapwood
(c) Heartwood (d) Cambium

45. In order to "girdle a tree" which of the folowing must be completely severed?

(a) Phloem
(b) Xylem
(c) Cambium
(d) Phloem and Cambium

46. Which of the following statements best describes "heartwood"?

(a) Lighter colored wood just inside the bark
(b) Part of the tree stem which is always live wood
(c) Carries nutrients and water from the roots to the crown
(d) None of the above

47. Both the inner bark and sapwood are formed by cell divisions of the _____ layer.

(a) Phloem (b) Xylem
(c) Cambium
(d) All of the above

48. Which of the following is a direct product of photosynthesis?
(a) Oxygen
(b) Carbohydrates
(c) Fats and protein
(d) All of the above

49. Annual rings on a tree are composed of:
(a) Summerwood (b) Spring wood
(c) Both a and b (d) None

50. The only living part of the tree trunk is the ______.
(a) Heartwood (b) Sapwood
(c) Cambium layer (d) Outer bark

51. _____ carries the sap from the roots to the leaves.
(a) Heartwood (b) Xylem
(c) Phloem (d) Cambium layer

52. Closely spaced annual rings indicate ___.___
(a) Fast growth
(b) Slow growth
(c) Very tall trees
(d) Very short trees

53. _____ carries plant food manufactured in the leaves down to the branches - trunk and roots.
(a) Heartwood (b) Sapwood
(c) Phloem (d) Xylem

54. The ____________ builds wood tissue in the trunk.
(a) Cambium layer (b) Heartwood
(c) The leaf (d) Phloem

55. __________ gives support and strength to the tree.
(a) Outer bark (b) Sapwood
(c) Heartwood (d) Cambium layer

56. __________ helps to protect the tree trunk from injury.
(a) Cambium layer (b) Sapwood
(c) Heartwood (d) Outer bark

57. The manufacture of plant food in the leaves of a plant is called ________.
(a) Respiration (b) Transpiration
(c) Photosynthesis (d) Respiration

58. The center of a wood is called the ______.
(a) Cambium layer
(b) Sapwood
(c) Pith (d) Bole

59. The word "buck" is often used in the discussion of forest harvesting activities. What does this word mean?
(a) To put logs in a pile
(b) To cut trees into logs
(c) To push a log truck out of the mud
(d) To move logs from the stump to the landing

60. The three "E's" associated with the Forest Stewardship Program are:
(a) Environmental, esthetics and economic
(b) Energy, esthetics and economics
(c) Esthetics, environmental and energy
(d) Everlasting, esthetics and energy

61. The profession dealing with maintenance of ornamental trees is called:
(a) Silviculture (b) Dendrology
(c) Arboriculture (d) Hydrology

62. In forestry the letters TSI stand for the words ___.
(a) Total saleable investment
(b) Timber stand improvement
(c) Total site income
(d) Taxable site index

63. The first step in forest harvesting is ___.___
(a) Skidding (b) Felling & bucking
(c) Loading (d) Hauling

64. Bucking is ______.
(a) Cutting a tree down
(b) Cutting a down tree into log lengths
(c) Cutting the limbs off a standing tree
(d) Cutting the limbs off a down tree

65. Debris left after a logging operation is called ______.

(a) Duff (b) Waste
(c) Slash (d) Stubble

66. Which among these preservatives is the most eco-friendly?

(a) Copper-Chrome-Arsenic (ASCU)
(b) Acid-Cupric-Chrome (ACC)
(c) Chromated Zinc Chloride (CZC)
(d) Copper Chrome Boric (CCB)

67. Kraft paper is made entirely from unbleached pulp produced from

(a) Sulphite pulp
(b) Mechanical pulp
(c) Soda pulp
(d) Sulphate pulp

68. Soft woods are found to be most suitable for the production of

(a) Chemical Process
(b) Mechanical Process
(c) Soda Process
(d) Chemical Process

69. Paper, which is more or less moisture content of the wood should be about

(a) Filler (b) Sizing
(c) Beating (d) Loading

70. The addition of non-fibrous mineral matter to stock prior to sheet formation is known as

(a) Ageing (b) Loading
(c) Opacity (d) Beating

71. For the manufacture of rayon, the pulp should be contain more

(a) Lignin (b) Cellulose
(c) Hemi cellulose (d) Acetic acid

72. The cross banded construction of wood is called as

(a) Laminated wood
(b) Laminae wood
(c) Soft wood
(d) Ply wood

73. The caustic soda treated celloulose is called as

(a) Cellulose
(b) Hemi cellulose
(c) Alkali cellulose
(d) Cellulose Xanthation

74. Method of assembling of wood components in ply wood manufacture is called as

(a) Cross banded construction
(b) Balanced construction
(c) Lamination
(d) None of these

75. For the construction of arched or curved assemblies, which of the following is used

(a) Ply wood
(b) Laminated wood
(c) core board
(d) Particle board

76. Most common commercial chemical agents used for bleaching are

(a) Sodium compounds
(b) Chlorine compounds
(c) Acids compounds
(d) None of these

77. For manufacture of news print paper, which of the following species is used

(a) *Bambusa arundinacea*
(b) *Pinus roxburghii*
(c) *Boswellia serrate*
(d) *Eulaliopsis binata*

78. Commonly used chemical pulping process in India is

(a) Sulphate process
(b) Sulphite process
(c) Acidic Process
(d) Semi-chemical process

79. Excellent material for Aero plane propeller blades is

(a) Impreg (b) Compreg
(c) Staypak (d) Staybord

80. ASCU is which of the following type of preservative

(a) Oil type

(b) Organic solvent type

(c) Tar type

(d) Fixed water soluble

81. The wood preservative Arsenic – Copper-Chromate in the ratio of 1:3:4 comes under the category of

(a) Oil type

(b) Water soluble leachable type

(c) Water soluble fixed type

(d) Organic solvent type

82. The specific gravity of world's lightest wood is

(a) 0.4 (b) 0.44

(c) 0.044 (d) 0.004

83. Heating of wood of these in absence of oxygen to produce methanol is called as

(a) Wood saccharification

(b) Wood distillation

(c) Wood hydrolysis

(d) None of these

84. Paper, which is more or less impervious to the penetration of inks, is said to be

(a) Filler (b) Sizing

(c) Beating (d) Loading

85. Which of the following glue is commonly used in ply wood industries

(a) Phenolic glue

(b) Animal glue

(c) Starch glue

(d) Casein glue

86. Yield in sulphate pulping processes is

(a) 90-95% (b) 45-50%

(c) 70 % (d) 50-60%

87. Mechanical treatment given to fibres to increase tensile strength, bursting strength etc is

(a) Loading (b) Beating

(c) Bleaching (d) Sizing

88. Wood that is specially treated to increase its strength is called as

(a) Improved wood

(b) composite wood

(c) soft wood

(d) ply wood

89. Glues once cured are not softened by heat are called as

(a) Thermosetting glues

(b) thermoplastic glue

(c) Synthetic glues

(d) None of these

90. For treatment of Bamboo poles commonly used wood preservation process is

(a) Bethal Process

(b) Boucherie Process

(c) Lowry Process

(d) Boulton Process

91. Estimating the amount of board feet, cords, or pieces in a timber stand is called ________.

(a) Scaling (b) Guessing

(c) Browsing (d) Cruising

92. Which is the most durable part of a wooden fence post?

(a) Sapwood

(b) Heartwood

(c) Both are equally durable

(d) None

93. Decay or wood rot is caused by which of the following organisms?

(a) Bacteria

(b) Fungi

(c) Soil insects

(d) All of the above

94. A butt log is:

(a) A log with excessive swelling at one end

(b) A log with many defects

(c) The first log cut above the stump

(d) The smallest log (in diameter) cut from the tree

95. Logging costs typically represent what percent of the total cost of saw timber delivered to a mill?

(a) 33% (b) 50%
(c) 75% (d) 80%

96. When timber buyers openly bid against one another at a timber sale it is called a (an):

(a) Auction sale
(b) Mill scale sale
(c) Sealed bid sale
(d) Stumpage sale

97. The length of time needed to air dry or season wood depends on:

(a) Climate (b) Method of stacking
(c) Wood size (d) All of the above

98. A cant hook is used to_______.

(a) Roll logs (b) Load logs
(c) Fell logs (d) Skid logs

99. The darker colored interior wood near the tree's center, often no longer living, is called

(a) Core wood (b) Heartwood
(c) Sapwood (d) Terminal wood

100. A Pulaski is used in forest firefighting for _____.

(a) Felling and clearing
(b) Grubbing and chopping
(c) Raking and sweeping
(d) Scraping and raking

101. The width of the cut made by a saw is called the ______.

(a) Cant (b) Gap
(c) Kerf (d) Notch

102. Paper was first made from ground mulberry bark by the _______.

(a) American (b) Chinese
(c) Egyptians (d) Indians

103. The tree which has more than one leader stem is called

(a) Forked (b) Buttressed
(c) Epicormic branches (d) Fluted

104. Species suitable for structural timber purpose (Pick the odd man out)

(a) *Pterocarpus santalinus*
(b) *Tectona grandis*
(c) *Dalbergia latifolia*
(d) *Ailanthus excelsa*

105. An example for ring porous wood is

(a) Dipterocarpus (b) Gmelina
(c) Hardwickia (d) Quercus

106. Pick out the tree which yields soft fibre

(a) *Sterculia villosa*
(b) *Albizia lebbek*
(c) *Azadirachta indica*
(d) *Acacia nilotica*

107. Root rot in sal is caused by

(a) Pseudomonas (b) Polyporus
(c) Ganoderma (d) Pythium

108. The species, which express rapid growth in the beginning of season, is called as

(a) Late wood (b) Early wood
(c) Spring wood (d) Summer wood

109. In broad-leaved species the mechanical function is performed by

(a) Vessels (b) Fibres
(c) Tracheids (d) Tyloses

110. In softwoods conduction and mechanical functions are performed by

(a) Vessels (b) Fibres
(c) Tracheids (d) Tyloses

111. In broad leaved species the conduction function is performed by

(a) Vessels (b) Fibres
(c) Tracheids (d) Tyloses

112. In ring porous wood the vessels are

(a) Larger (b) Smaller
(c) Medium (d) Intermediate

113. Pick the wavy grained wood

(a) *Pterocarpus santalinus*
(b) *Santalum album*
(c) *Pterocarpus marsupium*
(d) *Terminalia chebula*

114. The capacity of wood to bend out the shape without rupture is called
(a) Elasticity (b) Flexibility
(c) Fissibility (d) None of these

115. The capacity of wood to split along with fibres is termed as
(a) Elasticity (b) Flexiility
(c) Fissiility (d) None of these

116. The fibres saturation point for tropical timbers ranged between
(a) 0 – 15% (b) 15 – 30%
(c) 30 – 45% (d) 45 – 60%

117. Pick out the soft wood
(a) *Tectona grandis*
(b) *Albizzia lebbeck*
(c) *Chukrasia tabularis*
(d) *Toona ciliata*

118. Good fuel woods (Pick the odd man out)
(a) *Acacia nilotica*
(b) *Casuarina equisetifolia*
(c) *Chloroxylon swietenia*
(d) *Ailanthus excelsa*

119. Species suitable for cricket bat
(a) *Acacia nilotica*
(b) *Ailanthus excelsa*
(c) *Diospyros melanoxylon*
(d) *Salix alba*

120. Tyloses found only in the heart wood of .
(a) Soft wood (b) Hard wood
(c) Reaction wood (d) Ringporous wood

121. Ringporous wood species
(a) *Acacia nilotica*
(b) *Hardwickia binnata*
(c) *Azadiracta indica*
(d) *Tectona grandis*

122. Majority of Indian dicotyledonous timbers are
(a) Ring porous wood
(b) Semi ring porous wood
(c) Diffuse porous wood
(d) Reaction wood

123. Horizontally arranged cells, running radially from bark towards the centre of tree
(a) Vessels (b) Fibres
(c) Tracheids (d) Medullary rays

124. The layer that present between xylem and phloem is
(a) Cambium (b) Bark
(c) Meristem (d) None

125. Sal is a
(a) Ring porous wood
(b) Diffuse porous wood
(c) Intermediate wood
(d) None of these

126. The institute of wood science and technology is located in
(a) Thrissur (b) Jodhpur
(c) Jabalpur (d) Bangalore

127. The basic structural and physiological unit of a plant
(a) Cell (b) Fiber
(c) Tracheid (d) Cambium

128. Secondary thickening in wood is caused by
(a) Xylem (b) Phloem
(c) Cambium (d) Meristem

129. The lignin content of hard woods range between
(a) 5 and 10% (b) 8 and 18%
(c) 18 and 25% (d) 25 and 40%

130. The region in which cells divide repeatedly and form new cells is termed as
(a) Surface region
(b) Meristematic region
(c) Cambial region
(d) None of these

131. The water held by capillary attraction inside the empty spaces in and between the cell is termed as
(a) Free water
(b) Absorbed water
(c) Bound water
(d) None of these

132. The formula for estimating the moisture content of wood is
(a) (OD weight – Green wt / OD weight) X 100
(b) (Green wt– OD weight / Green wt) X 100
(c) (Green wt– OD weight / OD weight) X 100
(d) None

133. The density of wood is directly related to
(a) Stability
(b) Durability
(c) Flexibility
(d) Elasticity

134. Urea formaldehyde (UF) is used as
(a) Fire retardant (b) Wood adhesive
(c) Insect repellant
(d) Chemical preservative

135. Tree with high cellulose content
(a) Eucalyptus (b) Acacia
(c) Albizia (d) Anthocephalus

136. Tree with long fiber length
(a) Eucalyptus
(b) Bamboos
(c) Casuarina
(d) Leucaena

137. The ability of cell wall to reflect light is termed
(a) Colour (b) lusture
(c) Odour (d) None of these

138. Heavy wood (Tick odd man out)
(a) *Acacia nilotica*
(b) *Dalbergia sissoo*
(c) *Tectona grandis*
(d) *Abies pindrow*

139. Pick out most refractory wood
(a) *Sizygium cuminii*
(b) *Adina cordifolia*
(c) *Bombax ceiba*
(d) *Morus alba*

140. Choose the non refractory wood
(a) *Tectona grandis*
(b) *Veteria indica*
(c) *Mangifera indica*
(d) *Pinus roxburghi*

141. Moluscan borer (Tick the odd man out)
(a) Xylotria (b) Teredo
(c) Nausitoria (d) Limnoria

142. Perishable wood
(a) *Tectona grandis*
(b) *Albizia lebbeck*
(c) *Boswellia serrata*
(d) *Dalbergia paniculata*

143. The function of the phloem in the tree is to ____.
(a) Develop leaves from nutrients
(b) Conduct food manufactured in the leaves to the branches, trunk and roots
(c) Grow wood and bark cells
(d) Produce carbohydrates through photosynthesis

144. Epiblema of roots is equivalent to
(a) Pericycle (b) Endodermis
(c) Epidermis (d) Stele

145. Phellogen and Phellem respectively denote
(a) Cork and cork cambium
(b) Cork cambium and cork
(c) Secondary cortex and cork
(d) Cork and secondary cortex

146. In which of the following pairs of parts of a flowering plant is epidermis absent?
(a) Root tip and shoot tip
(b) Shoot bud and floral bud
(c) Ovule and seed
(d) Petiole and pedicel

147. A piece of wood having no vessels Trache(a) must be belong to
(a) Teak (b) Mango
(c) Pine (d) Palm

148. A plant tissue, when stained, showed the presence of hemicellulose and pectin in cell wall of its cells. The tissue represents
(a) Collenchyma (b) Sclerenchyma
(c) Xylem (d) Meristem

149. Fibres are likely to be absent in
(a) Secondary phloem
(b) Secondary Xylem
(c) Primary phloem
(d) Leaves

150. A vesselless piece of stem possessing prominent sieve tubes would belong to
(a) Pinus (b) Eucalyptus
(c) Grass (d) Trochodendron

151. Major components of wood
(a) Cellulose (b) Hemi cellulose
(c) Lignin (d) All the above

152. Vessels are present in
(a) Hard Wood (b) Heart Wood
(c) Sap wood (d) Soft Wood

153. In seasoning of wood, the rate of drying depends on
(a) Heat
(b) Humidity or dryness of the air
(c) Circulation of the air around the timber
(d) All the above

154. Based on seasoning behavior, the woods can be classified into
(a) 3 categories
(b) 4 categories
(c) 5 categories
(d) 6 categories

155. In Air Seasoning, the function of the crossers is
(a) Controls the dryness of air
(b) Allow uniform circulation of air around the timber
(c) Both a & b
(d) To maintain Relative Humidity

156. Serious wood destroying animals are
(a) Elephant
(b) Crustaceans and Molluscs
(c) Deers
(d) Borers

157. Ingredients of ASCU are
(a) Aresenic Penta Oxide + Copper Sulphate + Na or K di chromate
(b) Aresenic Penta Oxide + Zinc Sulphate + Na or K di chromate
(c) Aresenic Penta Oxide + Mercuric Chloride + Na or K di chromate
(d) Aresenic Penta Oxide + Boric Acid + Na or K di chromate

158. In plywood, the grains of each veneers arranged
(a) Parallel to adjacent veneer
(b) Right angle to adjacent veneer
(c) According to the end use
(d) None

159. Compregnated wood is
(a) Compressed wood
(b) Impregnated wood
(c) Compressed and impregnated wood
(d) None

160. In Laminated Veneer Lumber manufacturing, the adhesives used
(a) Phenol Formaldehyde
(b) Urea Formaldehyde
(c) Melamine Formaldehyde
(d) All the above

161. In pulp and paper manufacturing process, commonly used fillers
(a) Soap
(b) Talc
(c) Rosin
(d) Aluminum sulphate

162. Radial growth in trees due to the activity of
(a) Cambium
(b) Parenchyma cells
(c) Rays
(d) Sclerenchyma cells

163. The following microstructures affects the drying rate

(a) Tyloses

(b) Rays

(c) Sclerenchyma

(d) None

164. The lightest wood in the world is a

(a) Soft wood

(b) Hard wood

(c) Gymnosperms

(d) All the above

165. ------------------ is the reciprocal of resistivity and is evaluated in units mhos per centimeter

(a) Conductivity (b) Gravity

(c) Density (d) None

ANSWERSHEET

8. Wood Science And Technology

1. (a)	2. (a)	3. (d)	4. (a)	5. (b)	6. (a)	7. (b)	8. (b)	9. (c)	10. (a)
11. (c)	12. (c)	13. (b)	14. (b)	15. (c)	16. (b)	17. (b)	18. (c)	19. (c)	20. (d)
21. (a)	22. (a)	23. (b)	24. (b)	25. (d)	26. (a)	27. (a)	28. (b)	29. (b)	30. (b)
31. (b)	32. (b)	33. (b)	34. (c)	35. (c)	36. (c)	37. (b)	38. (a)	39. (c)	40. (d)
41. (d)	42. (d)	43. (a)	44. (a)	45. (d)	46. (d)	47. (c)	48. (b)	49. (c)	50. (c)
51. (b)	52. (b)	53. (c)	54. (a)	55. (c)	56. (d)	57. (c)	58. (c)	59. (b)	60. (a)
61. (c)	62. (b)	63. (b)	64. (b)	65. (c)	66. (d)	67. (b)	68. (b)	69. (b)	70. (b)
71. (b)	72. (d)	73. (d)	74. (a)	75. (a)	76. (b)	77. (d)	78. (a)	79. (b)	80. (d)
81. (c)	82. (c)	83. (b)	84. (b)	85. (a)	86. (b)	87. (b)	88. (a)	89. (a)	90. (b)
91. (d)	92. (b)	93. (b)	94. (c)	95. (a)	96. (a)	97. (d)	98. (a)	99. (b)	100. (b)
101. (c)	102. (b)	103. (a)	104. (a)	105. (d)	106. (a)	107. (b)	108. (b)	109. (b)	110. (c)
111. (a)	112. (a)	113. (a)	114. (b)	115. (c)	116. (b)	117. (d)	118. (d)	119. (d)	120. (b)
121. (d)	122. (d)	123. (d)	124. (a)	125. (b)	126. (d)	127. (a)	128. (c)	129. (c)	130. (b)
131. (a)	132. (c)	133. (b)	134. (b)	135. (a)	136. (b)	137. (b)	138. (d)	139. (a)	140. (a)
141. (d)	142. (d)	143. (b)	144. (d)	145. (b)	146. (a)	147. (c)	148. (a)	149. (c)	150. (d)
151. (d)	152. (a)	153. (d)	154. (a)	155. (b)	156. (b)	157. (a)	158. (b)	159. (c)	160. (d)
161. (b)	162. (c)	163. (a)	164. (a)	165. (a)					

9

Forest Botany

KEY POINTS

- Taxonomy is the science of systematically naming and organizing organisms into similar groups
- Taxonomic revolution and explosion was the era between 1960 and 1967
- Chemotaxonomy deals with biochemical profiles for various taxa and assessing phylogenetic relationships
- Numerical taxonomy deals with data from various sources and their analysis with the help of computers to develop quantitative methods of classification
- Phylogene – evolution of races, i.e., fish, amphibian, mammals, human
- Antageny – evolution/developments of the organism. It is a repetition of phytogeny
- The plant kingdom is divided into two taxa viz., bryophytes and vascular plants
- The seeded plants divides into two taxa, Gymnosperms and Angiosperms
- ICBN - International Code of Botanical Nomenclature (ICBN)
- Carolus Linnaeus – Father of Modern Botany
- Alberts Magnus (1193–1280), for the first time distinguished between monocotyledons and dicotyledons
- Species Plantarum book was published by Linnaeus
- Holotype - Anyone specimen designated by the author as nomenclature type
- Isotype – A duplicate specimen of the holotype
- Lectotype – Specimen selected from the original material to serve as a nomenclatural type when no holotype was designed earlier
- The system of naming every plant in two parts in Latin words is called "Binomial System of Nomenclature
- Binomial nomenclature was first adopted by Linnaeus published in his first edition of "Species Plantarum" in 1753
- Primary meristems are differentiated into three viz., Protoderm, Ground Meristem, Procambium
- Line is a sub-group of cultivar propagated by seed
- Strain is a sub-group of cultivar with specific characteristics
- Epidermis is the single exterior layer that protects the stems, leaves, flowers and roots
- Parenchyma tissues are made of simple, thin-walled cells

- The parenchyma cells become a storage unit called the cortex
- Sclerenchyma tissues are thick-walled support cells found throughout the plant as fiber
- Leaf epidermis produces a wax called cutin
- Stem and root epidermis make subrin
- Receptacle – base of the flower
- Flower containing sepals, petals, stamens and pistil called complete flowers
- Flower lacking sepals, petals, stamens, and/or pistils called incomplete flowers
- Flowers containing male and female parts called perfect flowers
- Flowers that lack either male or female parts called imperfect flowers
- Pistillate – flowers containing only female parts
- Staminate – flowers containing only male parts
- Monoecious – plants with separate male flowers and female flowers on the same plant
- Dioecious– plants with male flowers and female flowers on separate plants
- Gynoecious – plants with only female flowers
- Andromonoecious – plants with only male flowers
- *Liriodendron tulipfera* is known as the Tulip Tree in the north and as Yellow Poplar in the south
- Annuals - Plants that complete their life cycle in one year
- Biennials - Plants that complete their life cycle in more than one year
- A chromatid is one copy of a duplicated chromosome, which is generally joined to the other copy by a single centromere
- Coleoptile is the pointed protective sheath covering the emerging shoot in monocotyledons such as grasses
- Reproductive structure of gymnosperms are called cones
- Dicot stem, vascular bundles are arranged in a ring around the pith. This type of stele is called eustele
- Falgellum is a hair like structure acts as a primary organelle for locomotion in the cells
- A layer of polysaccharide slime produced by seeds upon imbibition is called Mucilage
- Absorption and Anchorage are the primary functions of root
- Anther and Filament are collectively called Stamen
- Central part of the ground tissue system is known as pith or Medulla
- Ginkgo biloba – a “living fossil”, male and female tree, used as a medicinal plant
- Chloroplasts are usually Absent in the epidermis of dicot stem
- Conjunctive tissue of dicot roots usually consists of Parenchymatous cells
- Conjunctive tissue of monocot roots is usually made up of Sclerenchymatous cells
- Cortex is made up of Parenchyma cells
- Cortex occurs between the epidermis and pericycle of ground tissue system
- Cryptogams is a plant that produces spores.
- Cuticle is made up of a waxy substance known as Cutin
- Cuticle is a thin layer impregnated with wax
- Cytoskeleton is absent in Prokaryotic cells

- Dehiscent cap of a seed or a fruit that opens during germination is called Operculum
- Dicot stem have open vascular bundle tissue system
- Endoplasmic reticulum is Absent in the Prokaryotic Cell
- Endoplasmic reticulum is Present in the Eukaryotic Cell
- Endosperm of a seed is the result of Double fertilization
- Epidermal tissue is made up of Epidermal cells
- Golgi apparatus is Absent in the Prokaryotic Cell
- In dicot roots the pith is generally Absent
- Lateral branches of the stem are Exogenous in origin.
- Leaf blade is known as lamina
- Leaf is the chief organ of Photosynthesis
- Seed is the mature and fertilized ovule
- Secondary growth is generally Absent in monocot roots
- Secondary growth is generally Present in dicot roots
- Suicide sacs of the cell is known as *Lysosomes*
- The epidermis is made up of Parenchyma cells
- The powerhouses of the cell is *Mitochondria*
- Cytology - Study of cytogenetics
- *Ethnoecology* is the scientific study of the way different groups of people in different locations understand ecosystems around them; the environments in which they live; and their relationship with these
- *Archaeoethnobotany* is the study of the ethnobotany of identification of the plant materials from archaeological sites for studies on migration of human cultures, and origin, dispersal and domestification of crops etc
- Study of Fungi is mycology
- Study of Algae - phycology
- Ethnobotany is the study of the relationships between plants and people
- Dendrology is the sub filed of botany which deals with the study of woody plants.

Anacardiaceae	Mango/Cashew family
Annonaceae	Custard apple family
Apocynaceae	Oleander/Nerium family
Betucaceae	Alder family
Bignoniaceae	Bignonia family
Bixaceae	Bixa family
Boraginaceae	Borage family
Burseraceae	Torch wood family
Caesalpinaceae	Cassia/Amaltus family
Casuraiaceae	Beef wood family
Glusiaceae/Guttiferae	Garcinia family

Combretaceae	Terminalia family
Compositae	Sunflower family
Dipterocarpaceae	Sal family
Ebenaceae	Tendu family
Ericaceae	Heath family
Euphorbiaceae	Euphorbia family
Fabaceae / `Papilionaceae	Pea family
Fagaceae	Beech/Oak family
Labiatae	Mint family
Lauraceae	Laurel family
Lythraceae	Lawsonia family
Magnoliaceae	Magnolia family
Malvaceae	Cotton family
Meliaceae	Neem/Mahogany family
Mimosaceae	Acacia family
Moraceae	Mulberry family
Myrtaceae	Eucalyptus family
Myristicaceae	Nutmeg family
Oleaceae	Olive family
Proteaceae	Grevillea
Ranunculaceae	Buttercup or crowfoot family
Rhamnaceae	Ber family
Rhizophoraceae	Mangrove family
Rubiaceae	Ixora family
Rutaceae	Citrus family
Salicaceae	Salix/ willow family
Santalaceae	Sandal family
Sapindaceae	Soapnut / soapberry
Sapotaceae	Madhuca, Manilkara
Simarubaceae	Quassia family
Solanaceae	Nigh shade family
Sterculiaceae	Sterculia family
Tamaricaceae	Tamarix spp
Tiliaceae	Linden/Jute family
Ulmaceae	Elm family

Species	Common Name
Pinus wallichiana	Indian blue pine / 5-leaved pine
Pinus roxburghii	Chir pine /3-leaved pine
Oxandra lanceolata	Lancewood
Adansonia spp	Bottle trees
Bombax ceiba	Cotton tree or tree cotton
Bixa orellana	Achiote / Aploppas/ lipstick tree
Liriodendron tulipifera	Tulip tree
Magnolia grandiflora	Southern magnolia or bull bay
Annona squamosa	Sugar apple, sweetsop
Annona reticulate	Custard apple
Theobroma cacao	Cacao, Food of the Gods
Sterculia foetida	Wild almond tree
Sterculia monosperma	China chestnut, Seven Sisters' Fruit
Abroma augusta	Devil's cotton
Cola nitida	Cola nut tree
Heritiera littorals	Sundri
Dipterocarpus turbinatus	Gurjun
Vateria indica	White Damar / Indian Copal Tree
Azadirachta indica	Mahogany/ Neem
Melia azedarach	Chinaberry tree, bead-tree /Cape lilac
Swietenia macrophylla	Honduras Mahogany
Swietenia mahagoni	West Indian Mahogany
Chloroxylon swietenia	Satinwood
Chikrassia tabularis	Chickrassy
Khaya senegalensis	African mahagony
Ailanthus excels	Tree of the gods or Tree of heaven
Simarouba glauca	Paradise tree
Aesculus indica	Indian Horse Chestnut
Dodonea viscose	Virali
Filicium decipiens	Fern tree
Litchi chinensis	Litchi
Sapindus emarginatus	Soapnut
S. mukorossi	Chinese soap berry
Schleichora oleosa	Lac tree / Kusum
Madhuca indica	Mahwa or mahua or Iluppai
Donella roxburghii	Star apple / Kattu Illuppai
Madhuca longifolia	Nattu Illuppai

Rauvolfia serpentine	Sarpagandi
Spathodea campanulata	Fountain Tree
Tectona grandis	Common teak
Tectona hamiltoniana	Dahat teak
Tectona philippinensis	Philippine teak
Pogostemon perilloides	Patchouli
Hevea brasiliensis	Para rubber tree
Manihot glaziovii	Manicoba rubber
Artocarpus altilis	Bread fruit
Artocarpus heterophyllus	Jack fruit
Artocarpus hirsutus	Aini
Broussonetia papyrifera	Paper mulberry
Castilla elastic	Mexican rubber tree
Ficus elastic	India rubber tree
Helianthus annuus	Sunflower
Carthamus tinctorius	Safflower
Buchanania angustifolia	Cuddappah almond
Feronia limonia	Elephant apple or wood apple
Murraya koenigii	Curry leaf tree
Garcinia indica	Kokam butter tree
G. mangostana	Mangusteen
Mesua ferrea	Iron wood
Ulmus wallichiana	Himalayan elm
Trema orientalis	Charcoal tree
Quercus alba	White oak
Taxus baccata	European Yew
Pinus mugo	Mountain Pine
Pinus roxburghii	Chir Pine
Pinus caribaea	Caribbean Pine
Pinus wallichiana	Blue Pine
Pinus gerardiana	Chilgoza Pine
Picea glauca	White Spruce
Picea mariana	Black Spruce
Picea rubens	Red Spruce
Picea purpurea	Purple Spruce
Abies pindrow	Pindrow Fir or West Himalayan Fir
Abies concolor	White Fir
Abies alba	Silver Fir

Juniperus indica	Black Juniper
Juniperus excels	Greek Juniper
Eucalyptus globulus	Blue gum
Eucalyptus grandis	Rose gum
Eucalyptus tereticornis	Red gum

MULTIPLE CHOICE QUESTION'S

1. The branchy of science deals with the naming and classifying of plants?
(a) Plant Taxonomy
(b) Plant Pathology
(c) Plant Morphology
(d) Plant Genetics

2. The classification used to name all living organisms is properly called ________.
(a) Binomial classification
(b) Latin names
(c) Scientific names
(d) All of the above

3. The basic unit of classification is
(a) Genus (b) Species
(c) Family (d) Taxon

4. Floating unicellular plants found in oceans and freshwater are called
(a) Algae
(b) Zooplanktons
(c) Phytoplanktons
(d) Epiphytes

5. Carolus Linnaeus proposed .... system of classification
(a) Phylogenetic (b) Two kingdoms
(c) Five Kingdoms (d) Natural

6. Who is known as the 'father of botany'?
(a) Theophrastus
(b) Linnaeus
(c) Aristotle
(d) Mendel

7. Thalloid and non flowering plants are known as-------
(a) Angiosperm (b) Thallophyta
(c) Gymnosperm (d) Dicotyledons

8. Artificial system of classification of plants was proposed by
(a) British botanist
(b) Swedish botanist
(c) German botanist
(d) Indian botanist

9. ________ is a sexual system of classification
(a) Artificial system
(b) Natural system
(c) Phylogenetic system
(d) Natural selection

10. The Binomial system was introduced by
(a) Carolus Linnaeus
(b) Gaspard Bauhin
(c) Sir Joseph Dalton Hooker
(d) Adolf Engler

11. Genera plantarum of Bentham and Hooker was published in a
(a) A single volume
(b) Two volumes
(c) Three volumes
(d) Four volumes

12. The present day 'orders' were referred in Bentham and Hooker classification as
(a) Series (b) Cohorts
(c) Orders (d) Families

13. Plants having flowers with free petals are placed under
(a) Monochlamydeae
(b) Monocotyledons
(c) Gamopetalae
(d) Polypetalae

14. Inferae includes
(a) 6 orders and 34 families
(b) 4 orders and 23 families
(c) 3 orders and 9 families
(d) 5 orders and 27 families

15. How many families were described by Bentham and Hooker in their classification?
(a) 204 (b) 212
(c) 202 (d) 102

16. In Bentham and Hooker's classification of plants, the present by "families" were referred to by them as
(a) Families (b) Cohorts
(c) Orders (d) Series

17. Thalamiflorae includes
(a) 4 orders and 23 families
(b) 6 orders and 34 families
(c) 5 orders and 27 families
(d) 3 orders and 12 families

18. Which one of the following series includes the epigynous flowers?
(a) Thalamiflorae (b) Disciflorae
(c) Inferae (d) Heteromerae

19. The family included under the series Unisexuales is
(a) Solanaceae (b) Euphorbiaceae
(c) Malvaceae (d) Musaceae

20. *Thespesia populnea* belongs to
(a) Solanaceae (b) Euphorbiaceae
(c) Malvaceae (d) Musaceae

21. Malvaceae is placed in the series
(a) Thalamiflorae (b) Inferae
(c) Heteromerae (d) Disciflorae

22. Anthers are monothecous in
(a) Solanaceae (b) Euphorbiaceae
(c) Malvaceae (d) Musaceae

23. In *Abelmoschus esculentus,* the fruit is
(a) Drupe (b) Schizocarp
(c) Regma (d) Loculicidal capsule

24. The change from meristematic tissue to permanent tissue is called
(a) Differentiation. (b) Self perpetuating
(c) Photosynthesis. (d) Cell division.

25. The type of tissue presents in the petioles of banana and Canna, is
(a) Stellate parenchyma
(b) Prosenchyma
(c) Aerenchyma
(d) Chlorenchyma.

26. The tissue generally present in all organs of plant is
(a) Parenchyma (b) Chlorenchyma
(c) Collenchyma (d) Sclerenchyma

27. The lamellar collenchyma is seen in the hypodermis of
(a) Datura (b) Helianthus
(c) Ipomoea (d) Nicotiana

28. The root hairs are produced from
(a) Rhizodermis
(b) Trichomes
(c) Accessory cells
(d) Trichoblasts

29. Pick the unicellular organism
(a) Fungi (b) Protozoa
(c) Plants (d) Animals

30. Bicollateral vascular bundles are seen in the family members of
(a) Malvaceae (b) Musaceae
(c) Solanaceae (d) Cucurbitaceae

31. Anther and Filament are collectively called
(a) Stamen (b) Androecium
(c) Androphore (d) Antephyllus

32. The casparian strips are found in the endodermis of
(a) Dicot stem (b) Dicot root
(c) Monocot stem (d) Dicot leaf.

33. The passage cells are found in endodermis of
(a) Dicot stem (b) Monocot stem
(c) Dicot root (d) Dicot leaf.

34. The polyarch condition is found in
(a) Monocot leaf (b) Dicot leaf
(c) Dicot stem (d) Monocot root

35. The inner most layer of the cortex is
(a) Epidermis (b) Hypodermis
(c) Endodermis (d) Pericycle

36. In angiosperms, the strucutres that produce male gametophyte are called
(a) Anthers (b) Sepals
(c) Pollen tubes (d) Stigmas

37. The outermost circle of flower is
(a) Sepals (b) Petals
(c) Carpels (d) Corollas

38. Pollination occurs when pollen lands on the
(a) Style (b) Stigma
(c) Filament (d) Anther

39. The seed leaves of flowering plants are known as
(a) Endosperm (b) Carpels
(c) Radicals (d) Cotyledons

40. In seed plants, the structure that encloses the male gametophyte and transports it to another plant is called
(a) Pollen grain (b) Seed
(c) Flower (d) Pollinator

41. An animal cell does not have _______
(a) Nucleus (b) Plastid
(c) Cell membrane (d) Centriole.

42. This layer of parenchyma cells protects the vulnerable root tip
(a) Cambium (b) Rootcap
(c) Endodermis (d) Pericycle.

43. Cell sap is found in ________
(a) Chloroplasts
(b) Mitochondrion
(c) Central vacuole
(d) Cytoplasm

44. This vascular tissue cell type is not found in gymnosperm
(a) Tracheid (b) Sieve tube element
(c) Vessels (d) Phloem.

45. This tissue is living at maturity and has unevenly thickened cell walls; acts to provide flexible support for growing stems.
(a) Parenchyma (b) Xylem
(c) Phloem (d) Collenchyma.

46. In monocot roots which types of vascular bundles are found?
(a) Collateral, conjoint and closed
(b) Radial with exarch xylem
(c) Bicollateral, conjoint and closed
(d) Radial with endarch xylem

47. The modified stem of Opuntia is:
(a) Phyllode (b) Phylloclade
(c) Cladode (d) Stamroinode

48. "Pheromones" are secreted by the cells for the following function
(a) All plants for growth an development
(b) Yeast for facilitating mating
(c) All fungi for sexual reproduction
(d) Rhizopus for formation of zygospore

49. What are the two types of tissues?
(a) Meristematic and permanent
(b) Meristematic and temporary
(c) Meristems and temporary
(d) None of the above

50. Parenchyma is a type of ____________.
(a) Complex tissue (b) Simple tissue
(c) Xylem (d) Phloem

51. Parenchyma: Simple:: Phloem: _______.
(a) Simple (b) Collenchyma
(c) Complex (d) Xylem

52. Transpiration and exchange of gases are the functions of
(a) Stomata
(b) Xylem
(c) Both (a) and (b)
(d) Neither (a) nor (b)

53. _______tissue forms new cells in plants.
(a) Meristematic (b) Permanent
(c) Simple (d) None of these

54. _____ tissues prevent loss of water in plants.
(a) Protective (b) Xylem
(c) Phloem (d) All of the above

55. A group of cells alike in form, function and origin are called _________.
(a) Tissue (b) Organ
(c) Organelle (d) None of these

56. Plant length is increased by _________.
(a) Apical meristems
(b) Lateral meristems
(c) Periblem
(d) Parenchyma

57. The cell division is restricted to ___.______
(a) Meristematic cells
(b) Permanent cells
(c) Secretory cells
(d) All the above

58. Active cell takes place in _.__________
(a) Xylem (b) Phloem
(c) Sclerenchyma (d) Cambium

59. Collenchyma mainly forms _________.
(a) Hypodermis
(b) Epidermis
(c) Phloem
(d) Inner cortex

60. Which of the following tissues is composed of mainly dead cells?
(a) Phloem (b) Epidermis
(c) Xylem (d) Endodermis

61. Phloem in the plants perform the function of __________.
(a) Conduction of food
(b) Conduction of water
(c) Providing support
(d) Photosynthesis

62. Which of the following characteristics is true of all protists?
(a) Contains a nucleus
(b) Photosynthetic
(c) Has a cell wall
(d) Very small

63. What organisms are most numerous on the Earth?
(a) Eukaryotes (b) Archaea
(c) Prokaryotes (d) Plants

64. In an aqueous environment microscopic animals and plants are collectively known as
(a) Herbivores (b) Fauna and Flora
(c) Planktons (d) Symbionts

65. Which of the following types of plants is not able to self-pollinate?
(a) Dioecious (b) Monoecious
(c) Ccomplete (d) Wind-pollinated

66. In flowering plants, pollen is released from the
(a) Anther (b) Stigma
(c) Carpel (d) Filament

67. The Embryonic root is called?
(a) Plumule (b) Hypocotyl
(c) Epicotyl (d) Radicle

68. Which of these structures is unique to the seed of a monocot?
(a) Cotyledon (b) Endosperm
(c) Coleoptile (d) Radicle

69. Fruits are developed from
(a) Microsporangia.
(b) Receptacles.
(c) Fertilized eggs
(d) Ovaries.

70. The process by which the micros-pores of a seed plant are transferred to another compatible plant is
(a) Pollination
(b) Fertilization
(c) Germination
(d) Reproduction

71. What is the botanical name for the male reproductive organ in flowering plants?
(a) Calyx (b) Frond
(c) Carpel (d) Stamen

72. Which of these terms is used to describe the means by which the male and female cells combine to form a new cell which then eventually develops into a new embryo?
(a) Fertilization (b) Fusion
(c) Respiration (d) Ptyxis

73. What is the name of the structure that is formed from the ovule of a plant after fertilization has taken place?
(a) Xylem (b) Seed
(c) Scion (d) Nut

74. What is the name of the ripened ovary which contains the seeds in many plants?
(a) Fruit (b) Leaf
(c) Flower (d) Cellulose

75. The plant called that sheds its leaves before winter, or during a dry spell is called
(a) Deciduous (b) Indehiscent
(c) Evergreen (d) Recessive

76. Which of the following item is a product of the possible product of the cambium of a plant?
(a) Cork (b) Wort
(c) Germ (d) Fruit

77. What substance produced from the plant is used to create its cell walls?
(a) Lignin (b) Cellulose
(c) Sap (d) Ligature

78. Which type of plant tissue is an active site of cell division?
(a) Permanent Tissue
(b) Meristematic Tissue
(c) Epithelial Tissue
(d) Connective Tissue

79. Which type of meristematic tissue is responsible for the increase in girth of the plant?
(a) Apical (b) Parenchyma
(c) Intercalary (d) Lateral

80. Which type of permanent tissue does carrot contain?
(a) Collenchyma (b) Aerenchyma
(c) Sclerenchyma (d) Parenchyma

81. Which type of parenchyma contains chlorophyll?
(a) Chlorenchyma
(b) Aerenchyma
(c) Chlorosome
(d) All types of parenchyma contain chlorophyll.

82. Which tissue does the husk of a coconut contain?
(a) Sclerenchyma (b) Collenchyma
(c) Parenchyma (d) Aerenchyma

83. Which substance is responsible for thickening of sclerenchyma walls?
(a) Suberin (b) Cutin
(c) Lignin (d) Pectin

84. Epidermis in desert plants has a thin waxy coating of a chemical substance. Name the chemical.
(a) Suberin (b) Cutin
(c) Pectin (d) Rennin

85. What is the tubular structures present in xylem?
(a) Xylem fibers
(b) Sieve tubes
(c) Xylem parenchyma
(d) Tracheids and vessels

86. Name the part of phloem with perforated walls.
(a) Phloem fibers
(b) Companion cells
(c) Sieve tubes
(d) Phloem parenchyma

87. Name of dead part of phloem is
(a) Companion cells
(b) Parenchyma
(c) Fibers
(d) Sieve tubes

88. What is a common function of modified roots?
(a) The storage of food.
(b) Photosynthesis.
(c) Help plants walk.
(d) Gaseous exchange.

89. Cytoskeleton is absent in
(a) Eukaryotic (b) Prokaryotic
(c) Mesokaryotic (d) None of the above

90. What are plants called if they grow one year, flower the next and die?
(a) Perennial (b) Annual
(c) Biennial (d) None

91. The leaves are connected to the stem by a thin stalk is called
(a) Stem (b) Vein
(c) Blade (d) Petiole

92. What does the word 'gymnosperm' mean?
(a) Naked seed
(b) Lose leaves in the fall
(c) Covered seed
(d) None of these

93. Angiosperm mean
(a) None of these
(b) Naked seed
(c) Covered seed
(d) Lose leaves in the fall

94. How many seed leaves does a dicotyledon have?
(a) 0 (b) 2
(c) 1 (d) 3

95. The four basic parts of a plant are all of these except?
(a) Roots (b) Leaves
(c) Sepals (d) Stems

96. Monocot pollen grains are generally
(a) Monocoplate (b) Bicolpate
(c) Tricolpate (d) Multicolpate

97. Exine of pollen grain is made up of
(a) Sporopollenin
(b) Cellulo-pectin
(c) Callose
(d) Cellulose

98. Ovule is
(a) Megasporangium
(b) Microsporangium
(c) Integumentedmegasporangium
(d) Megaspore

99. The tenuinucellate ovule has
(a) Small amount of nucellus
(b) Large amount of nucellus
(c) Chalazalnucellus
(d) Micropylenucllus

100. The point at which funicle is attached to body of ovule is called
(a) Raphe (b) Hilum
(c) Endothelium (d) Chalaza

101. A typical angiospermic embryo sac is usually
(a) One celled (b) Two celled
(c) Three celled (d) Seven celled

102. Continued self pollination results in
(a) Better progeny
(b) Weak progeny
(c) New varieties
(d) Both (a) and (c)

103. Endosperm is generally
(a) Haploid (b) Diploid
(c) Triploid (d) Polyploid

104. In monocot seeds, radicle is covered by a protective covering is called

(a) Coleoptile (b) Coleorrhiza

(c) Scutellum (d) Aleurone layer

105. Embryo axis is also called as

(a) Tegman (b) Tiller

(c) Tigellum (d) Scutellum

106. The first node indicates point of attachment of ----- to embryo axis

(a) Plumule (b) Radicle

(c) Cotyledons (d) All of the above

107. The branch which deals with study of fungi is known as--------

(a) Physiology (b) Ecology

(c) Mycology (d) Phycology

108. The type of seed in which endosperm is completely utilized by developing embryo

(a) Endospermic seed

(b) Non endospermic seed

(c) Albuminous seed

(d) None of these

109. Agamospermy produces new plant through the formation of

(a) Gamete (b) Bulbil

(c) Bud (d) Asexual embryo

110. Palynology deals with study of

(a) DNA

(b) Chromosomes

(c) Pollen grains

(d) Flowers

111. How many pollen mother cells should undergo meiotic division to produce 64 pollen grains?

(a) 64 (b) 32

(c) 16 (d) 8

112. In a fully developed male gametophyte the number of nuclei is

(a) One (b) Two

(c) Three (d) Four

113. If an endosperm cell of an angiosperm contains 24 chromosomes, the number of chromosomes in each cell of the root will be

(a) 8 (b) 4

(c) 16 (d) 24

114. Which of the following is necessary for seed formation in angiosperm?

(a) Ovule

(b) Pollination

(c) Double fertilization

(d) All the above

115. Anthesis is referred as

(a) Elongation of pollen tube is style

(b) Emergence of anthers from corolla tube

(c) Dehiscence of anthers

(d) The first opening of a flower

116. Karyogamy is called

(a) Fusion of zoospores

(b) Fusion of cytoplasm

(c) Fusion of spores

(d) Fusion of gametic nuclei

117. What is the simplest design of a plant body?

(a) Cytoplasm (b) Mud

(c) Cellulite (d) Single cell

118. Which of these trees is NOT an angiosperm?

(a) Siberian Crabapple

(b) Maidenhair Tree (Ginkgo Biloba)

(c) Crack Willow

(d) Common Laburnum

119. The first step in the germination of a seed is

(a) Pollination.

(b) Fertilization.

(c) Imbibition of water.

(d) Hydrolysis of starch and other food reserves.

120. In flowering plants, pollen is released from the
(a) Anther (b) Stigma
(c) Carpel (d) Filament

121. Which of the following types of plants is not able to self-pollinate?
(a) Dioecious (b) Monoecious
(c) Complete (d) Wind-pollinated

122. A perfect flower is fertile, but may be either complete or incomplete. Which of the following correctly describes a perfect flower?
(a) It has no sepals
(b) It has fused carpels.
(c) It is on a dioecious plant.
(d) It has both stamens and carpels.

123. Biogeochemical cycles are also known as
(a) Cycles of matter
(b) Sedimentary cycles
(c) Material cycling
(d) Gaseous cycles

124. The plant kingdom classification based on the presence and absence of embryo was proposed by
(a) Linnaeus (b) Eichler
(c) Tippo (d) Whittaker

125. Three kingdom classification was proposed by
(a) Haeckel (b) Copeland
(c) Whittaker (d) Gray

126. Grasses have ______ meristems.
(a) Intercalary (b) Apical
(c) Lateral. (d) None of these

127. This field of botany specializes in studying the internal structure of plant organs is
(a) Plant physiology
(b) Plant anatomy
(c) Plant morphology
(d) Plant taxonomy

128. Father of taxonomy
(a) Darwin (b) Linnaeus
(c) Hinterthuer (d) Copeland

129. The Kingdom in which mushrooms are found is:
(a) Monera (b) Bacteria
(c) Protista (d) Fungi

130. Maximum absorption of rainfall water is done by:
(a) Tropical deciduous forest
(b) Tropical evergreen forest
(c) Tropical savanna
(d) Scrub forest

131. Basidiomycetes produces ------------type of fruiting bodies
(a) Ascocarp (b) cleistothecium
(c) Basidiocarp (d) Basidiospore

132. The study of fungi is known as --------
(a) Mycology (b) Phycology
(c) Ecology (d) Taxonomy

133. The algae growing in seawater is known as -------
(a) Freshwater algae
(b) Terrestrial algae
(c) Marine algae
(d) Lithophytic algae

134. The study of algae is known as-- - -- - --
(a) Mycology (b) Phycology
(c) Taxonomy (d) Physiology

135. What is the scientific term for a leaf stalk?
(a) Petiole (b) Pedicel
(c) Peduncle (d) Stem

136. Which of the following is an edible 'Fungi'?
(a) Mucor (b) Penicillium
(c) Agaricus (d) Rhizopus

137. Energy transferred from one trophic level to another is
(a) 5% (b) 15%
(c) 10% (d) 20%

138. Pick the unicellular organism
(a) Fungi (b) Plants
(c) Protozoa (d) Animals

139. 'Pneumatophores' are present /common in:
(a) Xerophytes (b) Hygrophytes
(c) Mesophytes (d) Halophytes

140. Father of Indian Ethnobotany is
(a) Dr.R.K. Luna
(b) Dr. S. K. Jain
(c) Dr. S. S. Negi
(d) Dr. L. S. Khanna

141. In which form does the food transported in plants?
(a) Sucrose (b) Fructose
(c) Glucose (d) Lactose

142. Largest flower is:
(a) Rafflesia (b) Lotus
(c) Rose (d) Lily

143. Which group of cell possesses both animal and plant characteristics?
(a) Myxomycetes
(b) Basidiomycetes
(c) Phycomycetes
(d) Rhodomycetes

144. Central part of the ground tissue system is known as pith or
(a) Lamella
(b) Medulla
(c) Cambium
(d) Casparian strip

145. Meristematic tissue is responsible for the growth of the plant. Which type of meristematic tissue is responsible for the increase in girth of the plant?
(a) Apical
(b) Parenchyma
(c) Intercalary
(d) Lateral

146. Suicide sacs of the cell is known as
(a) Nucleus (b) Lysosomes
(c) Mytocontria (d) Ribosomes

147. Which type of parenchyma contains chlorophyll?
(a) Chlorenchyma
(b) Aerenchyma
(c) Chlorosome
(d) All types of parenchyma contain chlorophyll.

148. Which tissue does the husk of a coconut contain?
(a) Sclerenchyma (b) Collenchyma
(c) Parenchyma (d) Aerenchyma

149. Which substance is responsible for thickening of sclerenchyma walls?
(a) Suberin (b) Cutin
(c) Lignin (d) Pectin

150. During .... period first seed-forming plants had appeared on the earth
(a) Silurian
(b) Carboniferous
(c) Cambrian flora
(d) Ordovician flora

151. The exterior portion of the root is called
(a) Xylem fibers
(b) Sieve tubes
(c) Xylem parenchyma
(d) Epidermis

152. The membrane that delimits the vacuoles is called
(a) Mesoplast (b) Protoplast
(c) Tonoplast (d) Metaplast

153. Two or more types of tissues form
(a) Companion cells
(b) Parenchyma cells
(c) Tissue Systems
(d) Sieve tubes

154. What is a common function of modified roots?
(a) The storage of food
(b) Photosynthesis
(c) Help plants walk
(d) Gaseous exchange

155. The cell was discovered by
(a) Dioscorides
(b) Robert Hooke
(c) Rudolf Virchow
(d) Louis Pasteur

156. The ability of an organism to change in response to the environment is called
(a) Adaptation (b) Avoidance
(c) Response (d) None of the above

157. Scientific study on bryophytes is known as
(a) Mycology (b) Taxonomy
(c) Bryology (d) Pharmacology

158. Scientific study on lichens is known as
(a) Ecology (b) Lichenology
(c) Cytology (d) None of the above

159. Scientific study of algae is called
(a) Molecular biology
(b) Phycology
(c) Physiology
(d) None of the above

160. The structure of DNA was discovered by
(a) Theodor Schwann
(b) Matthias Schleiden
(c) James Watson and Francis Crick
(d) Charles Darwin

161. The study of fossil plants is defined as
(a) Palaeobotany (b) Taxonomy
(c) Ecology (d) Histology

162. The epidermis is made up of ___________ cells
(a) Collenchyma
(b) Meristematic
(c) Paranchymatus
(d) Sclerenchyma

163. Leaf is the chief organ of
(a) Photosynthesis
(b) Storage
(c) Conduction
(d) All the above

164. Leaves which have the stipules are called ___________
(a) Ex-stipulate (b) Stipulate
(c) Lanceolate (d) Palmate

165. Leaf blade is known as ___________
(a) Pulvinus (b) Stipules
(c) Pinnae (d) Lamina

166. ___________ are mature, fertilized ovules
(a) Seeds (b) Fruit
(c) Anther (d) All the above

167. Radicle is an embryonic ________ of the plant.
(a) Shoot (b) Root
(c) Leaves (d) None of the above

168. Stigma, Style, Ovary and Ovule collectively called as ___________
(a) Stamen (b) Callyx
(c) Pistil (d) Corella

169. The individual flowering units of the monocot type of inflorescence is referred as___________
(a) Spikelet (b) Leaflet
(c) Raceme (d) Cyme

170. Fungi which live within the living plant tissues are known as
(a) Parasite
(b) Epiphyte
(c) Endophyte
(d) Halophyte

171. Basidiocarpis formed from
(a) Primary mycelium
(b) Secondary mycelium
(c) Tertiary mycelium
(d) Quaternary mycelium

172. Anti-cancerous compound Lentinan and Eritadine are extracted from
(a) Fruit bodies of *Agaricus bisporus*
(b) Fruit bodies of *Lentinus edodeds*
(c) Fruit bodies of *Pleurotus ostreatus*
(d) Fruit bodies of *Volvariella volvacae*

173. Which of the following description is correct for anisogamy?

(a) Fusing gametes are morphologically and physiologically similar
(b) Fusing gametes are morphologically and physiologically dissimilar
(c) Motile male gametes are produced in large number and female gamete produced singly
(d) Fusing gametes are formed by different parental thallus

174. The number of cotyledon in a pinus seed is

(a) One (b) One or two
(c) Many (d) None of these

175. Who classified the Gymnosperms on the basis of wood structure

(a) Engler and Prantl
(b) Chamberlain
(c) Seward
(d) Pant

176. The seeds which of the following plants are consumed by human beings ?

(a) *Pinus insularis*
(b) *Pinus merkurii*
(c) *Pinus gerandiana*
(d) *Pinus roxburghii*

177. Which one among the gymnosperm plant is called a living fossil?

(a) Gnetum (b) Ginkgo
(c) Ginkgodium (d) Ephedra

178. Special type of inflorescence is

(a) Spadix (b) Cyathium
(c) Corymb (d) Panicle

179. Capsule is a

(a) Dry fruit
(b) Dry dehiscent fruit
(c) Loculicidal capsule
(d) septicidal capsule

180. Cohorts and orders of Bentham and Hooker's classification is today known as

(a) Orders and Families
(b) Families and Orders
(c) Families and Genera
(d) Genera and Species

181. Bentham and Hookers system of classification is based on

(a) de Condolle and Lindley's system
(b) de Condolle system
(c) Robert Brown's system
(d) Exclusive natural system

182. Presence of stomata on both surfaces of leaves is called

(a) Epistomatic (b) Amphistomatic
(c) Hypsotomatic (d) Multistomatic

183. Robert H. Whittaker classified the organisms into how many kingdoms ?

(a) Five (b) Four
(c) Three (d) Two

184. Gram positive cell wall is negatively charged because of the presence of

(a) D-alanine (b) Teichoic acid
(c) Peptidoglycan (d) Glucose

185. Meiosis occurs in the

(a) Somatic cells
(b) Germ cells
(c) Vegetative cells
(d) Meristem

186. Two or more crops grown togeather in the kitchen garden are called

(a) Succession crop
(b) Companion crop
(c) Annual crop
(d) Biennial crop

187. In Monocots, grafting is not possible because

(a) Cambium is absent
(b) Vascular bundles are absent
(c) Parallel venation
(d) Herbaceous

188. An example of the algal sulphated polysaccharide is

(a) Fucoidin (b) Sodium alginate
(c) Agar-Agar (d) Carrageenan

189. Calyptra is seen in
(a) Gametophyte
(b) Sporophyte
(c) Archegoniophore
(d) Antheridophore

190. The layer which is protective in function of the sporangium of ferns is called as
(a) Annulus (b) False indusium
(c) Indusium (d) Stomium

191. Aggregate fruits are:
(a) Acorns (b) Borne in clusters
(c) Cones (d) Winged seeds

192. Gymnosperms bear naked seeds. They are the:
(a) Broad-leafed trees
(b) Cone bearing trees
(c) Fruit trees
(d) Oaks

193. The genus of the tree noted for its great size and long life is _____.
(a) Alnus (b) Betula
(c) Salix (d) Sequoia

194. Achene, samara, drupe and pome are types of _________.
(a) Bark (b) Flowers
(c) Fruit (d) Leaves

195. .... is the embryonic **Shoot** of the plant
(a) Radicle
(b) Mesocotyl
(c) Hypocotyl
(d) None of the above

196. An ovate leaf is ____________.
(a) Egg Shaped
(b) Heart shaped
(c) Kidney shaped
(d) Triangular in shape

197. A leaf margin described as entire is
(a) Coarsely toothed (b) Lobed
(c) Smooth (d) Wavy

198. The plant kingdom division Magnoliophyta includes ____________.
(a) Angiosperms
(b) Broadleaf trees
(c) Oaks and hickories
(d) All of the above

199. Open grown softwood trees crown take ____________ form.
(a) Triangular (b) Bulb like
(c) Columnar (d) Square

200. Open grown hardwood trees crown become ____________ in form.
(a) Triangular (b) Bulb like
(c) Columnar (d) Square

201. Xylem and phloem are the components of ....tissue system
(a) Vascular (b) Vessel
(c) Cambial (d) None of the above

202. If the male and female flower parts occur in separate flowers on the same tree, the species is called
(a) Monoecious
(b) Dioecious
(c) All of the above
(d) None of these

203. Ovate, deltoid, lanceolate, star shaped are examples of:
(a) Leaf margins
(b) Leaf apexes
(c) Leaf bases (d) All of the above

204. Entire, serrate, dentate, lobed and incised are examples of:
(a) Leaf margins (b) Fruit types
(c) Bud types (d) All of the above

205. Major characteristics of twigs used in tree identification are:
(a) Buds (b) pith
(c) Lenticels (d) All of the above

206. Which of the following are female parts of a tree flower?
(a) Sepal (b) Pistil
(c) Pollen (d) All of the above

207. The outside edge of a leaf is called
(a) Lobe (b) Margin
(c) Petiole (d) Leaflet

208. The leaf stem is also called
(a) Petiole (b) Attach
(c) Joiner (d) Midrib

209. When a single leaf is attached to the twig it is said to be _____.
(a) Truncate (b) Pinnate
(c) Palmate (d) Simple

210. When many small leaflets are attached to the petiole to form a larger leaf is called _____.
(a) Multiple (b) Compound
(c) Acute (d) Oblique

211. Which is not a type of leaf arrangement?
(a) Alternate (b) Opposite
(c) Whorled (d) Oblique

212. As per Benthem and Hookers' classification, there are _________ non-standard exceptions to the family name ending 'aceae'.
(a) 8 (b) 6
(c) 4 (d) 2

213. The natural system of classification is based on
(a) Numerical (b) Habit
(c) Form (d) None of these

214. A dry and nut-like fruit is called
(a) Caryopsis (b) Drupe
(c) Berry (d) Samara

215. The largest family among the Indian flora is
(a) Fabaceae (b) Meliaceae
(c) Asteraceae (d) Myrtaceae

216. The origin of teak is
(a) India (b) Burma
(c) Indonesia (d) Nepal

217. The arrangement of ovules within the ovary is
(a) Aestivation (b) Phyllotaxy
(c) Placentation (d) None of these

218. Ergrostisergrostis is an example for
(a) Basionym (b) Homonym
(c) Autonym (d) Tautonym

219. A pair of appendages that present at the point of attachment of a leaf to the stem is
(a) Bracts (b) Bracteoles
(c) Stipules (d) None of these

220. Thick and fleshy root without any particular shape is
(a) Tuberous (b) Rhizome
(c) Bulbs (d) None of these

221. Sucking roots of parasites are
(a) Prop roots (b) Haustoria
(c) Hanging roots (d) Tilt roots

222. The specimen or other element designated by the author or used by him as the nomenclatural type
(a) Lectotype (b) Holotype
(c) Isotype (d) None of these

223. Father of Botany is
(a) Linnaeus (b) Aristotle
(c) Theophrastus (d) John Ray

224. If all the petals are free, the condition is called
(a) Polypetale (b) Polysepalae
(c) Gamopetalae (d) Gamosepalae

225. The alternative name of Guttiferae is
(a) Clusiaceae (b) Asteraceae
(c) Poaceae (d) None

226. *Acacia nilotica* belongs to the family
(a) Papilionaceae (b) Mimoceae
(c) Caesalpinaceae (d) None of these

227. Peeling of bark is a characteristic feature of
(a) Magnoliaceae (b) Meliaceae
(c) Moraceae (d) Myrtaceae

228. *Santalum album* is found in _________ forests
(a) Dry deciduous (b) Evergreen
(c) Semi-evergreen (d) Moist deciduous

229. Plants with male and female flowers separate are

(a) Monoecious (b) Dioecious

(c) Polygamous (d) None of these

230. Pellucid dots are characteristic of the family

(a) Magnoliaceae (b) Rutaceae

(c) Myrtaceae (d) Asteraceae

231. __________ is an exotic

(a) Eucalyptus (b) Casuarina

(c) Simaruba (d) All of the above

ANSWERSHEET
9. Forest Botany

---

1. (a)	2. (a)	3. (b)	4. (c)	5. (b)	6. (a)	7. (b)	8. (b)	9. (a)	10. (a)
11. (c)	12. (b)	13. (d)	14. (c)	15. (c)	16. (c)	17. (b)	18. (c)	19. (b)	20. (c)
21. (a)	22. (c)	23. (d)	24. (a)	25. (a)	26. (a)	27. (b)	28. (d)	29. (b)	30. (d)
31. (a)	32. (b)	33. (c)	34. (d)	35. (c)	36. (d)	37. (a)	38. (b)	39. (c)	40. (a)
41. (b)	42. (b)	43. (c)	44. (c)	45. (d)	46. (b)	47. (b)	48. (b)	49. (a)	50. (b)
51. (c)	52. (a)	53. (a)	54. (a)	55. (a)	56. (a)	57. (a)	58. (d)	59. (a)	60. (c)
61. (a)	62. (c)	63. (c)	64. (c)	65. (a)	66. (a)	67. (d)	68. (c)	69. (d)	70. (a)
71. (d)	72. (a)	73. (b)	74. (a)	75. (a)	76. (a)	77. (b)	78. (b)	79. (d)	80. (b)
81. (a)	82. (a)	83. (c)	84. (b)	85. (d)	86. (c)	87. (c)	88. (a)	89. (b)	90. (c)
91. (d)	92. (a)	93. (c)	94. (b)	95. (c)	96. (a)	97. (a)	98. (d)	99. (a)	100. (b)
101. (d)	102. (b)	103. (c)	104. (b)	105. (c)	106. (c)	107. (c)	108. (b)	109. (d)	110. (c)
111. (c)	112. (c)	113. (c)	114. (d)	115. (d)	116. (d)	117. (d)	118. (b)	119. (c)	120. (a)
121. (a)	122. (d)	123. (a)	124. (c)	125. (a)	126. (a)	127. (b)	128. (b)	129. (d)	130. (b)
131. (d)	132. (a)	133. (c)	134. (b)	135. (a)	136. (c)	137. (b)	138. (c)	139. (d)	140. (b)
141. (a)	142. (a)	143. (a)	144. (b)	145. (d)	146. (b)	147. (a)	148. (a)	149. (c)	150. (b)
151. (d)	152. (c)	153. (c)	154. (a)	155. (b)	156. (a)	157. (c)	158. (b)	159. (b)	160. (c)
161. (a)	162. (c)	163. (a)	164. (b)	165. (d)	166. (a)	167. (b)	168. (c)	169. (a)	170. (c)
171. (b)	172. (b)	173. (b)	174. (a)	175. (c)	176. (c)	177. (b)	178. (b)	179. (b)	180. (a)
181. (a)	182. (b)	183. (a)	184. (b)	185. (b)	186. (b)	187. (a)	188. (d)	189. (b)	190. (c)
191. (b)	192. (b)	193. (d)	194. (c)	195. (c)	196. (a)	197. (c)	198. (d)	199. (a)	200. (b)
201. (a)	202. (a)	203. (c)	204. (a)	205. (d)	206. (b)	207. (b)	208. (a)	209. (d)	210. (b)
211. (d)	212. (a)	213. (c)	214. (a)	215. (c)	216. (b)	217. (c)	218. (d)	219. (c)	220. (a)
221. (b)	222. (b)	223. (c)	224. (a)	225. (a)	226. (b)	227. (d)	228. (a)	229. (b)	230. (c)
231. (d)									

10

Tree Seed Technology

KEY POINTS

- Seed is a fertilized matured ovule
- Seed the basic input of plant regeneration
- Seed is matured ovule that consists of an embryonic plant together with a store of food, all surrounded by a protective coat
- Seeds should have optimum moisture content (low (6 – 10%) - orthodex seed, high (17 – 30%) – recalcitrant seed)
- Plant growth originates within the buds in regions known as meristems
- The fusion of male gamete of pollen grain to the egg cell of female ovule is known as fertilization, which results in formation of fruit and seed
- Testa is a seed coat. It is derived from the integuments and its several layers thick.
- Testa is the outer most layer of the seed coat
- Tegmen is an inner coat of testa.
- Pericarp - The body of a fruit developed from the ovary well and enclosing the seeds.
- Raphe- A cord or ridge of fibrovasculer tissue connecting the base of the nucleus with placenta, the adhesent funicle of anatropous ovule (Type of ovule).
- Micropyle - The operture in the ovule that makes the position of the radials.
- Hilum - A scar left on the seed where it was formerly attached to the funicle or placenta.
- Ceruncle - A wart or protuberance near the hilum of the seed.
- Legenostome- The tree apex of the nucellus.
- Lens - The contracted tissue the tree position of nucellus frequently attached to the base of the lagenostone.
- Coleorhiza - The steath of the monocotyledonous embryo, which is piereed by the true radicle.
- Aril (Arillus) - An expansion of the funicle, arising in the placenta and enveloping the seed.
- Areole - A place marked out on a surface such as on the flat sides on the seed coats in seeds of sub family Mimosaceae and Caesalpinaceae.
- Appendages - External attachment on the seed that favours dispersal of seed

- Endosperm - Storage tissue of seeds and is formed by fussion of pollen cell with 2 polar nuclei. It is nonviable in nature.
- Embryo - It consists of embryonic axis and cotyledon either one or two.
- Embryonic axis - It is growing part of seed the miniolize plant consisting of plumule and radial.
- Cotyledon - It is the storage tissue of dicots (mostly) and is the part of embryo. It is viable.
- Seutellcum - It is the colyledon of mancot viable.
- Coleoptile - Covering of minialize plumule.
- Plumule - It is miniature part of shoot region of plant
- Radicle - It is the miniature part of root region of plant.
- Hypocotyl - It is the region below the colyledon above the radicle of the seed.
- Perisperm - it is unutilized nucellus region of seed normally seen as patery growth outside the endosperm and inside the seed coat.
- Epicotye - It is region above the cotyledon and below the primary leaf of seed (plumule).
- In crop plants in which cotyledons come above the soil surface are called epigeal germination
- Plants in which cotyledons do not come above the soil surface are called hypogeal germination
- National seeds corporation of India was established in 1960
- As per the Indian seed Act (1966) seeds are multiplied adopting 3 tier system as breeder seed to foundation seed to certified seed
- Breeder seed in the progeny of nucleus seed
- Breeder seed is used for the production of foundation seed
- Certification is not required for breeding seed
- Certified seed is the progeny of either foundation or registered or certified seed
- Foundation seed is the progeny of breeder seed
- Genetic purity is determined by grow out test
- Study of pollen grain is termed as Palynology
- Pollen grain is a double walled structure, with hard outer cover (exine) and thin inner layer (intine)
- Cryopreservation – seed storage at -196°C the temperature of liquid nitrogen
- Cultigens – any variety of crops under cultivation with no known wild state
- Growth and development of embryo is called embryogeny
- Exocarp – outermost layer of fruit wall
- Mentor pollen – ineffective compatible pollen
- Myreaochory – Dispersal of seeds by ants
- The wall of the ovary or fruit is called pericarp
- An ovule with more than one embryo is called polyembryony

MULTIPLE CHOICE QUESTION'S

1. Seed are formed, but the embryos developed without fertilization
(a) Parthenogenesis
(b) Apomixis
(c) Sporogenesis
(d) Gametogenesis

2. Embryos develop from the embryo sac without pollination
(a) Parthenogenesis
(b) Pseudogamy
(c) Androgenesis (d) Apomixis

3. Chromosome number of *Jatropha curcus* is
(a) 2n=22 (b) 2n=24
(c) n=46 (d) None of the above

4. Which one of the following tree is unisexual
(a) Golden shower tree
(b) Dung oil tree
(c) Paradise tree
(d) None of the above

5. The most important factor which determines stability of seed is
(a) Light (b) Temperature
(c) Moisture (d) Nutrient

6. In seed testing, the grow out test is conducted to determine
(a) Physical purity (b) Seed viability
(c) Genetic purity (d) None of these

7. The stalk of the ovules is called
(a) Pedicel (b) Petiole
(c) Funicle (d) Micropyle

8. A fertilized mature ovule, which possesses am embryonic plant, food reserve and protective coating in a plant species is
(a) Embryo Sac (b) Embryonic callus
(c) Embryo (d) Seed

9. The visible scar (spot) observed on a seed at the place of the detachment from seed
(a) Micropyle (b) Stome
(c) Pole (d) Hillum

10. Triploids are normally
(a) Sterile (b) Fertile
(c) Male sterile (d) None of these

11. Genetically most pure seed is
(a) Breeder seed (b) Nucleus seed
(c) Foundation seed (d) Certified seed

12. Usually, seed production areas should contain a minimum area (ha) of
(a) 1 ha (b) 2 ha
(c) 3 ha (d) 5 ha

13. Notification of seed control order by Government of India in the year
(a) 1985 (b) 1989
(c) 1983 (d) 1954

14. The seed Act was enacted in the year
(a) 1965 (b) 1966
(c) 1968 (d) 1969

15. The original tree (Donor tree) from which parts to be propagated are taken are called
(a) Clone (b) Ramet
(c) Ortet (d) Progeny

16. Orthotropic growth refers to
(a) Tree (b) Shrub
(c) Herb (d) None of these

17. The most common plant growth regulators compound used in rooting of tree species is
(a) IAA (b) IBA
(c) NAA (d) 2, 4-D

18. Pure Live Seed is related to
(a) Physical purity (b) Genetic purity
(c) Germination percentage
(d) Defective purity

19. Seed lot certificate colour is
(a) Orange (b) Green
(c) Purple (d) Orange and Green

20. Seed sample certificate colour is
(a) Blue (b) Orange
(c) Purple (d) White

21. Difference between seed and grain is
(a) Seed processing (b) Genetic purity
(c) Seed quality (d) Cost

22. Embryo exised test ------------------used for surface disinfestation
(a) NaOCl (b) Lactophenal
(c) $AlK(SO_4)$ (d) All of these

23. Factors that affect tree seed demand and supply
(a) Plantation area
(b) Plantation types
(c) Species selection
(d) All the above

24. Among the gymnosperm, the seed demand is more for
(a) Pinus spp. (b) Quercus spp.
(c) Podocarpus spp. (d) Araucaria spp.

25. Seeds are unique in natural regeneration and propagation because
(a) Seed constitute unique genetic composition
(b) Seeds are produced in large number
(c) Seeds can be stored for long period
(d) All the above

26. The "Embryon" is published by
(a) ISTA
(b) AOSA
(c) ASEAN Forest Tree Seed Centre
(d) CAB International

27. The principle mode of pollination in Gymnosperm is by
(a) Insects (b) Wind
(c) Birds (d) Bats

28. Seeds which maintain high moisture content during maturation
(a) Recalcitrant seeds
(b) Orthodox seeds
(c) Hard seeds
(d) All the above

29. The scar on the seed coat left by the funiculus is
(a) Micropyle (b) Hilum
(c) Chalaza (d) Caruncle

30. Dalbergia seeds are dispersed by
(a) Wind (b) Animals
(c) Water (d) None of the Above

31. Ovule integument gets transformed into
(a) Seed (b) Fruit wall
(c) Seed coat (d) Cotyledons

32. Innate seed dormancy is
(a) Caused due to internal factors
(b) Different from quiescence
(c) Called primary dormancy
(d) All the above

33. A seed which is just waiting for favourable environmental condition to germinate is called
(a) Dormant seed
(b) Quiescent seed
(c) Non-viable seed
(d) Dead seed

34. Seed are subjected to hydraulic pressure at 2000 atm. to
(a) Increase permeability of seed coat to water
(b) Break seed coat
(c) Completion of after ripening period
(d) Leach out the germination inhibitors

35. Failure in pollination and fertilization in tree crops is due to
(a) Shortage of pollen production
(b) Closed flowers or cones
(c) Inbreeding barrier
(d) All the above

36. If most of the trees in the stand have abundant flowers, the reproduction capacity is
(a) Very good (b) Good
(c) Poor (d) Very poor

37. Collection of immature fruits may
(a) Minimize seed loss
(b) Avoid dormancy
(c) Reduce pre-processing damage
(d) All the above

38. Labelling of seed lot in tree crops is
(a) Not important
(b) Very difficult
(c) Important for maintaining the identity
(d) Not followed elsewhere

39. Speed of seed deterioration depends on
(a) Species
(b) Condition of seed at collection
(c) Environment
(d) All the above

40. Bulk reduction in fleshly fruits can be achieved by
(a) Drying
(b) Removing the fruit pulp
(c) Soaking
(d) Threshing

41. Which of the following is recommended to maintain the viability in dry fruits of recalcitrant species
(a) Moist storage (b) Depulping
(c) Sun drying (d) Hot air drying

42. Removal of wings from seed/fruit during processing is called
(a) Grading (b) Dewinging
(c) Pre-cleaning (d) Pre-curing

43. Keeping the fruits at moist condition for prolonged period before extraction is called
(a) Bulking (b) Pre-cleaning
(c) Pre-curing (d) None of the above

44. Seeds of tree species can be graded based on its density or weight by using
(a) Specific gravity separator
(b) Indented cylinder
(c) Hand sieves
(d) All the above

45. Orthodox seeds can be stored safely at a seed moisture content of
(a) 0-4% (b) 8-10%
(c) 18-20% (d) 28-30%

46. While processing of tree seeds, seed quality may be affected by
(a) Mechanical damage
(b) Heat damage
(c) Chemical damage
(d) All the above

47. The Colour of breeder seed tag is
(a) Golden yellow (b) Blue
(c) Green (d) White

48. Which of the following is a recalcitrant seed
(a) *Accacia nilotica*
(b) *Melia dubia*
(c) *Syzygium cuminii*
(d) *Delonix regia*

49. Characteristics of Good quality seed includes
(a) Physical quality
(b) Physiological quality
(c) Genetic quality
(d) All the three

50. The stage at which the maximum dry matter accumulation in seed is
(a) Harvestable maturity
(b) Physical maturity
(c) Physiological maturity
(d) Full maturity

51. Foundation seed is the progeny of
(a) Breeder seed
(b) Certified seed
(c) Labeled seed
(d) Registered seed

52. Breeder seed is the progeny of
(a) Foundation seed
(b) Registered seed
(c) Nucleus seed
(d) Certified seed

53. Certification is not required for
(a) Nucleus seed
(b) Breeder seed
(c) Foundation seed
(d) Certified seed

54. Improved seed includes
(a) Nucleus seed
(b) Breeder seed
(c) Foundation seed
(d) All the above

55. Middle layer of fruit wall is called
(a) Mesocarp (b) Mericarp
(c) Meristem (d) Microbyle

56. The Rudimentary root of the seed or seedling that forms the primary root of the young plant is known as
(a) Rachis (b) Radicle
(c) Rachilla (d) Racerne

57. New genes are created by
(a) Mutation (b) Recombination
(c) Inversion (d) None

58. Mode of pollination in maize is
(a) Self-pollination
(b) Vegetative propagation
(c) Cross-pollination
(d) None

59. The genome contained by bread wheat are
(a) ABC (b) XYZ
(c) ABD (d) None

60. Seed drying is very important to maintain its
(a) Viability and vigour
(b) Protein content
(c) Oil content
(d) Chemical composition

61. The two most important factors influencing the life span of seeds under storage are
(a) Relative humidity and temperature
(b) Rainfall and RH
(c) Climate and temperature
(d) Seed size and moisture content

62. The chemical seed treatment is given to control
(a) Fungi & bacteria
(b) Soil insects
(c) Storage insects
(d) All of these

63. In India, normally how many generation system seed are produced?
(a) 1 (b) 2
(c) 3 (d) 4

64. Off types are removed by
(a) +ve selection
(b) -ve selection
(c) Pure line selection
(d) Mass selection

65. Seed certification involves
(a) 1 phase (b) 3 phase
(c) 6 phase (d) 7 phase

66. Interaction between alleles of two different loci is called as
(a) Epistasis
(b) Dominance
(c) Overdominance
(d) Noneof these

67. Seedlessnes in fruits is called as
(a) Parthenogensis (b) Parthenocarpy
(c) Apomixis (d) None of these

68. Solar treatment is used for
(a) Loose smut (b) Stem rust
(c) Powery mildew (d) Blast

69. Which of the following is not the part of seed embryo
(a) Scutellum (b) Endosperm
(c) Cotyledon (d) Radicle

70. Presence of inhibitors in pericarp leeds to
(a) Mechanical dormancy
(b) Physiological dormancy
(c) Chemical dormancy
(d) Morphological dormancy

71. The area of seed stand treated and managed for abundant seed production is
(a) Seed production area
(b) Seed orchard
(c) Seed source
(d) Provenance

72. Minimum arca suggested to establish seed production area is
(a) 20 Ha (b) 15 Ha
(c) 10 Ha (d) 4 Ha

73. Pollen dilution zone recommended for seed production area is
(a) 400 m (b) 300 m
(c) 200 m (d) 100 m

74. Flower induction can be enhanced by
(a) Girdling
(b) Hormone application
(c) Inducing stress
(d) All the three

75. Seeds can be graded according to its length using
(a) Specific gravity separator
(b) Spiral separator
(c) Indented cylinder
(d) Magnetic separator

76. Allowing the seed to attain moisture equilibrium with in the seed during artificial drying process is
(a) Cooling (b) Aeration
(c) Stratification (d) Tempering

77. Which of the following is a pre-storage seed treatment
(a) Dormancy breaking
(b) Seed coating
(c) Hydration-dehydration
(d) Halogenation

78. Seed certification requires
(a) Improved variety
(b) Genetic purity
(c) Physical purity
(d) All of the above

79. Plants which cotyledons do not come above the soil surface are called
(a) Hypogeal germination
(b) Epigeal germination
(c) a and b
(d) None

80. The first private seed company came into existence in
(a) Mumbai (b) Lucknow
(c) New Delhi (d) Kolkata

81. The flowering plants a second seed coat is known as
(a) Integument
(b) Aleurone layer
(c) Tegamen
(d) Inner ventral scale

82. In India, foundation seed is produced by
(a) Registered growers
(b) National Seed Corporation
(c) SSCA
(d) All the above

83. Failure in pollination and fertilization in tree crops is due to
(a) Shortage of pollen production
(b) Closed flowers or cones
(c) Inbreeding barrier
(d) All the above

84. If most of the trees in the stand have abundant flowers, the reproduction capacity is
(a) Very good (b) Good
(c) Poor (d) Very poor

85. Collection of immature fruits may
(a) Minimize seed loss
(b) Avoid dormancy
(c) Reduce pre-processing damage
(d) All the above

86. Labelling of seed lot in tree crops is
(a) Not important
(b) Very difficult
(c) Important for maintaining the identity
(d) Not followed elsewhere

87. Speed of seed deterioration depends on
(a) Species
(b) Condition of seed at collection
(c) Environment
(d) All the above

88. Bulk reduction in fleshly fruits can be achieved by
(a) Drying
(b) Removing the fruit pulp
(c) Soaking (d) Threshing

89. Which of the following is recommended to maintain the viability in dry fruits of recalcitrant species
(a) Moist storage
(b) Depulping
(c) Sun drying (d) Hot air drying

90. A set standards used to regulate own or community activity in relation to biological world is
(a) Biopotency (b) Biopiracy
(c) Biowar (d) Bioethics

91. In most of monocot plants, germination is
(a) Epigeal (b) Hypogeal
(c) Viviparous (d) Hydrogeal

92. Emergence of radicle and plumule from seed is called
(a) Pregemination
(b) Post germination
(c) Germination
(d) Dormancy

93. In stratification, dormant seeds are kept at
(a) Very high temp
(b) Low temperature (0-100°C)
(c) Under aerated condition
(d) Optimum temperature

94. Longevity of seed can be prolonged by keeping them at
(a) Dry condition and low temperature
(b) Humidity and low temperature
(c) Low temperature and dryness
(d) High temperature

95. The resumption of active growth of embryo is called as
(a) Germination (b) Growth
(c) Quiescence (d) Viability

96. Biopiracy means
(a) Use of biopatents
(b) Thefts of plants and animals
(c) Stealing of bioresources
(d) Exploitation of bioresources without authentic permission

97. Bioethics is related to
(a) Preventing biopiracy
(b) Regulation of unethical activities likegene cloning in animals
(c) Preventing theft of living materials
(d) Moral guidance to the problems in biology

98. Tetrazolium test for testing viability of seeds is attributable to
(a) Clement (b) Lakon
(c) Bonner (d) Skoog

99. Pick the odd man out – Orthodox seed
(a) Acacia (b) Eucalyptus
(c) Terminalia (d) Shorea

100. The growth hormones used for shoot multiplication under in vitro condition are collectively called as
(a) Auxins (b) Cytokinins
(c) Giberellins (d) Vitamins

101. Type of pre treatment required for teak is
(a) Acid scarification
(b) Hot water treatment
(c) Soaking in ambient for 24 hrs
(d) Alternate wetting and drying for 15 days

102. Which among the following tree seeds has the lowest viability period?

(a) *Acacia nilotica*

(b) *Cassia fistula*

(c) *Casuarina equisetifolia*

(d) *Albizzia lebbek*

103. Initial seed of an improved variety is called

(a) Nucleus seed

(b) Breeder seed

(c) Foundation seed

(d) Certified seed

104. The seed which farms generally used are called

(a) Breeder seed (b) Foundation seed

(c) Certified seed (d) Nucleus Seed

ANSWERSHEET
10. Tree Seed Technology

1. (b)	2. (a)	3. (a)	4. (c)	5. (c)	6. (c)	7. (c)	8. (b)	9. (d)	10. (a)
11. (b)	12. (d)	13. (c)	14. (b)	15. (c)	16. (a)	17. (b)	18. (c)	19. (d)	20. (a)
21. (b)	22. (a)	23. (d)	24. (a)	25. (d)	26. (c)	27. (b)	28. (a)	29. (b)	30. (a)
31. (c)	32. (d)	33. (b)	34. (a)	35. (d)	36. (a)	37. (d)	38. (c)	39. (d)	40. (b)
41. (a)	42. (b)	43. (c)	44. (a)	45. (b)	46. (d)	47. (a)	48. (c)	49. (d)	50. (c)
51. (a)	52. (c)	53. (b)	54. (d)	55. (a)	56. (b)	57. (a)	58. (c)	59. (c)	60. (a)
61. (a)	62. (d)	63. (c)	64. (b)	65. (c)	66. (a)	67. (b)	68. (a)	69. (b)	70. (c)
71. (a)	72. (d)	73. (c)	74. (d)	75. (c)	76. (d)	77. (d)	78. (d)	79. (a)	80. (c)
81. (c)	82. (b)	83. (d)	84. (a)	85. (d)	86. (c)	87. (d)	88. (b)	89. (a)	90. (d)
91. (b)	92. (c)	93. (b)	94. (b)	95. (a)	96. (d)	97. (b)	98. (b)	99. (d)	100. (b)
101. (d)	102. (c)	103. (a)	104. (c)						

11

Forest Genetics and Tree Breeding

KEY POINTS

- Genetics is the study of heredity and variation
- Heredity is the transmission of characters from parents to off springs.
- Cytogenetics is the study of various aspects of chromosomes and their effects on the development of characters of organism
- Darwin (1809 – 1882) proposed theory of natural selection
- When expression of one gene depends on the expression or absence of another gene in an individual, it is called gene interaction
- The interaction of genes at different loci that affects the same character is called Epistasis.
- The term Epistasis was first used by Bateson in 1909
- The gene that masks the expression of another gene is called Epistatic gene and the gene whose expression is masked is termed Hypostatic gene
- Epistasis is also referred to as intergenic or interallelic gene interaction
- Allele is the short term for allelomorph
- The term allelomorph was coined by Bateson
- Allele is the alternative form of a gene
- Alleles are of two types viz., either dominant or recessive and wild and mutant type
- The ability of a gene or gene combination to be expressed phenotypically to any degree is called penetrance
- Individuals with one set of chromsomes in their somatic cells are called haploids or monoploids
- Autopolyploid is a polyploid arising through multiplication of a single genome
- Autopolyploids are generally lower in fertility than the diploids
- Polyploidy has a highly variable role in the evolution of plant genera
- An aneuploid is an organism whose somatic chromosome number is not an exact multiple of the monoploid number
- A monosomic is an individual that lacks one chromosome of the normal complement of somatic cells (2n - 1)
- A nullisomic is an individual that lacks both members of one specific pair of chromosomes (2n - 2)

- A trisomic is an individual with one chromosome more than the normal complement of the somatic cells (2n + 1)
- A tetrasomic is an individual with two chromosomes more than the normal complement of the somatic cell (2n + 2)
- The genetic material in an organism may be described by four elements: deoxyribonucleic acid (DNA), codons, genes and chromosomes.
- DNA is the main carrier of genetic information
- Codon is a group of three nucleotides, it forms the unit of genetic code.
- Gene is the basic unit of heredity, which controls a particular inherited character.
- Chromosomes are the carriers of genes and are rod-like or thread-like structure, which in great majority of sexually reproduced organisms occurs, in homologous pairs in the nucleus of animal and plant cells.
- The organisms or cells having chromosomes in homologous pairs are known as diploids, and the condition of being diploid is known as diploidy.
- Mutation is a sudden heritable change in the genetic material.
- Recombination is any process that gives rise to new combination of genes.
- Introgression is the spreading of genes of one species into the gene pool of another by hybridization and back crossing
- Phenotypic variation is the variation in form, function and behaviour that results from the interplay of limiting factors on genotypes during the development of individuals
- The resultant offspring are often more vigorous than either parent and is known as hybrid vigour
- Cline is a gradient in a measurable characteristic.
- Ecotype is a group of plants of similar genotypes that occupy a specific ecological niche.
- No. of characteristics Cline is one
- No. of characteristics Ecotype is many
- A land race is a population of individuals that has become adapted to specific environment in which it has been planted.
- The populations from which the land race trees will be selected must have a broad genetic base.
- The plantation from which selections will be made should be reasonably large, generally in the order of 400 ha (1000 acres) or more.
- Pollen viability % = $-\dfrac{\text{Number of stained round pollen}}{\text{Total number of polen counted}} \times 100$

Pollen Viability %	Viability Status
60-100.00	Fertile
30-59.99	Partially fertile
1-29.99	Partially sterile
0-0.99	Completely sterile

- The ratio of genotypic variance to the phenotypic variance or total variance is known as heritability and expressed in percent.
- The heritability is the heritable portion of phenotypic variance.
- Broad-sense Heritability is the proportion of the phenotypic variance among the individuals to total genotypic variance among individuals.
- Broad sense heritability can range from 0 to 1
- Narrow-sense heritability is the proportion of the phenotypic variance to additive genetic variance
- Genetic Gain is the expected gain from selection (G) and is a product of heritability of the characteristic(h^2) and the selection differential
- $GA = h^2 \times \alpha p \times K$

	Low	Moderate	High
Heritability	< 30 %	30-60 %	> 60 %
Genetic Advance	< 10%	10- 20 %	> 20%

- High heritability, high genetic advance indicates the additive effect of the traits and forms this base for selection
- High heritability with low genetic advance indicates the presence of number of Non-additive gene.
- Low heritability with high genetic advance indicates the presence of number of Non-additive gene.
- Low heritability and low genetic advance – environmental influence is very high generally.
- A positive correlation between any two characteristics indicates that the two tend to rise or fall together
- A negative correlation indicates that a change in one tends to lead to a change in the other in the opposite direction
- Environmental correlation is a measure of the effect environment on the relationship between characteristics irrespective of genotypic differences
- Phenotypic correlation is the association of two characteristics that are directly measured and is called correlation of phenotypic values.
- Colchicine treatment is the most effective and the most widely used treatment for chromosome doubling
- Progeny testing - Selection of parent trees based upon the performance of their progeny.
- Progeny is the offspring of any specific mating or of any one individual with an apomictic reproduction.
- Half sibs are made up of individuals sharing the same one parent in common , which is usually the female parent.
- The polycross designs produce half-sib progeny.
- The half-sib progeny test is a poly cross test or one parent progeny test.

- Both the male and female parents are known in full-sib progeny, more information can generally be obtained from full-sib progeny tests than from half-sib progeny tests.
- Clonal testing - To identify suitable clones for deployment
- In 1717 Thomas Fairchild produced the first artificial hybrid
- 1903 Johansen proposed the famous 'pure line theory'
- G.H.Shull work in maize is the fore runner for the present day hybrid maize programme.
- Majority of our breeding programmes aim at increased yield.
- Genetic erosion : Disappearance of land races due to introduction of high yielding varieties
- The place from where the plants evolved from wild area and showing great diversity is called centre of origin
- There are certain areas where some crops exhibit maximum diversity of forms but this may not be the centre of origin for that particular crop. Such centres are known as Secondary centres of origin
- N.I. Vavilov proposed the concept of 'centres of origin'.
- Law of homologous Series is proposed by N.I Vavilov. According to this law "the characters found in one species also observed in other related species".
- Self incompatibility and sterility are the two mechanisms which encourage cross pollination.
- In self incompatible plants, the flowers will produce functional or viable pollen grains which fail to fertilize the same flower or any other flower of the same plant.
- Male sterility is characterized by nonfunctional pollen grains, while female gametes function normally
- The process of using different lines (genotype) and producing hybrid is known as line breeding.
- Apomixis as "the substitution for sexual reproduction or another asexual reproductive process that does not involve nuclear or cellular fusion (i.e. fertilization)"
- IPGRI – Supervised by consultative group on international Agricultural Research
- CGIAR – 1972 by FAO, World Bank.
- IPGRI established by CGIAR in 1994.
- 1946 – Plant introduction started at IARI, New Delhi.
- NBPGR – by ICAR – 1976 – New Delhi
- The germplasm collection is a collection of large number of genotypes of a crop species and its wild relatives.
- Genetic advance - It is the improvement in the mean genotypic value of the selected families over the base population is known as genetic advance under selection.
- Transgressive breeding - Transgressive segregation is the production of plants in F_2 generation that are superior to both the parents for one or more characters.
- When crosses are made between two different species or between two different genera, they are generally termed as - Distant hybridization (or) Wide hybridization
- Thomas Fairchild 1717 was the first man to do distant hybridization.
- Chromosomes are gradually eliminated from the zygote.

- A group of individuals within a family are referred to as siblings
- The group of related individuals when only one parent is common is called a half sib family
- When both parents are common, they constitute full-sib family
- The study of allele frequency distribution and change under the influence of the four main evolutionary processes: natural selection, genetic drift, mutation and gene flow
- A mutation is a heritable change in the genetic constitution of an organism, usually at the level of a gene
- Natural selection is a strong force that usually reduces variability.
- It is the migration of allelles from one population or species into another
- A Complex mechanism that operates through chance fluctuation in allele frequencies within a population
- The Hardy-Weinberg law is the fundamental law of population genetics and provides the basis for studying mendelian populations
- Forest tree improvement is applied when control of parentage is combined with other forest management activities, such as site preparation or fertilization, to improve the overall yields and quality of products from forestlands
- Mass selection involves choosing individuals solely on the basis of their phenotypes, without regard to any information about performance of ancestors, siblings, offspring, or other relatives.
- A pureline is the progeny of a single, homozygous, self pollinated plant
- Selection: Choosing individuals with desired qualities to serve as parents for the next generation
- Selection differential: The average phenotypic value of the selected individuals, expressed as a deviation from the population mean
- Family selection involves the choice of entire families on the basis of average phenotypic values
- This is a form of selection in which individuals are chosen are basis of the performance of their siblings and not their own performance.
- Within-Family Selection - Here individuals are chosen on the basis of their deviation from the family mean, and family values per se are given no weight when selections are made.
- Breeding is for one trait at a time until a desired level of improvement is made for that trait.
- A method of multi-trait selection that involves setting minimum values for each trait of interest.
- The selection index is a form of multitrait selection that combines information on all traits of interest into a single index
- The selection procedure that involves many cycles of selection and breeding is known as recurrent selection
- Variation - Refers to the presence of differences among the individuals of plant population
- Variability results due to differences either in the genetic constitution of the individuals or in the environment in which they are grown

- Additive variance arises due to cumulative effect of alleles at all gene loci for a particular trait.
- Dominance: This refers to the interaction of specific alleles at gene loci.
- Epistasis - Interaction among gene loci.
- Genotypic variation = Additive + Non additive genetic variation
- Phenotypic variation = Additive + Non additive genetic variation + Environmental variation
- Candidate Tree - A tree that has be selected for grading because of its desirable phenotypic qualities but that has not yet been tested or graded
- Select, Superior or Plus Tree - A tree that has been recommended for production or breeding orchard use following grading.
- Elite Tree - Trees that have proven to be genetically superior by means of progeny testing. Winner from a selection program & is the kind of tree that is most desired for mass production of seeds or vegetative propagules
- Comparison or Check Tree - Trees that are located in the same stand are of nearly the same age, are growing on the same sites as the select tree & against which select tree is graded.
- Advanced – Generation selection - A tree selected from genetic test crosses among the parents from the previous generation. Some form of family and within family selection - choose advanced generation selections.
- A rule of thumb - 1 tree / hectare i.e . One /1000 tree
- Candidate tree – yellow band 5 cm width above breast height
- Plus tree - yellow band 5 cm width below breast height
- In between yellow band Plus tree number is written TN NS C2
- Incomplete pedigree design - only one parent (usually the mother tree) is known while the sources of pollen (male parent) is not known
- Complete pedigree design - Both the male and female parents are known
- Open pollinated mating design - Trees are allowed to mate random through open pollination
- Single-pair mating- Each parents is mated to one other member of the population; this creates the maximum no. of unrelated families in each generation with a minimum number of crosses
- In complete-diallel mating design, each parent is crossed with each of the other parent trees once as a male parent and once as a female parent, as well as to itself.
- Nested design - Groupers of parents of one sex are mated to members of other sex, therefore progeny composed of full sib families
- Selfing - Transfer of pollen grains from the anther to the stigma of same flower is known as autogamy or self pollination.
- Bisexuality. Presence of male and female organs in the same flower is known as bisexuality.
- All the self pollinated plants have hermaphrodite flowers.
- Homogamy - Maturation of anthers and stigma of a flower at the same time is called homogamy.

- Cleistogamy - When pollination and fertilization occur in unopened flower bud, it is known as cleistogamy.
- Chasmogamy -Opening of flowers only after the completion of pollination is known as chasmogamy.
- Position of Anthers - In some species, stigmas are surrounded by anthers in such a way that self pollination is ensured.
- Autogamy is the transfer of pollen from the anthers to the stigma in the same flower and may occur with or without the assistance of pollinators.
- Geitonogamy is the mode of selfing which involves transfer of pollen between the flowers of the same tree and requires a pollen agent.
- Crossing - Transfer of pollen grains from the anther of one plant to the stigma of another plant is called allogamy or cross pollination.
- Allogamy leads to heterozygosity.
- Dicliny - It refers to unisexual flowers.
- Monoecy - When male and female flowers are separate but present in the same plants, it is known as monoecy.
- Dioecy - When staminate and pistillate flowers are present on different plants, it is called dioecy.
- Dichogamy refers to maturation of anthers and stigma of the same flowers at different times.
- Dichogamy promotes cross pollination even in the hermaphrodite species.
- When pistil matures before anthers, it is called protogyny
- When anthers mature before pistil, it is known as protandry.
- When styles and filaments in a flower are of different lengths, it is called heterostyly.
- Hindrance to self-pollination due to some physical barriers such as presence of hyline membrane around the anther is known as herkogamy.
- Self incompatibility - The inability of fertile pollens to fertilize the same flower is referred to as self incompatibility.
- Self incompatibility prevents the self-pollination and promotes cross pollination.
- Male sterility - In some species, the pollen grains are non functional.
- Plant hybridization consists in producing new types of offspring by artificial pollination brought about between the flowers of two different species or varieties.
- This process called hybridization or crossing can help to produce offspring which have the desired characteristics of both the parents.
- Heterosis is defined as the superiority of F1 hybrid over both the parents in terms of yield or some other characters.
- The term heterosis was first used by Shull in 1914
- Mutation - Changes in genes and chromosomes.

MULTIPLE CHOICE QUESTION'S

1. The DNA molecule is usually made up of how many strands?
(a) 1 (b) 2
(c) 3 (d) 6

2. The RNA molecule is made up of how many strands?
(a) 1 (b) 2
(c) 3 (d) 6

3. How many different types of RNA molecules are needed to make protein?
(a) 1 (b) 2
(c) 3 (d) 4

4. What is the form of RNA that carries the code from the DNA to the site where the protein is assembled?
(a) Messenger RNA (b) Nuclear RNA
(c) Ribosomal RNA (d) Transfer RNA

5. Which of the following carries amino acids to ribosomes, where amino acids are linked into the primary structure of a polypeptide?
(a) mRNA (b) tRNA
(c) hnRNA (d) rRNA

6. Transfer RNA differs from other types of RNA because it
(a) Transfers genetic instructions from the cell nucleus to the cytoplasm.
(b) Specifies the amino acid sequence of a particular protein.
(c) Carries an amino acid at one end.
(d) Contains codons.

7. The nitrogenous base found in DNA but not in RNA is
(a) Adenine. (b) Cytosine.
(c) Uracil. (d) Thymine.

8. Which substance is found in RNA but not in DNA?
(a) Thymine (b) Deoxyribose
(c) Ribose (d) Guanine

9. The nitrogenous base found in RNA but not in DNA is
(a) Adenine (b) Cytosine
(c) Guanine (d) Uracil

10. Uracil will pair with
(a) Ribose (b) Adenine
(c) Cytosine (d) Thymine

11. The synthesis of an RNA molecule from a DNA template strand is
(a) Replication. (b) Translation.
(c) Transcription. (d) Dna synthesis.

12. The relationship between RNA and the DNA that served as the template is
(a) Antagonistic (b) Opposite.
(c) Complementary.
(d) An exact duplicate.

13. Transcription
(a) Occurs on the surface of the ribosome
(b) Is the final process in the assembly of a protein
(c) Is the synthesis of any type of rna from a dna template.
(d) Is catalyzed by dna polymerase.

14. Which of the following catalyzes the process of transcription?
(a) RNA polymerase
(b) DNA polymerase
(c) Ligase (d) Gyrase

15. Three dimensional shape of tRNA is
(a) L-shaped (b) Clover leaf-like
(c) X-shaped (d) Y-shaped

16. In transcription,
(a) Several RNA molecules can be transcribed from the same DNA molecule
(b) Promoters are needed so that RNA polymerase can bind to DNA
(c) DNA produces messenger RNA
(d) All of these

17. The portion of the DNA molecule that is translated is composed of
(a) Introns (b) Anticodons
(c) Exons (d) Transcriptons

18. The portion of the DNA molecule that is not translated and is a noncoding portion of DNA is composed of
(a) Introns. (b) Anticodons.
(c) Exons. (d) Transcriptons.

19. In transcription,
(a) Several amino acids are assembled by the messenger RNA molecules at one time.
(b) A special sequence called a promoter is necessary for transcription to begin.
(c) Certain polypeptide sequences are governed by one ribosome, while other sequences are produced by other ribosomes.
(d) The transfer RNA molecules arrange the messenger RNA codons into the appropriate sequence.

20. When a gene transcription occurs, which of the following is produced?
(a) More DNA
(b) Protein or polypeptide sequences
(c) Messenger RNA
(d) Enzymes

21. The genetic code is made up of units consisting of how many nucleotides?
(a) 2 (b) 3
(c) 5 (d) 6

22. There are how many different kinds of amino acids in proteins?
(a) 3 (b) 6
(c) 12 (d) 20

23. There are how many different RNA codons?
(a) 3 (b) 12
(c) 28 (d) 64

24. If the codon consisted of only two nucleotides, there would be how many different kinds of codons?
(a) 4 (b) 8
(c) 16 (d) 32

25. The concept that a set of three nucleotides specifies a particular amino acid provides the basis for
(a) The one gene, one enzyme hypothesis.
(b) The one gene, one polypeptide hypothesis.
(c) The genetic code.
(d) Biochemical reactions among nucleic acids.

26. Of all the different codons that exist, three of them
(a) Are involved in mutations.
(b) Do not specify a particular amino acid.
(c) Cannot be copied.
(d) Do not specify a particular amino acid but provide instructions such as "stop".

27. Each "word" in the mRNA language consists of how many letters?
(a) Three (b) Four
(c) Five (d) More than five

28. If each nucleotide coded for a single amino acid, how many different types of amino acids could be combined to form proteins?
(a) 4 (b) 16
(c) 20 (d) 64

29. If the DNA triplets were ATG–CGT, the mRNA codons would be
(a) AUGCGU. (b) ATGCGT.
(c) UACGCA. (d) UAGCGU.

30. If the DNA triplets were ATG–CGT, the tRNA anticodons would be
(a) AUGCGU. (b) ATGCGT.
(c) UACGCA. (d) UAGCGU.

31. Ribosomes function as
(a) A single unit
(b) Two-part units
(c) Three-part units
(d) Four-part units

32. Which of the following needs an anticodon to bind to the protein synthesizing machinery?
(a) DNA (b) mRNA
(c) rRNA (d) tRNA

33. The "start" codon is
(a) UUU. (b) CUC.
(c) GUC. (d) AUG.

34. Amino acids are joined together in proteins by
(a) Hydrogen bonds
(b) Peptide bonds.
(c) Anticodons
(d) Wobble effects.

35. Translation begins when the initiator tRNA binds to the
(a) Large ribosomal subunit.
(b) Small ribosomal subunit.
(c) The second trna.
(d) The terminator codon.

36. Termination of protein translation is due to
(a) The lack of amino acids.
(b) Wobble effects.
(c) The lack of messenger rna.
(d) A stop.

37. The difference between normal hemoglobin and sickle-cell hemoglobin is in the
(a) Heme portion of the molecules.
(b) Number of chains of amino acids.
(c) Substitution of a specific amino acid for another specific amino acid.
(d) Addition of one amino acid to the normal hemoglobin molecule.

38. A gene mutation
(a) Is a change in the nucleotide sequence of DNA.
(b) May arise spontaneously.
(c) Can occur in all organisms.
(d) All of these

39. Mutations can be
(a) Beneficial.
(b) Harmful.
(c) Heritable
(d) All of these

40. Which of the following can directly cause mutations?
(a) Ionizing radiation
(b) Nonionizing radiation
(c) Alkylating agents
(d) Both ionizing and nonionizing radiation

41. Anticodon
(a) Sites at which RNA polymerases can bind and initiate transcription
(b) Binds to small subunit platform of a ribosome
(c) Guided and catalyzed by RNA polymerases
(d) A tRNA triplet that binds to mRNA

42. Codon
(a) Sites at which RNA polymerases can bind and initiate transcription
(b) Binds to small subunit platform of a ribosome
(c) Guided and catalyzed by RNA polymerases
(d) A set of three nucleotides

43. Messenger RNA
(a) Sites at which RNA polymerases can bind and initiate transcription
(b) Binds to small subunit platform of a ribosome
(c) Guided and catalyzed by RNA polymerases
(d) A tRNA triplet that binds to mRNA

44. Promoter
(a) Sites at which RNA polymerases can bind and initiate transcription
(b) Binds to small subunit platform of a ribosome
(c) Guided and catalyzed by RNA polymerases
(d) A tRNA triplet that binds to mRNA

45. Transcription
(a) Sites at which RNA polymerases can bind and initiate transcription
(b) Binds to small subunit platform of a ribosome
(c) Guided and catalyzed by RNA polymerases
(d) A tRNA triplet that binds to mRNA

46. This codon terminates a coding region.
(a) AUG (b) UAA
(c) UUU (d) UUA

47. The anticodon AAA would pair with this codon.
(a) AUG (b) UAA
(c) UUU (d) UUA

48. A single mutation involving the second letter of codon AUA would convert it to this codon.
(a) AUG (b) UAA
(c) UUU (d) AAA

49. A DNA codon of ATT would be complementary to this RNA codon.
(a) AUG (b) UAA
(c) UUU (d) UUA

50. This codon specifies an amino acid and indicates the beginning of a coding region.
(a) AUG (b) UAA
(c) UUU (d) UUA

51. The process of making RNA from DNA is:
(a) Transcription
(b) Translation
(c) Conjunction
(d) Peptide synthesis

52. The process of making proteins from RNA is:
(a) Transcription (b) Translation
(c) Conjunction (d) Peptide synthesis

53. Which of the following is NOT one of the bases in a DNA molecule?
(a) Adenine (b) Cytosine
(c) Uracil (d) Thymine

54. Cytosine will form a base pair only with:
(a) Cytosine (b) Adenine
(c) Thymine (d) Guanine

55. What is a nucleotide?
(a) A phosphate, five-carbon sugar, and a nitrogen base
(b) A group of linked amino acids
(c) One half of a DNA molecule
(d) A protein, a sugar, and a phosphate

56. The weak bonds between complementary nitrogen bases involve:
(a) RNA
(b) DNA
(c) Hydrogen
(d) Five-carbon sugars

57. DNA assumes a form termed what?
(a) A triple-linked chain
(b) A six sided figure
(c) A step ladder
(d) A double helix

58. DNA may be described functionally as:
(a) An enzyme
(b) A source for RNA production
(c) An arrangement of mRNA molecules
(d) A sink for disposal of RNA

59. Where is the messenger RNA synthesized in a cell?
(a) Nucleus
(b) Mitochondria
(c) Ribosome
(d) Endoplasmic reticulum

60. Which of the following correctly describes messenger RNA?
(a) Single stranded
(b) Uracil replaces thymine
(c) Many sugar to phosphate bonds occur
(d) All of the above

61. Which of the following are correct types of RNA?
(a) mRNA (b) qRNA
(c) tRNA (d) a and c

62. The directions used in protein synthesis are provided by:
(a) mRNA
(b) Centromere
(c) RNA in the ribosomes
(d) tRNA

63. The coding of amino acids by multiple sets of nucleotides is referred to as:
(a) Base pairing specificity
(b) Hydrogen bonding
(c) Chemical coding
(d) Triplet coding

64. What is the role of ribosomes in protein synthesis?
(a) They provide a source of amino acids
(b) They provide a site for transfer RNA to link to messenger RNA
(c) They translate the basic DNA code using transfer rna
(d) They carry the proteins to their site of action

65. Which of the following is NOT a major grouping of proteins in cells?
(a) Enzymes
(b) Antibodies
(c) Energy storage units
(d) Structural units

66. __________ is responsible for replacing RNA primers with new DNA in Okazaki fragments.
(a) RNA exonuclease
(b) DNA polymerase III
(c) DNA ligase
(d) DNA polymerase I

67. All the amino acids have multiple codons except?
(a) Met (b) Arg
(c) Cys (d) Leu

68. The __________ separate during anaphase of Mitosis.
(a) Homologous chromosomes
(b) Alleles of a gene
(c) Daughter cells
(d) Sister chromatids

69. Which of the following will give cell-specific regulation of gene expression?
(a) Estrogen receptor
(b) TFIID
(c) TATA box
(d) AP-1

70. The codon is found in
(a) DNA (b) rRNA
(c) tRNA (d) mRNA

71. Who developed first electron microscope:
(a) Oparin & Haldane
(b) Watson & Crick
(c) Miller & Urey
(d) Knoll & Ruska

72. The stage of meiosis in which chromosomes pair and cross over is:
(a) Prophase I
(b) Metaphase I
(c) Prophase II
(d) Metaphase II

73. Polyploidy refers to:
(a) Extra copies of a gene adjacent to each other on a chromosome
(b) An individual with complete extra sets of chromosomes
(c) A chromosome which has replicated but not divided
(d) Multiple ribosomes present on a single mRNA

74. A gene showing co dominance
(a) Has both alleles independently expressed in the heterozygote
(b) Has one allele dominant to the other
(c) Has alleles tightly linked on the same chromosome
(d) Has alleles expressed at the same time in development

75. DNA ligase is:
(a) An enzyme that joins fragments in normal DNA replication
(b) An enzyme involved in protein synthesis
(c) An enzyme of bacterial origin which cuts DNA at defined base sequences
(d) An enzyme that facilitates transcription of specific genes

76. Replication of DNA:
(a) Takes place in a "conservative" manner
(b) Takes place in a "dispersive" manner
(c) Takes place in a "semi-conservative" manner
(d) Usually involves one origin of replication per chromosome in eukaryotes

77. A duplication is:
(a) An exchange between non-homologous chromosomes, resulting in chromosomes with new genes adjacent to each other.
(b) Loss of genes in part of a chromosome
(c) An extra copy of the genes on part of a chromosome
(d) A reversal of order of genes on a chromosome

78. A mutation in a codon leads to the substitution of one amino acid with another. What is the name for this type of mutation?
(a) Nonsense mutation
(b) Missense mutation
(c) Frameshift mutation
(d) Promoter muttion

79. Mapping of human chromosomes:
(a) Has been restricted to the sex chromosomes because of small family sizes
(b) Proceeded much more successfully as large numbers of DNA markers became available.
(c) Has determined that the number of linkage groups is about twice the number of chromosomes
(d) Has demonstrated that almost all of the DNA is involved in coding for genes

80. Germ Plasm theory was postulated by
(a) Lamark (b) Weisman
(c) Kolliker (d) Hertwig

81. The ability of a given gene or gene combination to be expressed phenotypically to any degree
(a) Penetrance (b) Expressivity
(c) Pleiotropism (d) Prepotency

82. Total number of genotypes in human ABO blood group system
(a) 3 (b) 12
(c) 6 (d) 4

83. Eye colour in drosophila is an example of
(a) Sex linked inheritance
(b) Sex limited inheritance
(c) Sex influenced inheritance
(d) Incomplete dominance

84. If the X/A ratio is 0.5, the individual will be
(a) Intersex
(b) Normal female
(c) Super female
(d) Normal male

85. In translocation, the exchange of chromosomal segments occurs
(a) Within the same chromosome
(b) Between homologous chromosomes
(c) Between non homologous chromosomes
(d) Both b and c

86. If a woman heterozygous for colour blindness marries a colour blind man, what is the probability that their first child will be colour blind daughter?

(a) 50% (b) 25%
(c) 75% (d) 100%

87. Shortest phase in mitosis is

(a) Prophase (b) Anaphase
(c) Metaphase (d) Telophase

88. Soluble RNA (sRNA) is

(a) tRNA
(b) rRNA
(c) mRNA
(d) mitochondrial RNA

89. The theory of epigenesis was proposed by

(a) K.F. Wolf (b) Charles Darwin
(c) Swammerdam (d) Kolliker

90. Phenotypic F2 ratio of duplicative recessive epistasis (complementary gene interaction) is

(a) 12:3:1 (b) 9:7
(c) 9:6:1 (d) 9:3:3:1

91. Source of energy in nucleus is

(a) Glycolysis only
(b) Tca cycle only
(c) Both a and b
(d) None of the above

92. Crossing over taking place at

(a) Mitosis (b) Meiosis ll
(c) Meiosis l (d) All of the above

93. Xanthine is deaminated product of

(a) Adenine (b) Thymine
(c) Cytosine (d) Guanine

94. The types of phenotypes in F2 generation is

(a) 2n (b) 3n
(c) n2 (d) n3

95. Sum total of genes in a population is

(a) Genotype (b) Karyotype
(c) Gene pool (d) Gene frequency

96. Rho factor is required for the termination of

(a) Replication (b) Transcription
(c) Translation (d) Transversion

97. The types of histone proteins present in eukaryotic chromosome are

(a) 4 (b) 5
(c) 6 (d) 7

98. The classical test cross ratio in dihybrid is

(a) 7:1:1:7 (b) 1:7:7:1
(c) 1:1:1:1 (d) both a and b

99. Chromatin consists of

(a) DNA & protein only
(b) RNA & protein only
(c) RNA & protein only
(d) DNA, RNA & protein

100. Epistasis is ....: type of interaction

(a) Inter allelic
(b) Intra allelic
(c) Allelic genetic
(d) Both a and c

101. If the number multiple alleles for a trait is 5, then what is the total number of genotypes

(a) 9 (b) 16
(c) 12 (d) 15

102. Chromosome theory of linkage was proposed by

(a) Beadle and Tatum
(b) Bateson and Punnett
(c) Morgan and Castle
(d) Sutton and Bovery

103. Quantitative traits are controlled by

(a) Major genes (b) Minor genes
(c) Poly genes (d) Both b and c

104. Breeding value (BV) is

(a) Equal to the TA
(b) Twice the TA
(c) 1/2 of TA
(d) 1/4 of TA

105. Heritability in narrow sense is
(a) VA/VP (b) VE/VP
(c) VG/VP (d) VA+VG/VP

106. Repeatability value set in ....of heritability
(a) Lower limit of h2
(b) Upper limit of h2
(c) Intermediate of h2
(d) Both a and b

107. The most effective method of selection
(a) Mass selection
(b) Pedigree selection
(c) Progeny testing
(d) Selection index

108. Quantitative traits shows
(a) Continuous variation
(b) Discontinuous variation
(c) Both a and b
(d) None of the above

109. Genotypic frequency of progeny depends on
(a) Genotypic freq. of parent
(b) Genotypic freq. of population
(c) Gene freq. of parent
(d) Both a and c

110. In a population, sex linked genes are carried by homogametic sex is ....of the total sex linked genes in the population
(a) 1/2 (b) 1/4
(c) 1/3 (d) 2/3

111. Type of gene action in general combining ability (GCA) is
(a) Additive
(b) Non additive
(c) Both a and b
(d) None of the above

112. Range of h^2 and fitness is
(a) 0 to 1 (b) -1 to +1
(c) 0 to infinity (d) 0 to xn

113. The proportion of population which shows genetic death is
(a) Genetic sterility
(b) Genetic linkage
(c) Genetic load (d) Genetic drift

114. For inbreeding, mated individuals should have common ancestors with in
(a) 2-3 generations
(b) 4-6 generations
(c) 6-7 generations
(d) More than 10 generations

115. Inbreeding coefficients of progenies produced by self mating, parent offspring mating, full sibs and half sibs are in the order
(a) 1,0.5,0.25,0.125
(b) 0.5,0.5,0.25,0.125
(c) 1,0.5,0.5,0.25
(d) 0.5,0.25,0.25,0.125

116. In the second generation, hybrid vigor will be
(a) Same as the first
(b) Doubled the first
(c) Half of the first
(d) None of the above

117. Crossing of two inbreed lines of the same breed is
(a) Incrossing
(b) Incross breeding
(c) Top incrossing
(d) Top incross breeding

118. The intensity of selection depends with number traits considered 'n' is
(a) 1/n (b) 1/n2
(c) n (d) 1/vn

119. Non additive gene action include,
(a) Epistasis (b) Dominance
(c) Interaction (d) All of the above

120. Mildest form of out breeding is
(a) Cross breeding
(b) Close breeding
(c) Out crossing
(d) Rotational crossing

121. The dispersive process mainly occurs in small population in which,
(a) Only direction can be predicted not amount
(b) Only amount can be predicted not direction
(c) Both amount and direction can be predicted
(d) Direction and amount cannot be predicted

122. The contribution of offspring to the next generation is called
(a) Fitness
(b) Adaptive value
(c) Selective value
(d) All of the above

123. Genotypic correlation is due to
(a) Polymorphism
(b) Linkage
(c) Pleiotropism (d) Both b and c

124. A new breed can be evolved by
(a) Out crossing (b) Cross breeding
(c) Grading up (d) None of the above

125. If the coefficient of selection is 0.25, then the fitness is
(a) 1 (b) 0.25
(c) 0.75 (d) 0.5

126. Grading up produces pure breed in how many generations
(a) 2-4 (b) 4-5
(c) 10-12 (d) 7-8

127. Which of the following would represent a mutation?
(a) A change in the base sequence of DNA
(b) Crossing over in meiosis
(c) Both (a) and (b)
(d) Neither (a) nor (b)

128. Arrange the following structures/ chemicals in order of size from smallest to largest.
1. codon 2. gene 3. base 4. chromosome
(a) 1, 2, 3 & 4 (b) 2, 1, 3 & 4
(c) 3, 1, 2 & 4 (d) 1, 2, 4 & 3

129. A change in which of the following would result in a mutation?
(a) Order of bases in DNA
(b) Gene
(c) Chromosome
(d) All of The Above

130. The production of a protein from mRNA is called ______.
(a) Replication
(b) Transformation
(c) Transcription
(d) Translation

131. The role of mRNA is to
(a) Carry the Genetic Instructions to the Cytoplasm.
(b) Identify and Transport Amino Acids.
(c) Provide Nucleotides for RNA Production.
(d) Produce Amino Acids For Protein.

132. Transcription refers to
(a) DNA replication.
(b) Protein Synthesis Using mRNA and tRNA
(c) mRNA synthesis from DNA.
(d) Converting an Inactive Protein into its Active Form.

133. Which of the following molecules has an amino acid attached to it?
(a) DNA (b) tRNA
(c) mRNA (d) All of the above

134. Codons may
(a) Act as Start Codons.
(b) Act as Stop Codons.
(c) Code for a Specific Amino Acid.
(d) All of the above.

135. How many amino acids are coded for in this mRNA sequence: GGUACCUUU-ACU
(a) 6 (b) 4
(c) 1 (d) 12

136. Thirty percent of the bases in DNA extracted from a prokaryotic cell is adenine. What percentage of cytosine is present in this DNA?

(a) 20 (b) 30

(c) 40 (d) 10

137. All cells of a species (except for the sex cells) contain the same amount of DNA. However, the cells of multicellular organisms vary in both structure and function from one tissue type to another. This specialization is evidence that most of the DNA in a cell is

(a) Inactive

(b) Removed As The Cell Matures

(c) Converted Into Protein

(d) Converted Into Glycoprotein

138. Which of the following is the role of the structural genes in an inducible enzyme system?

(a) To Code for Proteins that Metabolize an Inducer Substance

(b) To Produce A Repressor that Binds to The Operator

(c) To Stop The Transcription of The Operator

(d) To Recognize and form a Complex with the Inducible Enzyme

139. Substances that cause cells to increase production of an enzyme (or group of enzymes) are known as

(a) Introns (b) Repressors

(c) Inducers (d) Promoters

140. New alleles arise by

(a) Mutation (b) Migration

(c) Genetic drift (d) Random mating

141. New variations of genes in a population may be produced by

(a) mutation

(b) crossing over

(c) sexual reproduction

(d) all of these

142. The Hardy-Weinberg formula

(a) is useful in determining the extent to which a sexually reproducing population is evolving

(b) is used to predict when genetic drift will occur in a sexually reproducing population

(c) is useful in determining the extent to which polyploidy is occurring in specific plant populations

(d) is used to predict when specific groups of organisms will become extinct

143. The maintenance of Hardy-Weinberg equilibrium is encouraged

(a) when sexual selection occurs

(b) when mutations occur

(c) in small populations

(d) when there is no gene flow among different populations

144. In the Hardy-Weinberg equation, the term q2 refers to the frequency of

(a) a recessive allele of a given locus

(b) the homozygous recessive genotype at a given locus

(c) recessive alleles in a population

(d) heterozygotes in a population

145. The influence of genetic drift on allele frequencies increases as

(a) Gene flow increases.

(b) Population size decreases.

(c) Mutation rate decreases.

(d) The number of heterozygous loci increases.

146. Two individuals are members of the same species if they

(a) possess the same number of chromosomes

(b) breed at the same time

(c) are phenotypically indistinguishable

(d) can mate and produce fertile offspring

147. __________ is the practice of rapidly multiplying stock plant material to produce a large number of plants
(a) Micro propagation
(b) Macro propagation
(c) Seed propagation
(d) All the above

148. Self incompatibility (SI) uses first reported by
(a) Breese (b) Koelreuter
(c) Davis (d) Riddle

149. __________ is somewhat similar to runners but with shorter internodes
(a) Suckers (b) Rhizomes
(c) Corm (d) Bulbs

150. The pollen sterility is caused by nuclear genes is termed as
(a) Cytoplasmic male sterility
(b) Cytoplasmic Genetic male sterility
(c) Genetic male sterility
(d) None of the above

151. The respective stigma of Eucalyptus tereticornis shows exudation and change in colour blue to
(a) Reddish white
(b) White
(c) Creamish white
(d) None of the above

152. Physical separation of anther and stigma is called
(a) Cleistogamy (b) Allogamy
(c) Autogamy (d) Herkogamy

153. Gametophytic system is described by
(a) Hygher
(b) East & Manglsdorf
(c) Koelreuter
(d) Hayward

154. Tropical trees are predominantly __________ trees
(a) Selfed (b) Apomixis
(c) Bisexual (d) Outcrossed

155. Mechanism of self incompatibility
(a) Pollen –Stigma interaction
(b) Pollen tube –Style interaction
(c) Pollen tube –Ovule interaction
(d) All the above

156. Transfer of pollen grains from the anther to the stigma of same flower is
(a) Cleistogamy (b) Allogamy
(c) Autogamy (d) Homogamy

157. Eucalyptus is mainly pollinated through
(a) Ants (b) Honey bee
(c) Water (d) Air

158. Plants which are not flowering plants
(a) Green algae
(b) Ferns
(c) Gymnosperms
(d) All the above

159. The development of embryo sac from a megaspore is known as
(a) Megagametogenesis
(b) Microgametogenesis
(c) Microphyle
(d) Sporogenesis

160. Maturation of anthers and stigma of a flower at the same time is called
(a) Cleistogamy (b) Allogamy
(c) Autogamy (d) Homogamy

161. Pollination occurs through wind is known as
(a) Entomophylly (b) Zoophylly
(c) Anemophylly (d) Hydrophylly

162. Development of phenotypically stable population called
(a) Vybrid (b) Hybrid
(c) Zybrid (d) None of the above

163. An organism which has acquired a heritable variation as a result of sudden change in the hereditary material is called as a:
(a) Hybrid (b) Biotype
(c) Mutant (d) Albino

164. A Polyploid having chromosome sets from different sources such as different species is known as

(a) Autopolyploid (b) Aneuploid

(c) Tetraploid (d) Allopolypoid

165. The increased performance of F1 hybrid over its parents is called

(a) Dominance (b) Heterosis

(c) Crossing over (d) None of these

166. Ratio of the genetic variance to the total variance which determines the relationship between phenotype and genotype is known as

(a) Genetic advance

(b) Penetrance

(c) Hetritability

(d) Dominance

167. Mass pedigree selection method of breeding is particularly, useful for improvement of

(a) Self pollinated crops

(b) Self compatible cross pollinated crops

(c) Self incompatible crops

(d) Diocious crops

168. Emasculation of flower is necessary for hybridization in

(a) Male sterile plants

(b) Self pollinated plants

(c) Self incompatible plants

(d) Dioecious plants

169. Mendel did not face the problem of linkage, however, seven characters studied by him were located on how many chromosomes?

(a) Seven (b) Four

(c) Five (d) Three

170. Test cross ratio in case of duplicate gene action will be

(a) 1:1 (b) 1:1:1:1

(c) 3:1 (d) 1:3

171. A:T ratio in one strand of DNA duplex is 2, what will be the ratio in the complementary strand?

(a) 1.0 (b) 0.5

(c) 2.5 (d) 2.0

172. The term gene was coined by

(a) Mendel (b) Bateson

(c) Shull (d) Johannsen

173. Multiple effects of a single gene is known as

(a) Pleiotropy (b) Polyploidy

(c) Heteroploidy (d) None of these

174. An example of a heterozygous but homogeneous population is

(a) Pureline

(b) Open Pollinated variety

(c) Hybrid variety

(d) Synthetic variety

175. The cross of an inbred line with an open pollinated variety is known as

(a) Back cross (b) Top cross

(c) Three way cross (d) Double cross

176. The first artificial hybrid was made by

(a) Mendel (b) Thomas Fairchild

(c) Koelreuter (d) Beal

177. Polyploid derived from two different species is called

(a) Autopolyploid (b) Triploid

(c) Allopolyploid (d) Monoploid

178. DNA synthesis takes place in which stage?

(a) G1 phase (b) G2 Stage

(c) S Phase (d) None of these

179. The number of bases in the single turn of the DNA helix is

(a) 5 (b) 7

(c) 10 (d) 12

180. The common form of DNA present in the living organisms is the

(a) A form (b) Z form

(c) C form (d) B form

181. Okazaki fragments are formed due to synthesis of DNA in direction
(a) 5'-3' (b) 3'-5'
(c) Both (d) None of these

182. An amino acid which is related to drought tolerance in plants is
(a) Glutamic acid
(b) Proline
(c) Glycine
(d) Hydroxyproline

183. The variance which is fixable is :
(a) Additive (b) Dominance
(c) Epistatic (d) All above

184. The term cybrid refers to the
(a) Somatic hybrids
(b) Gametic hybrids
(c) Both (a) and (b)
(d) None of these

185. During mitosis, chromosomal DNA duplication takes place in
(a) Interphase (b) Prophase
(c) Metaphase (d) Anaphase

186. A mating among the following cannot lead to heterosis
(a) AAbb x aaBB
(b) aaBBcc x AABBCC
(c) AABBccdd x AABBccdd
(d) AABBccdd x aabbCCDD

187. IPGRI headquater is situated in
(a) New Delhi, India
(b) New York, USA
(c) Sydney, Australia
(d) Rome, Italy

188. Law of homologus series is given by
(a) Biffen (b) Debary
(c) Darwin (d) Vavilov

189. Land races are
(a) Advanced strains
(b) Hybrid varieties
(c) Local varieties
(d) All of the above

190. A condition in which F1 plants of an interspecific cross are vigorous and fertile but their F2 progeny is weak and sterile is called
(a) Hybrid sterility
(b) Hybrid in viability
(c) Hybrid breakdown
(d) All of the above

191. Term heterosis was introduced by
(a) Davenport (b) Shull
(c) Hull (d) East

192. Highest uniformity is present in
(a) Double cross (b) Multiple cross
(c) Single cross (d) Three way cross

193. Progeny of a single plant obtained by asexual reproduction is
(a) Pure line (b) Inbred
(c) Clone (d) Strain

194. The progeny of a single homozygous self pollinated plant is known as
(a) Pure line (b) Inbred
(c) Clone (d) Land race

195. The term genetics is coined by
(a) Johannsen (b) Bateson
(c) Punnet (d) Morgan

196. An individual with gametic chromosome number is
(a) Haploid (b) Monoploid
(c) Dihaploid (d) Polyploid

197. The individual with 2n+1 chromosome are commonly known as
(a) Euploid (b) Aneuploid
(c) Heteroploids (d) None of these

198. Pollination and fertilization in an unopened flower bud is the situation
(a) Cleistogamy (b) Chasmogamy
(c) Homogamy (d) Dichogamy

199. The situation in which one gene hides the effect of a second gene when both are present in a chromosome is called
(a) Dominance (b) Epistasis
(c) Both a and b (d) None of these

200. An individual with 2n-1-1 chromosome constitution is called
(a) Monosomic (b) Double monosomic
(c) Disomic (d) Double trisomic

201. The smallest unit of genetic material that cannot be divided but is interchangeable among units refers to
(a) Cistron (b) Muton
(c) Recon (d) Plasmon

202. In apomixes, the progeny closely resembles
(a) Male parent
(b) Female parent
(c) Both male and female parent
(d) None of these

203. Superiority of F1 hybrid over a standard commercial variety is called
(a) Heterobeltiosis
(b) Luxuriance
(c) Economic heterosis
(d) Relative heterosis

204. NBPGR is located at
(a) Lucknow (b) Karnal
(c) Bikaner (d) New Delhi

205. Mutation due to change in gene because of
(a) Loss (b) Degeneration
(c) Addition (d) All of the above

206. The chemical mutagens usually induce
(a) Chromosomal breakages
(b) Gene mutation
(c) Chromosomal translocation
(d) None of these

207. A cybrid is a
(a) Sexual hybrid
(b) Hybrid formed by cell division
(c) Plant formed by tissue culture
(d) Synonymous to hybrid

208. Reversion occurs due to
(a) Gene interaction
(b) Mutation
(c) Gene interaction and mutation
(d) None of these

209. Physical basis of life is
(a) Nucleus (b) Cell
(c) Protoplasm (d) Food

210. The capacity of a cell to develop into a complete plant is called
(a) Tissue culture (b) Fecundity
(c) Totipotency (d) Haploidy

211. Colchicine interferes with
(a) Replication
(b) Recombination
(c) Formation of spindle fibres
(d) None of these

212. Colchicine is generally used for
(a) Chromosome doubling
(b) Hybridization
(c) Mutation
(d) All the above

213. Chromosomal theory of inheritance was given by
(a) Beadle and Tatum
(b) Watson and Crick
(c) Bateson and Punnet
(d) Sutton and Boveri

214. RFLP is a
(a) Biochemical marker
(b) Molecular marker
(c) Morphological marker
(d) Physiological marker

215. Chromosome other than sex chromosomes are called
(a) Allosomes (b) Autosomes
(c) Heterosomes (d) Autoallosomes

216. Improved seed distributed to farmers for commercial cultivation is
(a) Nucleus (b) Breeder
(c) Foundation (d) Certified

217. Recommended concentration of EMS for induced mutagenesis is
(a) 0.03 to 1.5 % (b) 0.30 to 3.0 %
(c) 1.00 to 3.5 % (d) >4 %

218. DNA can be denatured by heating at a temperature of
(a) 720c (b) 860c
(c) 900c (d) 940c

219. The following amongst the molecular markers is codominant in nature
(a) RAPD (b) SSR and RFLP
(c) VNTR (d) None of these

220. Genomic imprinting occurs in
(a) Plants (b) Animals
(c) Humans (d) All of the above

221. In mitosis, longest phase is
(a) Anaphase (b) Telophase
(c) Prophase (d) Metaphase

222. Collection of Germplasm from foreign countries is called
(a) Indigenous species
(b) Introduction
(c) Exotic species
(d) Invasive species

223. The best method for developing disease resistant plants is
(a) Mutation (b) Selection
(c) Hybridization (d) Back crossing

224. The growth of tissues of living plants on a suitable culture medium is called
(a) Anther culture
(b) Plant tissue culture
(c) Genetic engineering
(d) None of these

225. Clonal selection is useful in
(a) Flowering species
(b) Non flowering species
(c) Seeds propagating species
(d) None of these

226. Population which intermate freely are called as
(a) Random mating population
(b) Mendalian population
(c) Panmicitic population
(d) All the above

227. What is the most effective method for transfer of on oligogenic character in an otherwise agronomically superior genotype?
(a) Pedigree
(b) Bulk population
(c) Recurrent selection
(d) Back cross

228. Which of the following will NOT promote speciation?
(a) gamete differences
(b) gene flow
(c) seasonal fertility
(d) natural selection

229. When germination of all seeds occurs over a relatively brief period it is called
(a) Continuos
(b) Intermittent
(c) Quasi simultaneous
(d) Periodic

230. Physical dormancy is found in
(a) *Eucalyptus delegatensis*
(b) *Betula pendula*
(c) *Acacia nilotica*
(d) *Fraxinus* spp

231. Water soaking seed treatment is best suited for
(a) *Cassia fistula*
(b) *Santalum album*
(c) *Terminalia arjuna*
(d) *Grevillea robusta*

232. Vertical antenna like branches often occurring in conifers outside the normal branch whorl
(a) Epicormic branch
(b) Lateral branch
(c) Ramicorn
(d) None of the above

233. For a wind pollinated species, the width of the pollen dilution zone is
(a) 500-600m (b) 100-150m
(c) 200-400m (d) 450-500m

234. Clonal seed orchard has
(a) Broader genetic base and high selection differential
(b) Broader genetic base and less selection differential
(c) Narrow genetic base and less selection differential
(d) Narrow genetic base and high selection differential

235. Pollen contamination in seed orchard can be reduced by
(a) geographic isolation
(b) supplemental mass pollination
(c) both a & b
(d) none of the above

236. Example for closed panmictic population
(a) Seed stand
(b) Seed production area
(c) Seed orchard
(d) None of the above

237. Minimum number of clones to be planted in the first generation seed orchard is
(a) 10-20 (b) 15-25
(c) 25-35 (d) 25-40

238. Tumbling is the method for
(a) Dry extraction of fruits
(b) Wet extraction of pulpy fruits
(c) Extraction of cones
(d) None of the above

239. Dry strobile is seen in
(a) Dogwood (b) Redwood
(c) Casuarina (d) Fir

240. A tree that has been selected for grading because of the desirable character is
(a) Candidate tree
(b) Check tree
(c) Elite tree
(d) Plus tree

241. Factorial and diallel designs are commonly used in the seed orchards of
(a) Broad leaved spp
(b) Conifers
(c) Both a & b
(d) None of the above

242. Systematic removal of individuals in orchards
(a) Rouging (b) Thinning
(c) Clearing (d) None of the above

243. The species with special orchard problems are
(a) Monoecious species
(b) Dioecious species
(c) Both a & b
(d) None of the above

244. The unique functional position of organisms is
(a) Niche
(b) Habitat
(c) Species interaction
(d) Ecotone

245. Speciation occurs through physical separation of population.
(a) Ecotype (b) Sympatry
(c) Cline (d) Allopatry

246. CVP index on forest productivity does not include the following
(a) Climatic (b) Biotic
(c) edaphic (d) None of these

247. Oxygen sag relates to
(a) Soil pollution (b) Water pollution
(c) Air pollution (d) Noise

248. Trees that have proven to be genetically superior by means of progeny testing
(a) Candidate tree
(b) Plus tree
(c) Elite tree
(d) Comparison tree

249. Most frequently used seed orchard design is
(a) Completely randomized design
(b) Randomized block design
(c) Pure rows
(d) Chess board/ Rotating block

250. Plus tree selection method used in an even aged stand
(a) Baseline selection
(b) Comparison tree selection
(c) Regression selection
(d) Mother tree system

251. The original geographic area from which seed or other propagules were obtained is known as
(a) Seed orchard
(b) Seed production area
(c) Seed source
(d) Provenance

252. In tree improvement programme, many cycles of selection and breeding refers to
(a) Tandem selection
(b) Independent culling
(c) Selection index
(d) Recurrent selection

253. The book on applied tree improvement was written by
(a) B.D.Singh
(b) Namkoong
(c) M.Gibson
(d) Bruce Zobel and John Talbert

254. The pioneering tree improvement studies were attempted by
(a) Japanese (b) Chinese
(c) Swedish (d) None of these

255. Successful tree improvement programme depends on
(a) Parentage
(b) Efficient Management
(c) a and b
(d) None of these

256. Variation within the clonal progenies is due to
(a) Environment (b) Genetics
(c) Both a & b (d) Biotic

257. Selection differential can be calculated using
(a) Selection intensity and phenotypic variance
(b) selection intensity and genotypic variance
(c) Phenotypic and genotypic variance
(d) None of these

258. The following mating designs are termed as incomplete mating design
(a) Full diallel (b) Polymix
(c) Nested (d) Partial diallel

259. More genetic gain can be achieved in the following seed orchard
(a) Seedling seed
(b) Clonal
(c) Extensive seedling seed
(d) all the above

260. Pureline theory was proposed by
(a) Jinks (b) Vavilov
(c) Muller (d) Johannsen

261. Homoclime approach suited to
(a) Hybrids (b) exotics
(c) clone (d) Genetic depression

262. Self incompatibility hampers the development of
(a) Inbreds (b) Clone
(c) Hybrids (d) exotics

263. Progeny having both parent identities is known as
(a) Halfsib
(b) Fullsib
(c) One parent family
(d) None of these

264. Selection can be resorted for
(a) Non heritable trait
(b) Environmental influenced trait
(c) heritable trait
(d) All of these

265. Provenance - hybrid seed orchard is used for
(a) Hybridising diverse and compatible provenance
(b) Hybridising genetically similar provenance
(c) Hybridising two different but compatible species
(d) Hybridising diverse and non compatible provenances

266. The presence of male and female flowers in different plants is called
(a) Dioecious (b) Monecious
(c) Halfsib (d) None of these

267. Emasculation by suction method is more suitable
(a) Big flower (b) Medium flower
(c) Small flower (d) Noneof these

268. In trees, normally the width of pollen dilution zone is used
(a) 100-200m (b) 200-300m
(c) 300-400m (d) 400-500m

269. IBPGR refers to
(a) Indian Bureau of Plant Genetic Resources
(b) International Bureau of Plant Genetic Resources
(c) Indian Botanical Plant Genetic Resources
(d) International Botanical Plant Genetic Resources

270. __________ proposed the concept of hormonal control of organ formation.
(a) Skoog and Miller
(b) Gautherect
(c) Caplin and Steward
(d) Kohlenbeck

271. The virus free plant was first produced by
(a) Morel and Martin
(b) White
(c) Cocking & White
(d) Stanford & Muir

272. __________ is a major carbon source in plant tissue culture medium
(a) Glucose (b) Sucrose
(c) Mannose (d) Fructose

273. ________ is a method of sterilizing with water vapour under high pressure.
(a) Hot sterilization
(b) Autoclaving
(c) Flame sterilization
(d) Filter sterilization

274. Cybrid is a ______________
(a) Somatic embryo
(b) Somatic hybrid
(c) Somo clonal variant
(d) Virus free plant

275. __________ mediated transformation is an important indirect gene transfer technique
(a) Agrobacterium
(b) Protoplast transformation
(c) Pollen
(d) Particle bombardment

276. The two poly nucleotide strands of the DNA double helix are
(a) Parallel
(b) Complementary
(c) Non complementary
(d) Parallel and Complementary

277. Tree improvement consists of
(a) Silviculture (b) Tree breeding
(c) Both a & b (d) None of the above

278. The unorganized mass of tissues are called as
(a) Callus
(b) Totipotency
(c) Dedifferentiation
(d) Shooting

279. The original tree (Donor tree) from which parts to be propagated are taken are called
(a) Clone (b) Ramet
(c) Ortet (d) Progeny

280. Ramet is
(a) Population of genetically similar plant
(b) Individual of clone
(c) Callus
(d) Clone

281. Cleistogamous flowers are always
(a) Insect pollinated
(b) Wind pollinated
(c) Cross pollinated
(d) Self pollinated

282. Orthotropic growth refers to
(a) Tree (b) Shrub
(c) Herb (d) None of these

283. The crossing among randomly selected plants in F_2 refers to...............
(a) Recurrent selection
(b) Disruptive mating
(c) Diallele selective mating
(d) Biparental mating

284. The trees produced from the seed of a parent called
(a) Selection
(b) Population
(c) Recurrent selection
(d) Progeny

285. Population of genetically similar plants obtained from same invidual by vegatative method
(a) Propagule (b) Bud
(c) Clone (d) Callus

286. Area where superior phenotypes or genotypes are established and managed intensively and entirely for seed production is referred
(a) Provenance
(b) Family
(c) Seed Orchard
(d) Seed source

287. A condition in which F1 plants of an interspecific cross are vigorous and fertile and their F1 progeny is weak and sterile
(a) Sterility
(b) Fully Incompatibility
(c) Hybrid breakdown
(d) Partial Incompatibility

288. An individual having more than two basic sets of chromosomes
(a) Autopolyploids
(b) Allopolyploids
(c) Polyploids
(d) Amphidiploids

289. Ability of a gene to affect more than one phenotypic characteristic
(a) Pseudoalleles
(b) Modifying factors
(c) Pleiotrophism
(d) Penetrance

290. In cross pollinated species a true breeding line obtained by continuous inbreeding
(a) Variety (b) Hybrid
(c) Inbred (d) Synthetic variety

291. A plant with functional pollen fails to set seed when self pollinated
(a) Sterility
(b) Incompatibility
(c) Gamete abortion
(d) Genetic barrier

292. Regeneration of whole plant from tissues of an actively dividing plant part such as stem tip, is referred
(a) Anther culture
(b) Callus culture
(c) Meristem culture
(d) Ovule culture

293. Variation that is observed among the plants which are regenerated from callus cultures of somatic explants such as meristem

(a) Gametoclonal variation

(b) Protoclonal variation

(c) Somoclonal variation

(d) Zygotic variation

294. Regeneration of plants from isolated meristematic cell or tissues

(a) Embryogenesis

(b) Macro-propagation

(c) Micro-propagation

(d) Protoplast culture

295. The plant part used in regeneration in tissue culture termed as

(a) Callus (b) Explant

(c) Ovule (d) Embryo

296. Isolation, introduction and expression of foreign DNA in plants and animals

(a) Gene cloning

(b) Genetic Engineering

(c) Transgenic breeding

(d) Gene splicing

297. Mating between closely related individuals such as selfing and sib mating is called

(a) Inbred (b) Hybrid

(c) Inbreeding (d) None of these

298. Trees that have proven to be genetically superior by means of progeny testing is

(a) Candidate tree

(b) Plus tree

(c) Elite tree

(d) Comparison tree

299. The term overdominance was given by

(a) Collins

(b) Shell and East

(c) Hull

(d) Bruce

300. Plus tree selection method used in an even aged stand is

(a) Baseline selection

(b) Comparison tree selection

(c) Regression selection

(d) Mother tree system

301. The genotypes having single locus difference is known as.........

(a)Heterosis

(b) Multilines

(c) Pureline

(d) Isogenic lines

302. Usually, seed production areas should contain a minimum area (ha) of

(a) 1 ha (b) 2 ha

(c) 3 ha (d) 4 ha

303. Breeding value (BV) is

(a) Equal to the TA

(b) Twice the TA

(c) 1/2 of TA

(d) 1/4 of TA

304. Heritability in narrow sense is

(a) VA/VP (b) VE/VP

(c) VG/VP (d) VA+VG/VP

305. Repeatability value set in ....of heritability

(a) Lower limit of h2

(b) Upper limit of h2

(c) Intermediate of h2

(d) Both a and b

306. Pseudoheterosis is also known as

(a) Average heterosis

(b) Economic heterosis

(c) Luxuriance

(d) None of these

307. Pureline selection method commonly used in

(a) Self pollinated species

(b) Cross pollinated species

(c) Both a and b

(d) None of the above

308. Male sterility promotes
(a) Inbreeding
(b) Out breeding
(c) Cross breeding
(d) None of the above.

309. Wide crossing is also known as
(a) Distant hybridization.
(b) Intergeneric hybridization.
(c) Interspecific hybridization.
(d) None of the above

310. Character showing continuous variation and considerable effect of the environment is
(a) Quantitative character
(b) Qualitative character
(c) Heritable character
(d) None of the above.

311. Non-additive variation accounts for ------------phenotypic variation
(a) 10 % (b) 25 %
(c) 75 % (d) 100 %

312. Genetic gain depends upon
(a) Hertability
(b) Phenotypic variance
(c) Selection intensity
(d) All of the above

313. Larger units of similar vegetation are referred as
(a) Community (b) Population
(c) Ecosystem (d) Biome

314. Lack of seed set on self pollination is
(a) Male sterility
(b) Self incompatibility
(c) Self fertility
(d) None of the above.

315. If all variation was due to genetics, the broad sense heritability would be equal to
(a) 0 (b) 1
(c) -1 (d) 2

316. Emasculation means
(a) Removal of female flowers/stigma
(b) Removal of male flowers /anthers
(c) Dusting of pollen grains on female flowers
(d) None of the above.

317. Genetic gain in tree improvement programme is estimated by the formula
(a) $G = g \times h^2$
(b) $G = S \times h$
(c) $G = S \times h^2$
(d) $G = E + GE$

318. Which studies are performed in order to describe genetic variation within and between population?
(a) Population variation
(b)Genetic variation
(c) Population dynamics
(d) Genetic gene

319. Where from the best quality seeds would be obtained for producing quality plating stock?
(a) Seed stand
(b) Seed production area
(c) Seed orchard
(d) Clonal orchard

320. Selection in evenaged forest
(a) Regression method
(b) Check tree method
(c) Mass selection
(d) Family solution

321. The dispersive process mainly occurs in small population in which,
(a) Only direction can be predicted not amount
(b) Only amount can be predicted not direction
(c) Both amount and direction can be predicted
(d) Direction and amount cannot be predicted

322. The contribution of offspring to the next generation is called
(a) Fitness
(b) Adaptive value
(c) Selective value
(d) All of the above

323. Genotypic correlation is due to
(a) Polymorphism
(b) Linkage
(c) Pleiotropism
(d) Both b and c

324. The subdivision of gene which is the site of mutation is
(a) Mutant (b) Muton
(c) Mutation (d) None of the above

325. If the coefficient of selection is 0.25, then the fitness is
(a) 1 (b) 0.25
(c) 0.75 (d) 0.5

326. Which markers are related to variation in chromosome morphology ?
(a) DNA markers
(b) Morphological markers
(c) Biochemical markers
(d)Cytological markers

327. Which is the only gaseous hormone?
(a) Abscisic acid (b) Auxins
(c) Cytokinins (d) Ethylene

328. Which hormone is primarily responsible for enhancing apical dominance?
(a) Gibberellins (b) Auxins
(c) Abscisic acid (d) Ethylene

329. Which of the following is responsible for promoting seed dormancy?
(a) Abscisic acid (b) Auxins
(c) Cytokinins (d) Ethylene

330. Which Hormone promotes the ripening of fruit?
(a) Gibberellins (b) Auxins
(c) Cytokinins (d) Ethylene

331. Which of the hormones regulates cell division and differentiation?
(a) Gibberellins (b) Auxins
(c) Cytokinins (d) Ethylene

332. Which hormone instigates "the triple response?"
(a) Abscisic acid (b) Auxins
(c) Cytokinins (d) Ethylene

333. Which hormone is located in every major organ and living tissue?
(a) Abscisic acid (b) Auxins
(c) Cytokinins (d) Gibberellins

334. Which hormone has anti-aging effects?
(a) Gibberellins (b) Auxins
(c) Cytokinins (d) Ethylene

335. Which hormone is responsible to stem elongation?
(a) Gibberellins (b) Auxins
(c) Abscisic acid (d) Ethylene

336. Which hormone allows seeds to ignore environmental conditions & germinate?
(a) abscisic acid (b) auxins
(c) cytokinins (d) gibberellins

337. Which of the following is the only known naturally occurring auxin?
(a) 2, 4-D (b) IAA
(c) GA (d) ABA

338. Which of the following hormones would be most useful in causing the rooting of plant cuttings?
(a) Abscisic acid (b) Cytokinins
(c) Gibberellins (d) Auxins

339. Which plant hormone(s) is (are) most closely associated with cell division?
(a) Ethylene (b) Cytokinin
(c) Abscisic acid (d) Phytochrome

340. The application of which of the following hormones would be a logical first choice in an attempt to produce normal growth in mutant dwarf plants?
(a) Cytokinin (b) Gibberellin
(c) Abscisic acid (d) Ethylene

341. One effect of gibberellins is to stimulate cereal seeds to produce
(a) Lipids. (b) Abscisic acid.
(c) Starch. (d) Amylase.

342. While responses to plant hormones are normally slow, which of the following hormones has been shown to be involved in the rapid opening and closing of stomata?
(a) Auxin (b) Cytokinin
(c) Ethylene (d) Abscisic acid

343. If you were shipping green bananas to a supermarket in Boston, which of the following chemicals would you want to eliminate from the plants' environment?
(a) CO_2 (b) Cytokinins
(c) Ethylene (d) Auxin

344. Auxin is responsible for many plant growth responses, including all of the following except
(a) Phototropism.
(b) Formation of adventitious roots.
(c) Apical dominance.
(d) Promotion of abscission.

345. Plants synthesize auxin from the amino acid
(a) Cystine. (b) Phenylalanine.
(c) Ornithine. (d) Tryptophan.

346. Preserving germplasm in frozen state is
(a) Cryopreservation
(b) Cold storage
(c) In situ preservation
(d) Vernalisation

347. An example of ex-situ conservation is
(a) Seed bank
(b) Sacred groves
(c) National parks
(d) Wildlife Sanctuary

348. Ex situ conservation is carried out in
(a) Sanctuary (b) National park
(c) Biosphere reserve
(d) Zoo.

349. A condition in which pollination and fertilization take place in an unopened flower is termed as
(a) Homogamy (b) Apogamy
(c) Polygamy (d) Cleistogamy

350. Cleistogamy encourages
(a) Self pollination
(b) Cross pollination
(c) Apomixis
(d) Inbreeding depression

351. Clone in general is
(a) Homozygous (b) Heterogenous
(c) Heterozygous (d) None of these

352. Cross of F1 hybrid with either parents is called
(a) Test cross (b) Single cross
(c) Back cross (d) None of these

353. Development and formation of pollen grains in anther of the stamen is known as
(a) Pollination
(b) Fertilization
(c) Microsporogensis
(d) Megasporogensis

354. Development of a fruit without fertilization is known as
(a) Parthenogensis (b) Polyembryony
(c) Budding (d) Apomixis

355. Development of seed without-sexual process is
(a) Apomixis (b) Polyembryony
(c) a and b (d) None of these

356. Development of embryos without fertilization
(a) Meiosis (b) Apomixis
(c) Sporogenesis (d) Gametogenesis

357. Heterosis is commercially exploited in
(a) Cross pollinated species
(b) Often cross pollinated species
(c) Self pollinated species
(d) Vegetatively propagated species

358. Heterosis over the better parent is called as

(a) Relative heterosis

(b) Standard heterosis

(c) Pseudoheterosis

(d) Heterobeltiosis

359. In self incompatibility system

(a) Pollen not produced

(b) Self pollination takes place

(c) Pollen does not germinate on self pollination

(d) All the above

360. The basic difference between a synthetic and composite variety lies in

(a) The difference in synthesis

(b) Testing for picking ability

(c) The mode of pollination

(d) The number of components

361. The best method to break the unfavourable linkage is

(a) Single seed descent method

(b) Diallel selective mating

(c) Pedigree method

(d) Bulk method

362. The condition in which the anthers and stigmas of a bisexual flower mature at the same time is called

(a) Cleistogamy (b) Heterogamy

(c) Syngamy (d) Homogamy

363. The condition in which the pollen of a flower has no fertilizing effect on the stigma of the same flower is called

(a) Herkogamy (b) Dichogamy

(c) Dicliny (d) Self-sterility

364. The effect of the genotype of the pollen grain on the phenotype of the seed is termed as

(a) Apospory

(b) Pollinia

(c) Endopolyploidy

(d) Xenia

365. When the male gametes are shed into embryo sac through micropyle is called

(a) Porogamy (b) Chasmogamy

(c) Mesogamy (d) Syngamy

366. The period from when the flower open to when seed is set called as

(a) Anthesis (b) Apomixis

(c) Herkogamy (d) Dichogamy

367. Which is a common method used in cross pollinated crops?

(a) Mass selection

(b) Bulk method

(c) Pure line selection

(d) Pedigree method

368. Which of the following procedures are not used for producing new varieties of self pollinated crop?

(a) Recurrent selection

(b) Mass selection

(c) Pure line selection

(d) Hybridization

369. The best method to carry forward the maximum number of genotypes / genes to the advanced generations against the force of natural selection is

(a) Pedigree method

(b) Bulk method

(c) Single seed descent method

(d) Back cross method

370. Pollen grains falling on the stigma of another flower of the same kind but on a different plant is called as

(a) Cross-Pollination

(b) Self-Pollination

(c) Often Cross pollination

(d) Mixed Pollination

371. Pollen is dispersed by

(a) Wind

(b) Water

(c) Animal and Birds

(d) All of the above

372. The diversity of place at the ecosystem level is termed as
(a) Morphological Diversity
(b) Ecosystem Diversity
(c) Species Diversity
(d) Genetic Diversity

373. Pollen viability test can be done by
(a) Baker's procedure
(b) p-Phenylenediamine test
(c) None of the above
(d) Both a & b

374. A measure of the possible choices of information provided by a gene is called as
(a) Genetic diversity
(b) Species diversity
(c) Alpha diversity
(d) Morphological diversity

375. Assessment of Genetic Diversity can be done by
(a) Species Richness
(b) Simpson Index
(c) Shannon-Wiener index
(d) All of the above

376. The Shannon-Wiener index is defined as:
(a) E = H/log (S)
(b) D = sum (Pi2)
(c) H = -sum (Pilog[Pi])
(d) 1 – D

377. A measure that accounts for both richness and proportion (percent) of each species
(a) Simpson diversity index
(b) Shannon-Wiener index
(c) All of the above
(d) None of the above

378. Conservation of species in their natural habitat
(a) In situ Conservation
(b) Ex situ Conservation
(c) Habitat Conservation
(d) Invitro Conservation

379. Example for Ex situ Conservation
(a) Natural parks
(b) Nature reserves
(c) Sanctuary
(d) Botanical gardens

380. Protection & maintenance of forest diversity through legal means is known as...
(a) Protected areas
(b) Reserved areas
(c) Botanical gardens
(d) None of the above

381. First preservation plot in India
(a) Bihar in 1906
(b) Gujarat in 1906
(c) Delhi in 1908
(d) Maharashtra in 1909

382. Preservation plot is maintained for conservation of ______
(a) Phenotypic diversity
(b) Genetic diversity
(c) Animal diversity
(d) Morphology diversity

383. Tree showing the inferred evolutionary relationships among various biological species based upon similarities and differences in their physical and/or genetic characteristics.
(a) Phylogenetic tree
(b) Evolutionary tree
(c) Commercial tree
(d) Both a & b

384. The current method of choice to infer phylogenetic trees is
(a) Cladistics (b) osteometric data
(c) Phenetics (d) Synapomorphies

385. A__________. is a broad term for the diagrammatic representation of a phylogenetic tree.
(a) Chronogram (b) Cladogram
(c) Dendrogram (d) Dendrology

386. Phylogenetics is richly informed....
(a) Taxonomy of organisms
(b) Classification of organisms
(c) Identification of organisms
(d) All of the above

387. A directed tree with a unique node corresponding to the (usually imputed) most recent common ancestor of all the entities at the leaves of the tree is called as.....
(a) Rooted phylogenetic tree
(b) Un rooted phylogenetic tree
(c) Bifurcated phylogenetic tree
(d) None of above

388. Interactions between two different species as selective forces on each other, resulting in adaptations that increase their interdependency.
(a) Covaluation (b) Coevolution
(c) Pollination (d) Reproduction

389. Most pollen falls within ________ from its sources.
(a) 150m (b) 250m
(c) 350m (d) 140m

390. Control pollen was killed by placing in an incubator at
(a) 56°C for up to 2 week
(b) 50°C for up to 1 week
(c) 60°C for up to 1 week
(d) 66°C for up to 2 week

391. A sugary solution produced in special flower glands is called as ________
(a) Nectar
(b) Honey
(c) Pollen
(d) None of the above

392. Pollen was stored for Periods ranging from 1 day to ...
(a) 3 years (b) 10 years
(c) 5 years (d) 15 years

393. Some plants take advantage of the sex drive of certain insects, it is called as________
(a) Plant diversion
(b) Plant duplication
(c) Plant Mimicry
(d) Plant modification

394. ________ in which only one parent is known for any given progeny.
(a) Incomplete pedigree designs
(b) Complete pedigree designs
(c) All of the above
(d) None of the above

395. Poly Cross Mating Designs is otherwise known as________
(a) Pollen Mix Mating Designs
(b) Total cross Mating Designs
(c) Open Pollinated Mating Design
(d) Cross Mating Design

396. Members of one sex are crossed in all combinations with several members of the other sex is called as¬________
(a) Nested design
(b) Factorial design
(c) Disconnected Factorial design
(d) Random design

397. Each parent is mated to another member of the population only once, it is known as ______
(a) Single pair mating design
(b) Double pair mating design
(c) Three pair mating design
(d) None of the above

398. Groups of parents of one sex are mated to members of the other sex is known as________
(a) Nested design
(b) Group Design
(c) Diallele Design
(d) Factorial Design

399. A__________ is a molecular unit of heredity of a living organism.
(a) Allele (b) Gene
(c) Chromosome (d) Cell

400. The first interspecific hybridization between sweet William and Carnation was developed by
(a) Thomas Fairchild
(b) Knight
(c) John Le Couteur
(d) De Vilmorin

401. Genes can be detected by__________
(a) Transgene Detection methods
(b) Ligase-mediated gene detection technique
(c) Apoptotic detectionmethods
(d) All of the above

402. Who discovered the Principles of Inheritance?
(a) Rimpu (b) Mendel
(c) Neilsson (d) Shull

403. The term heterosis for hybrid vigour is attributable to
(a) Jones (b) Bruce
(c) Hardy (d) Shull

404. How many Centers of Origin are identified?
(a) 8 (b) 9
(c) 7 (d) 10

405. Auxins (Pick the odd man out)
(a) IAA (b) BAP
(c) IBA (d) NAA

406. Auto Vegetative Propagation (Pick the odd man out)
(a) Rooting of cuttings
(b) Layering
(c) Budding
(d) Grafting

407. Heterosis is confined to
(a) F1 (b) F2
(c) BC1 (d) BC2

408. The average performance of an individual in a particular series of crosses is termed as
(a) General combining ability
(b) Specific combining ability
(c) Epistasis
(d) None of these

409. The selection method predominantly followed in plantations
(a) Regression method
(b) Comparison tree method
(c) Base line method
(d) All of these

410. The term provenance means
(a) Land race
(b) Geographical source
(c) Seed source
(d) None of the above

411. Ecotype refers to
(a) Variation in the phenotypic characters
(b) Phenotypic variation of different species
(c) Differential response of different species to environmental factors
(d) Differential response of the same species to environmental factors

412. The selection method used for improvement of traits for disease resistance and cold hardiness is
(a) Tandem selection
(b) Independent culling
(c) Selection index
(d) Recurrent selection

413. The seed orchard established using grafted plant is called
(a) Clonal seed orchard
(b) Seedling seed orchard
(c) Extensive seedling seed orchard
(d) None of the above

414. The growth regulators used for rooting in vegetative propagation are collectively called as

(a) Vitamins (b) Cytokinins
(c) Gibberllins (d) Auxins

415. The exotic tree which has natural hybrids in India is

(a) Casuarina (b) Eucalyptus
(c) Silver oak (d) Acacia

416. A plus tree whose genetic worth has been established is called

(a) Candidate tree
(b) Comparison tree
(c) Elite tree
(d) Superior tree

417. The pistil matures before anthers is called

(a) Protandry (b) Protogyny
(c) Herkogamy (d) Heterostyly

418. The stamens mature before pistil is called

(a) Protandry (b) Protogyny
(c) Herkogamy (d) Heterostyly

419. The Staminate and pistillate flowers occurs in the same plant is called

(a) Monoecy (b) Dioecy
(c) Protogyny (d) Protandry

420. The male and female flowers are present on different plants is called

(a) Monoecy (b) Dioecy
(c) Protogyny (d) Protandry

421. The transfer of pollen from the anthers to the stigma in the same flower is

(a) Autogamy (b) Allogamy
(c) Geitonogamy (d) None of the above

422. The transfer of pollen from a flower of one plant falls on the stigmas of other flowers of the same plant is called

(a) Autogamy
(b) Allogamy
(c) Geitonogamy
(d) None of the above

423. Single seed descent (SSD) method is the modified form of -------

(a) Bulk method
(b) Mass selection method
(c) Clonal selection
(d) Pedigree Method

424. ---------- is the Mating between phenotypic dissimilar individuals belonging to same populations

(a) Phenotypic Assortative Mating System
(b) Phenotypic Disassortative Mating System
(c) Genotypic Assortative Mating System
(d) Genotypic Disassortative Mating System

425. ---------- mating systems reduces frequency of other alleles

(a) Random mating system
(b) Phenotypic Assortative Mating System
(c) Phenotypic Disassortative Mating System
(d) Genotypic Assortative Mating System

426. The average phenotypic value of the selected individuals, expressed as a deviation from the population mean is

(a) Selection index
(b) Selection differential
(c) Half sib selection
(d) Full sib selection

427. Selection of parent trees based upon the performance of their progeny

(a) Mass selection
(b) sib selection
(c) Recurrent selection
(d) Progeny testing

428. In cryopreservation, cells are stored at temperature of

(a) –196°C (b) –112°C
(c) –120°C (d) –100°C

429. Synthetic seed in sandal wood is attributable to
(a) Nadgir (b) Bapat & Rao
(c) Lakshmi Sita (d) Mascarenhas

430. Tick the antiauxin
(a) Azoindole (b) Indole acetic acid
(c) Pyridoxine (d) None of these

431. The chemical basis of heredity is
(a) Protein
(b) DNA
(c) Carbohydrates
(d) Amino acids

432. PCR invented by
(a) Kary Muller (b) William benting
(c) Karl mullis (d) Totenzef

433. EDTA stands for
(a) Enzyme derived testing analysis
(b) Ethylene diamine tetra acetic acid
(c) Ethylene dichloro tetra acetic acid
(d) Ethylene dinitro triacetic acid

434. The buffer used in Electrophoresis is
(a) Tris acetate (b) Tricloride
(c) Ethanol (d) Bromin

435. RAPD stands for
(a) Randomly amplified Polymorphic DNA
(b) Restricted amplified polymorphic DNA
(c) Rapid Denaturation of Protein test
(d) Restriction and Polymerization of DNA

436. Father of Plant Tissue culture
(a) Murashige (b) Nitsch
(c) Skoog (d) Haberlandt

437. Anther culture technique was developed by
(a) Guha & Maheswari
(b) Bajaj
(c) Haberlandt
(d) H.J. Paikar

438. Single cell culture first developed by
(a) Muir (b) Gautheret
(c) Van overbeek (d) Morel

439. Cryopreservation technique was first developed by
(a) Zimmerman (b) Scibert
(c) Melchers (d) Bergamann

440. Protoplast fusion was attributable to
(a) Zimmerman (b) Scibert
(c) Melchers (d) Bergamann

441. The optimum temperature and Humidity in the culture room is normally
(a) 250c and 50% (b) 200c and 40%
(c) 200c and 80% (d) 250c and 80%

442. The membrane filter used in tissue culture should contain pore size of
(a) Less than 0.45 micro meter
(b) More than 0.45 micro meter
(c) Less than 1.00 micro meter
(d) More than 0.95 micro meter

443. The main energy source in Plant tissue culture technique is
(a) Agarose (b) Sucrose
(c) Fats (d) Proteins

444. During culture process Activated charcoal is added to adsorb
(a) Sucrose (b) Agarose
(c) Phenols (d) Fungi bodies

445. Agar is basically a
(a) Poly saccharide
(b) Lipid
(c) Protein (d) Phenolic compound

446. Pick the odd man out
(a) Cytokinins (b) Giberellins
(c) Absciscic acid (d) Thiamine

447. Auxins can dissolve in
(a) Ethanol (b) Water
(c) Oil (d) Petrol

448. In tissue culture IAA induces
(a) Cell division (b) Root induction
(c) Apical dominance
(d) All the above

449. Pick the odd man out
(a) Benzyl aminopurine
(b) Benzyl adenine
(c) Kinetin
(d) NAA

450. The potential to development of an adult organism from a single cell is referred as
(a) Totipotency (b) Apomixis
(c) Parthenocarpy (d) Callus induction

451. From a single cell, development of reproducible non zygotic embryos capable of germinating to form complete plants is called
(a) Somatic embryogenesis
(b) Shoot tip culture
(c) Meristem culture
(d) Zygote differentiation

452. Pick the odd man out
(a) Suspension culture
(b) Batch culture
(c) Continuous culture
(d) Anther culture

453. Haemocytometer is used to
(a) Cell counting
(b) Electrolyting callus culture
(c) Differentiation inducer
(d) Measure O_2 level in growing media

454. Morel and Martin (1952) developed techniques for
(a) Induced somatic embryogenesis
(b) Disease free plants
(c) Isolation of specific pathogens
(d) Callogenesis

455. The formation of Callus from explants is called
(a) Redifferentiation
(b) Differentiation
(c) Regeneration
(d) Culturing

456. Caulogenesis refers to
(a) Root formation
(b) Shoot formation
(c) Brach formation
(d) Callus formation

457. Rhizogenesis in plant tissue culture involves
(a) Root formation
(b) Shoot formation
(c) Brach formation
(d) Callus formation

458. NOA stands for
(a) Napthoxy acetic acid
(b) Nicotinamide oxalo acetate
(c) Non oxalic acetones
(d) None of these

459. What is HEPA?
(a) A polyurathene sheet
(b) A particulate media
(c) A high density aluminium sheet
(d) A filter used in Laminar chamber

460. The most commonly used medium in tissue culture (Pick the odd man out)
(a) Rose Bengal agar medium
(b) MS medium
(c) LS medium
(d) N6 medium

461. The lignification of cell wall is a characteristic feature of
(a) Non – Embryogenic calli
(b) Embryogenic calli
(c) Both of the above
(d) None of the above

462. Aseptic isolation and growth of sexually produced embryos in vitro to produce viable plant is called
(a) Somatic embryogenesis
(b) Embryo culture
(c) Anther culture
(d) Suspension culture

463. TMV stands for
(a) Tomographic molecular virulent
(b) Tobacco mosaic virus
(c) Tissues modified for vitality
(d) Tomato monomeric virus

464. FDA stands for
(a) Fluorescein di acetate
(b) Flavin di acetate
(c) Ferridoxin dissolved auxin
(d) None of the above

465. PEG used in tissue culture for
(a) Separation of tissues
(b) Fusion of protoplast
(c) Organ development
(d) Organ dissociation

466. Protoplast fusion is to develop the following (pick the odd man out)
(a) Aneuploids
(b) Cybrids
(c) Somatic Hybrids
(d) Transfer of genomes

467. National Chemical Laboratory is located
(a) Chennai (b) Delhi
(c) Pune (d) Mumbai

468. The genotypic clusters separated by largest distance show
(a) Maximum divergence
(b) Minimum divergence
(c) Equal divergence
(d) None of the above

469. D^2 statistics measures the forces of differentiation
(a) Inter-cluster
(b) Intra-cluster
(c) Both A & B
(d) Intermediate cluster

470. Base population includes
(a) Natural stands
(b) Unimproved plantations
(c) Both A & B
(d) Segregated population

471. The concept of D^2 statistics was proposed by
(a) Mahalanobis (b) Rao
(c) Peterson (d) Zobel

472. In recurrent mass selection the genetic gain for the traits of medium heritability is
(a) 12.0% (b) 12.5%
(c) 13.0% (d) 13.5%

473. Heterosis is
(a) Hybrid vigour
(b) Increased expression
(c) Back cross
(d) Inbreeding depression

474. Nucleus population is
(a) Highly productive individuals
(b) Control crossed
(c) Both A & B
(d) Assertive mating population

475. Selection in clonal orchard from forwards and backwards selection is executed
(a) Within the open pollinated breeding population
(b) Best trees from best families
(c) Both A & B
(d) Recurrent selection

476. In Clonal orchard from forwards and backwards selection gain achieved is
(a) Equal to seedling orchard
(b) Greater than seedling orchard
(c) Less than seedling orchard
(d) None of the above

477. The methods by which trees are combined to produce progeny is called
(a) Progeny testing
(b) Seeding seed orchard
(c) Mating design
(d) Provenances

478. Incomplete pedigree design include
(a) Open pollinated matting design
(b) Polymix matting design
(c) Both A & B
(d) Single pair matting design

479. General combining ability is
(a) Average performance of the progeny of individual parent
(b) Average performance of the progeny of multiple parents
(c) Average performance of the parents
(d) Both A & B

480. In open pollinated mating design estimates of genetic parameters is biased because of
(a) Random mating
(b) Non random mating
(c) Both A & B
(d) Not biased

481. Singe pair mating design ensures complete avoidance of
(a) In breeding (b) Out breeding
(c) Back cross (d) Multiple cross

482. In polymix mating design the number of male parents selected should not be less than
(a) 5 (b) 10
(c) 15 (d) 20

483. An operon is most closely associated with
(a) DNA replication
(b) Point mutations
(c) Translation
(d) Transcription

ANSWERSHEET
11. Forest Genetics and Tree Breeding

1. (b)	2. (a)	3. (c)	4. (a)	5. (b)	6. (c)	7. (d)	8. (c)	9. (d)	10. (b)
11. (c)	12. (c)	13. (c)	14. (a)	15. (b)	16. (d)	17. (c)	18. (a)	19. (b)	20. (c)
21. (b)	22. (d)	23. (d)	24. (c)	25. (c)	26. (d)	27. (a)	28. (a)	29. (c)	30. (a)
31. (b)	32. (d)	33. (d)	34. (b)	35. (b)	36. (d)	37. (c)	38. (d)	39. (d)	40. (b)
41. (d)	42. (d)	43. (b)	44. (a)	45. (c)	46. (b)	47. (c)	48. (d)	49. (b)	50. (a)
51. (a)	52. (b)	53. (c)	54. (d)	55. (a)	56. (c)	57. (d)	58. (b)	59. (a)	60. (d)
61. (d)	62. (a)	63. (d)	64. (b)	65. (c)	66. (d)	67. (a)	68. (d)	69. (a)	70. (d)
71. (d)	72. (a)	73. (b)	74. (a)	75. (a)	76. (c)	77. (c)	78. (b)	79. (b)	80. (b)
81. (a)	82. (c)	83. (a)	84. (d)	85. (a)	86. (b)	87. (b)	88. (a)	89. (a)	90. (b)
91. (a)	92. (c)	93. (d)	94. (a)	95. (c)	96. (b)	97. (b)	98. (c)	99. (d)	100. (a)
101. (d)	102. (c)	103. (d)	104. (b)	105. (a)	106. (b)	107. (c)	108. (a)	109. (c)	110. (d)
111. (a)	112. (a)	113. (c)	114. (b)	115. (d)	116. (c)	117. (a)	118. (d)	119. (d)	120. (c)
121. (b)	122. (d)	123. (d)	124. (b)	125. (c)	126. (d)	127. (a)	128. (c)	129. (d)	130. (d)
131. (a)	132. (c)	133. (b)	134. (d)	135. (b)	136. (a)	137. (a)	138. (a)	139. (c)	140. (a)
141. (d)	142. (a)	143. (d)	144. (b)	145. (b)	146. (d)	147. (a)	148. (b)	149. (a)	150. (a)

151. **(c)** 152. **(d)** 153. **(b)** 154. **(d)** 155. **(d)** 156. **(c)** 157. **(b)** 158. **(d)** 159. **(a)** 160. **(d)**
161. **(c)** 162. **(b)** 163. **(c)** 164. **(d)** 165. **(b)** 166. **(c)** 167. **(b)** 168. **(b)** 169. **(b)** 170. **(c)**
171. **(b)** 172. **(d)** 173. **(a)** 174. **(c)** 175. **(b)** 176. **(b)** 177. **(c)** 178. **(c)** 179. **(c)** 180. **(d)**
181. **(a)** 182. **(b)** 183. **(a)** 184. **(a)** 185. **(a)** 186. **(c)** 187. **(d)** 188. **(d)** 189. **(c)** 190. **(a)**
191. **(b)** 192. **(a)** 193. **(c)** 194. **(a)** 195. **(b)** 196. **(a)** 197. **(b)** 198. **(a)** 199. **(b)** 200. **(b)**
201. **(c)** 202. **(b)** 203. **(c)** 204. **(d)** 205. **(d)** 206. **(b)** 207. **(b)** 208. **(b)** 209. **(c)** 210. **(c)**
211. **(c)** 212. **(a)** 213. **(d)** 214. **(b)** 215. **(b)** 216. **(c)** 217. **(a)** 218. **(d)** 219. **(b)** 220. **(d)**
221. **(c)** 222. **(c)** 223. **(d)** 224. **(b)** 225. **(b)** 226. **(d)** 227. **(d)** 228. **(b)** 229. **(c)** 230. **(c)**
231. **(d)** 232 **(c)** 233. **(c)** 234. **(b)** 235. **(c)** 236. **(c)** 237. **(d)** 238. **(c)** 239. **(d)** 240. **(a)**
241. **(b)** 242. **(a)** 243. **(b)** 244. **(a)** 245. **(d)** 246. **(c)** 247. **(b)** 248. **(c)** 249. **(b)** 250. **(b)**
251. **(d)** 252. **(d)** 253. **(d)** 254. **(a)** 255. **(c)** 256. **(a)** 257. **(a)** 258. **(b)** 259. **(b)** 260. **(d)**
261. **(b)** 262. **(a)** 263. **(b)** 264. **(c)** 265. **(a)** 266. **(a)** 267. **(c)** 268. **(d)** 269. **(b)** 270. **(a)**
271. **(c)** 272. **(a)** 273. **(b)** 274. **(a)** 275. **(a)** 276. **(d)** 277. **(c)** 278. **(a)** 279. **(c)** 280. **(b)**
281. **(d)** 282. **(a)** 283. **(d)** 284. **(d)** 285. **(c)** 286. **(c)** 287. **(c)** 288. **(c)** 289. **(c)** 290. **(c)**
291. **(b)** 292. **(c)** 293. **(c)** 294. **(c)** 295. **(b)** 296. **(c)** 297. **(c)** 298. **(c)** 299. **(b)** 300. **(b)**
301. **(d)** 302. **(d)** 303. **(b)** 304. **(a)** 305. **(b)** 306. **(c)** 307. **(a)** 308. **(b)** 309. **(a)** 310. **(a)**
311. **(b)** 312. **(d)** 313. **(d)** 314. **(b)** 315. **(b)** 316. **(b)** 317. **(c)** 318. **(a)** 319. **(d)** 320. **(b)**
321. **(b)** 322. **(d)** 323. **(d)** 324. **(b)** 325. **(c)** 326. **(d)** 327. **(d)** 328. **(b)** 329. **(a)** 330. **(d)**
331. **(c)** 332. **(a)** 333. **(c)** 334. **(c)** 335. **(a)** 336. **(d)** 337. **(b)** 338. **(d)** 339. **(b)** 340. **(b)**
341. **(d)** 342. **(d)** 343. **(c)** 344. **(d)** 345. **(d)** 346. **(a)** 347. **(a)** 348. **(d)** 349. **(d)** 350. **(a)**
351. **(c)** 352. **(c)** 353. **(c)** 354. **(a)** 355. **(a)** 356. **(b)** 357. **(a)** 358. **(d)** 359. **(c)** 360. **(d)**
361. **(b)** 362. **(d)** 363. **(d)** 364. **(d)** 365. **(a)** 366. **(a)** 367. **(d)** 368. **(b)** 369. **(c)** 370. **(a)**
371 **(d)** 372. **(b)** 373. **(d)** 374. **(a)** 375. **(a)** 376. **(c)** 377. **(a)** 378. **(a)** 379. **(d)** 380. **(a)**
381. **(a)** 382. **(b)** 383. **(d)** 384. **(a)** 385. **(c)** 386. **(d)** 387. **(a)** 388. **(b)** 389. **(a)** 390. **(c)**
391. **(a)** 392. **(a)** 393. **(c)** 394 **(a)** 395. **(a)** 396. **(b)** 397. **(a)** 398. **(a)** 399. **(b)** 400. **(a)**
401. **(d)** 402. **(b)** 403. **(d)** 404. **(a)** 405. **(b)** 406. **(c)** 407. **(a)** 408. **(a)** 409. **(b)** 410. **(b)**
411. **(d)** 412. **(a)** 413. **(a)** 414. **(d)** 415. **(b)** 416. **(c)** 417. **(b)** 418. **(a)** 419. **(a)** 420. **(b)**
421. **(a)** 422. **(c)** 423. **(a)** 424. **(b)** 425. **(a)** 426. **(b)** 427. **(d)** 428. **(a)** 429. **(b)** 430. **(a)**
431. **(b)** 432. **(c)** 433. **(b)** 434. **(a)** 435. **(a)** 436. **(d)** 437. **(a)** 438. **(a)** 439. **(b)** 440. **(a)**
441. **(d)** 442. **(a)** 443. **(a)** 444. **(c)** 445. **(a)** 446. **(d)** 447. **(a)** 448. **(d)** 449. **(d)** 450. **(a)**
451. **(a)** 452. **(a)** 453. **(a)** 454. **(b)** 455. **(b)** 456. **(b)** 457. **(a)** 458. **(a)** 459. **(d)** 460. **(a)**
461. **(a)** 462. **(b)** 463. **(b)** 464. **(a)** 465. **(b)** 466. **(a)** 467. **(c)** 468. **(a)** 469. **(c)** 470. **(c)**
471. **(a)** 472. **(b)** 473. **(a)** 474. **(c)** 475. **(c)** 476. **(b)** 477. **(c)** 478. **(c)** 479. **(a)** 480. **(b)**
481. **(a)** 482. **(b)** 483. **(d)**

12

Physiology

KEY POINTS

- Plant physiology is a science to study the law of the life activity of plant.
- Epidermis bears root hairs which is known as piliferous layer
- Water potential is determined by three internal factors i.e. $\psi_w = \psi_w + \psi_s + \psi_p$
- Osmotic pressure has positive sign
- Osmotic potential has negative sign
- The diffusion of solvent molecules into the solution through a semi permeable membrane is called osmosis
- Most organisms are comprised of at least 70% or more water. Some plants, like a head of lettuce, are made up of nearly 95% water
- Water molecule has a positively-charged (hydrogen side) and negatively-charged side (oxygen)
- Water has a high boiling point and a high melting point when compared to other similar-sized molecules such as ammonia, carbon dioxide, hydrogen sulfide
- Red-orange and violet colours of light are those most greatly absorbed by green plants.
- The excess of water is lost from the aerial parts of plants in the form of water vapours is called transpiration
- Three types of transpiration viz., Stomatal transpiration, Cuticular transpiration and Lenticular transpiration
- The epidermis is a colourless layer of cells that protect the upper and lower sides of the leaf. The lower epidermis is interspersed with stomata.
- Guard cells are found on each side of the stoma and control the exchange of gases and the release of water by opening and closing the stomata.
- Transpiration is reduced by stomates being positioned on only the underside of the leaf.
- Long night" ("short day") plants - the photo-period is more than 9 hours but less than 14 hours per day.
- "Short night" ("long day") plants - the photo-period is more than 14 hours per day.
- "Indeterminate" plants such as carnations, cotton and sunflowers are relatively unaffected by the amount of daylight per day.
- Phytochrome is a blue pigment found in the leaf which allows plants to measure the amount of daylight
- Cohesion and adhesion are responsible for capillary action, the movement of water up a thin tube.

- Tropisms are movements caused by external stimulus.
- Phototropism - tropism in response to light
- Geotropism - tropism in response to the earth gravity
- Thigmotropism - tropism in response to mechanical stimulus.
- Skototropism - growth of vines toward a darkened region of the environment.
- Plants that are sensitive to relatively high salt concentrations as glycophytes
- Plants that are able to grow in the presence of high salts as Halophytes
- LAI = Total Leaf Area/ Land Area
- Leaf Area: L X W X A
  Where, L: leaf length; W: leaf maximum width; A: Constant
- Based on the mobility, elements are also classified into three types.
  Mobile elements : N, P, K, S and Mg
  Immobile elements : Ca, Fe and B
  Intermediate in mobility : Zn, Mn, Cu, Mo

MULTIPLE CHOICE QUESTION'S

1. Wilting of plants results from excessive :
(a) Absorption (b) Photosynthesis
(c) Respiration (d) Transpiration

2. Process of transpiration plants helps in a :
(a) Absorption of CO_2
(b) Upward conduction
(c) Opening of stomata
(d) None of these

3. Plasmolysis will occur when the cell is placed in a :
(a) Acidic solution
(b) Isotonic solution
(c) Hypotonic solution
(d) Hypertonic solution

4. Blue green algae are
(a) Actinomycetes (b) Acellular
(c) Prokaryotes (d) Eukaryotes

5. The solute present into the cells is strongly dependent on and ability of cells to respire aerobically and to produce it
(a) ADP (b) ATP
(c) RNA (d) DNA

6. Simplest amino acid is
(a) Alanine (b) Glycine
(c) Leucine (d) Aspartic acid

7. Fermentation is conducted by
(a) All fungi
(b) All bacteria
(c) Some fungi and some bacteria
(d) All micro-organisms

8. Peptide linkages are formed among :
(a) Amino acids
(b) Gluco molecules
(c) Surcose molecules
(d) Glycerol molecules

9. Proteins are made up of :
(a) Amino aids (b) Nucleic acids
(c) Sugars (d) Fatty acids

10. Carbon dioxide is liberated during
(a) Ascent of sap
(b) Transpiration
(c) Photosynthesis
(d) Respiration

11. Which is a micronutrient?
(a) Ca (b) Zn
(c) P (d) Mg

12. Which of the following effects of auxins on plants is the basis of commercial application?
(a) Callus formation
(b) Curvature of stem
(c) Induction of root formation
(d) In stem cuttings
(e) All of these

13. During the photosynthesis the oxygen in glucose comes from:
(a) Carbon dioxide
(b) Water
(c) Both from Carbon dioxide and water
(d) Oxygen in the air

14. When yeast and certain bacteria carry on aerobic respiration they produce :
(a) Carbon dioxide
(b) Nitrogen
(c) Water
(d) Oxygen

15. One of the following is a water soluble pigment :
(a) Carotene (b) Chlorophyll
(c) Anthocyanin (d) Xanthophyll

16. Photosynthesis is :
(a) Catabolic process
(b) Anabolic process
(c) Exothermic process
(d) Exergonic Process

17. Which of the following structures controls the transport of the material into and out of living cells?
(a) Centrosome (b) Cell membrane
(c) Cell wall (d) Ribosome

18. Functions and vital activities of the plants are studied under :
(a) Ecology (b) Physiology
(c) Cytology (d) Morphology

19. Excessive use of fertilizer causes death of plants due to;
(a) Endosmosis (b) Exosmosis
(c) Imbibition (d) Turgidity

20. Swelling of wooden door during rainy season due to :
(a) Endosmosis (b) Imbibition
(c) Capillary (d) Deplasmolysis

21. Stomata are bound by a pair of guard cells and open guard cells are :
(a) Flaccid (b) Large
(c) Turgid (d) Green

22. Osmosis means :
(a) Flow of solute from low concentration to higher concentration
(b) Flow on solute from higher concentration to low concentration
(c) Flow of solvent from low concentration to high concentration
(d) Flow of solvent from high concentration to low concentration

23. The rate of transpiration will greatly depend upon :
(a) Frequency of stomata
(b) Position of stomata
(c) State of stomata
(d) None

24. Photochemical reactions of photosynthesis take place in:
(a) Frets (b) Stroma
(c) Thylakoids (d) All of these

25. Which of the following is an essential part of photosynthesis?
(a) Glycolsis
(b) Photorespiration
(c) Plasmolysis
(d) Photolysis

26. Hydroponics are :
(a) Growing of plants in water culture
(b) Growing of aquatic plants
(c) Growing of floating aquatic plants
(d) Soil less cultivation of plants

27. When a seed is placed in a suitable medium for germination, by which of the following first process water enter into the seed coat?
(a) Imbibition (b) Absorption
(c) Osmosis (d) Root pressure

28. Plasma membrane :
(a) Impermeable
(b) A selective barrier
(c) A non-selective barrier
(d) Made of up cellulose

29. Anaerobic respiration is also known as :
(a) Intra-molecular respiration
(b) Inter-molecular respiration
(c) Extra-molecular respiration
(d) Molecular respiration

30. Which of the following is not a part of chlorophyll?
(a) Iron (b) Nitrogen
(c) Carbon (d) Hydrogen

31. Wilting of leaves is hot weather is due to:
(a) Iron and calcium
(b) Sodium and copper
(c) Calcium and potassium
(d) Iron and Magnesium

32. Leaves with yellow or colorless patches are termed as :
(a) Lecorine (b) Decolourized
(c) Etiolated (d) Variegated

33. Oxygen evolved in photosynthesis is :
(a) Less than CO_2 consumed
(b) More than CO_2 consumed
(c) Equal to CO_2 consumed
(d) None

34. The dry weight of a green leaf will be
(a) Lesser in evening
(b) Greater in the afternoon
(c) Greater in the morning
(d) Almost same throughout the whole day

35. Etiolated plants are formed due to lack of :
(a) Light (b) Hg
(c) Fe (d) Mg

36. For photosynthesis (i.e. for the synthesis of organic matter) the green plants need only :
(a) Light (b) Chlorophyll
(c) CO_2 & water (d) All of these

37. Growth is primarily affected by two climatic factors which are
(a) Light and temperature
(b) Temperature and relative humidity
(c) Light and wind
(d) Rainfall and temperature

38. One of the following is not found in intact cells:
(a) Lomasomes (b) Dictyosomes
(c) Microsomes (d) Ribosomes

39. A plant which needs not less than 10 hours of light to flower is
(a) Long day plant
(b) Short day plant
(c) Day neutral plant
(d) None of these

40. If cells are broken up and sedimented by centrifugation, the new structure formed in one of the fraction is :
(a) Lysosome (b) Ribosome
(c) Microsome (d) Centrosome

41. Probably the most useful measurable property in the soil-plant-air system is :
(a) Matric potential
(b) Water potential
(c) Osmotic potential
(d) Redox potential

42. DPD is equal to :
(a) OP X TP (b) OP-TP
(c) TP-OP (d) OP+TP

43. Seeds swell when placed in water is due to :
(a) Osmosis (b) Imbibition
(c) Hydrolysis (d) Plasmolysis

44. The loss of water in the form of vapour from aerial plant parts is known as :
(a) Osmosis (b) Respiration
(c) Photosynthesis (d) Transpiration

45. Semi-permeability is the property of :
(a) Plasma membrane
(b) Cell wall
(c) Golgibody
(d) Endoplasmic recticulum

46. Protoplasm is
(a) Polyphasic colloidal system
(b) Collodial solution
(c) Suspension
(d) True solution

47. Potometer is use to measure :
(a) Rate of transpiration
(b) Rate of respiration
(c) Growth
(d) Osmosis

48. Photosynthesis is a :
(a) Reductive, endermic, catabolic process
(b) Reductive, endergonic, anabolic process
(c) Reductive, exergonic, anabolic process
(d) Oxidative. Exergonic, catabolic process

49. Plants synthesis protein from :
(a) Starch (b) Sugar
(c) Amino acids (d) Fatty acids

50. The end products of respiration in plants are :
(a) Starch and O_2
(b) H_2O and energy
(c) CO_2, H_2O and energy
(d) Sugar and O_2

51. The process by which proteins are synthesized in a cell is called
(a) Translation (b) Transduction
(c) Translocation (d) Transcription

52. Process of endosmosis stops :
(a) When water concentration becomes equal
(b) When solution becomes isotonic
(c) When the leaves fall
(d) When there is no light

53. Pigment system is concerned with :
(a) Photolysis of water
(b) Reduction of CO_2
(c) Flowering
(d) None of these

54. Respiration
(a) Exothermic process
(b) Endothermic process
(c) Endergonic process
(d) Anabolic process

55. Photosynthesis is to the chloroplast, as ________ is to the mitochondrion.
(a) Respiration
(b) Microtubule
(c) Extracellular matrix
(d) Microfilament

56. The smooth endoplasmic reticulum functions in synthesis of
(a) DNA
(b) Lipids
(c) Polysaccharides
(d) Proteins

57. Which of the following is an eukaryotic characteristic?
(a) Cell fission
(b) Little processing of rna
(c) 70s ribosomes
(d) Membrane-bound nucleus

58. Lignin is an important component in
(a) Bacterial cell walls
(b) Plasmodesmata
(c) Tight junctions
(d) Secondary plant cell walls

59. Which of the following is true of a plant cell?

(a) Plant cells are often much larger than animal cells

(b) Plant cells may contain chloroplasts

(c) Only plant cells have a cell membrane.

(d) Chromosomes are found only in plant cells

60. Which organelle has only a single membrane around it?

(a) Peroxisome (b) Nucleus

(c) Chloroplast (d) Ribosome

61. Which of the following is a phototroph?

(a) A mushroom

(b) Bacteria growing on decomposing matter

(c) An earthworm

(d) A tree

62. To enter or leave a cell, substances must pass through

(a) A microtubule

(b) The golgi apparatus

(c) A ribosome

(d) The plasma membrane

63. Which of the following correctly matches an organelle with its function?

(a) Mitochondrion . . . Photosynthesis

(b) Ribosome . . . Manufacture of lipids

(c) Lysosome . . . Movement

(d) Central vacuole . . . Storage

64. Oxidation in biological systems is accompanied by

(a) Hydrogenation

(b) Resonance stabilization

(c) Repulsion

(d) Dehydrogenation

65. During the first phase of glycolysis, ATP is used to form what type of bond with glucose?

(a) Dehydrogenation

(b) Hydrogenation

(c) Substrate-level phosphorylation

(d) Phosphoester

66. The enzyme used to break down fructose-1,6-bisphosphate into two trioses is called

(a) Phosphoglucoisomerase

(b) Enolase

(c) Aldolase

(d) Triose phosphate isomerase

67. Beta oxidation is

(a) Process used to oxidize fatty acids.

(b) The second step in the oxidation of glucose.

(c) Process used to digest some polysaccharides.

(d) Occurs in the cytoplasm.

68. The sequential process of fatty acid catabolism to acetyl CoA is called

(a) Oxidative deamination

(b) B oxidation

(c) The kerbcycle

(d) The glyoxylate cycle

69. Transamination

(a) Can involve only a small number of amino acids

(b) Involves the amination of an alpha keto acid

(c) Liberates a free ammonia molecule

(d) Both B and C

70. Which scientist is credited with coining the term "cell" from the Latin cellulae?

(a) Robert Hooke

(b) Robert Brown

(c) Theodor Schwann

(d) Antonie van Leeuwenhoek

71. The Hatch-Slack cycle occurs in which of the following types of plants?

(a) C3 plants

(b) tropical grasses

(c) C4 plants

(d) b and c

72. The relationship between photosystem I (PSI) and photosystem II (PSII) is correctly described by which of the following statements?
(a) PSI absorbs only short wavelengths of light.
(b) Electrons are excited by PSI or PSII, but not both.
(c) The absorption maximum of PSII is 700 nm.
(d) PSI is designated P700, whereas PSII is designated P680

73. Which organelle is round and derives its name from the Latin word for "kernel"?
(a) Nucleus
(b) Lysosome
(c) Golgi complex
(d) Mitochondrion

74. Photolysis is ______________.
(a) The absorption of light by chlorophyll
(b) The assimilation of carbon dioxide
(c) The splitting of water
(d) None of the above

75. The molecules known as the energy currency of the cell are ______________.
(a) NAD (b) NADP
(c) ATP (d) ADP

76. In addition to ATP, what are the end products of glycolysis?
(a) CO2 and ethyl alcohol
(b) CO2 and H2O
(c) NADH and pyruvate
(d) H2O and ethyl alcohol

77. All of the following are functions of the Krebs cycle except
(a) Production of NADH
(b) Production of $FADH_2$
(c) Adding electrons and protons to oxygen to form water
(d) Production of ATP

78. Which of the following are products of the light reactions of photosynthesis that are utilized in the Calvin cycle?
(a) Electrons and H+
(b) ADP, Pi, and NADP+
(c) CO_2 and glucose
(d) ATP and NADP

79. The color of light least effective in driving photosynthesis is
(a) Yellow (b) Red
(c) Green (d) Blue

80. What is the primary function of the light reactions of photosynthesis?
(a) to use ATP to make glucose
(b) to produce ATP and NADPH
(c) to convert light energy to the chemical energy of PGAL
(d) to produce NADPH used in respiration

81. Where do the microtubules of the spindle originate during mitosis in both plant and animal cells?
(a) Centriole
(b) Centrosome
(c)Centromere
(d) Kinetochore

82. Plants appear green because
(a) They contain carotenoids
(b) They undergo photosynthesis
(c) They reflect green light
(d) They absorb green light

83. The primary source of energy for Earth is
(a) Solar energy
(b) Thermal energy
(c) Hydroelectric energy
(d) None of these

84. Land plants are composed of all the following tissue types except
(a) Vascular (b) Mesoderm
(c) Epidermal (d) Meristematic

85. Which functional plant cells lack a nucleus?
(a) Xylem only
(b) Sieve-tube cell only
(c) Both companion and parenchyma cells
(d) Both xylem and sieve-tube cells

86. ______ light is the most effective in producing a phototropic response.
(a) Green (b) Yellow
(c) Orange (d) Blue

87. Which of the following factors causes an increase in transpiration?
(a) A hot dry windy day
(b) A cool damp windy day
(c) A hot damp windy day
(d) None of the above

88. An apparatus for measuring transpiration in plants is called :
(a) Auxanometer (b) Potometer
(c) Hygrometer (d) Lactometer

89. Sialic acid is a constituent of :
(a) Cell membrane
(b) Cell wall
(c) Chromosomes
(d) Dictyosomes

90. Which pair of structures are usually found in both plant and animal cell
(a) Cell membrane and nucleolus
(b) Cell membrane and cell wall
(c) Nucleolus and chloroplast
(d) Nucleolus and cell wall

91. Unit membrane or (Plasma membrane) is made of :
(a) Cellulose
(b) Lipoproteins
(c) Phosphor proteins
(d) Phospholipids

92. The role of molybdenum is :
(a) Chromosome contraction
(b) Carbon assimilation
(c) Nitrogen fixation
(d) Flowering induction

93. The real force causing entry of water into a cell from other cell is :
(a) Diffusion pressure deficit
(b) Tugor pressure
(c) Osmotic pressure
(d) Wall pressure

94. Turgidity of a cell is maintained by :
(a) Diffusion pressure
(b) Wall pressure
(c) Osmotic pressure
(d) Turgor pressure

95. Mineral salt which are absorbed by the roots for the soil are in the form of
(a) Dilute solution
(b) Vary concentrated solution
(c) Concentrated solution
(d) Very dilute solution

96. When a cell is fully turgid, which of the following will be zero?
(a) Osmotic pressure
(b) Turgor pressure
(c) Wall pressure
(d) Suction pressure

97. In aerobic respiration in plants:
(a) Oxygen is taken in
(b) Oxygen is given out
(c) Carbon dioxide is given out
(d) Carbon dioxide is taken out

98. Water potential concept was formulated by
(a) Peter ray 1960
(b) M.H. Mathew 1954
(c) Otto Renner 1915
(d) W.C Hoffer 1920

99. Alcohol is produced during the process of
(a) Photosynthesis
(b) Aerobic respiration
(c) Anaerobic respiration
(d) None of the above

100. Auxins promote growth :

(a) Rarely (b) Always

(c) Mostly (d) Never

101. The first event in photosynthesis is :

(a) Synthesis in ATP

(b) Photoexcitation of cholorophyll all ejection of electron

(c) Photolysis

(d) Release of oxygen

102. Carbohydrates are present in the plasmalemma in the form of :

(a) Starch

(b) Cellulose

(c) Hemicellulose

(d) Phospholipids and phosphoproteins

ANSWERSHEET
12. Physiology

1. **(d)**	2. **(b)**	3. **(d)**	4. **(c)**	5. **(b)**	6. **(b)**	7. **(c)**	8. **(a)**	9. **(a)**	10. **(d)**
11. **(b)**	12. (d)	13. **(a)**	14. **(a)**	15. **(b)**	16. **(b)**	17. **(b)**	18. **(c)**	19. **(b)**	20. **(b)**
21. **(c)**	22. **(d)**	23. **(b)**	24. **(c)**	25. **(d)**	26. **(a)**	27. **(a)**	28. **(b)**	29. **(c)**	30. **(a)**
31. **(c)**	32. **(d)**	33. **(c)**	34. **(b)**	35. **(a)**	36. **(d)**	37. **(a)**	38. **(b)**	39. **(a)**	40. **(b)**
41. **(b)**	42. **(b)**	43. **(b)**	44. **(d)**	45. **(a)**	46. **(a)**	47. **(a)**	48. **(b)**	49. **(c)**	50. **(c)**
51. **(a)**	52. **(b)**	53. **(a)**	54. **(a)**	55. **(a)**	56. **(b)**	57. **(d)**	58. **(d)**	59. **(b)**	60. **(a)**
61. **(d)**	62. **(d)**	63. **(d)**	64. **(d)**	65. **(d)**	66. **(c)**	67. **(a)**	68. **(b)**	69. **(d)**	70. **(a)**
71. **(d)**	72. **(d)**	73. **(a)**	74. **(c)**	75. **(c)**	76. **(c)**	77. **(c)**	78. **(d)**	79. **(c)**	80. **(b)**
81. **(b)**	82. **(c)**	83. **(a)**	84. **(b)**	85. **(d)**	86. **(d)**	87. **(a)**	88. **(b)**	89. **(a)**	90. **(a)**
91. **(b)**	92. **(c)**	93. **(a)**	94. **(d)**	95. **(c)**	96. **(d)**	97. **(c)**	98. **(c)**	99. **(c)**	100. **(c)**
101. **(b)**	102. **(d)**								

13

Soil Science

KEY POINTS

- The term soil is derived from the Latin word 'Solum' means floor
- Soil is the uppermost loose layer of the earth suitable for plant growth
- The soil consists of four major components viz., mineral matter (45%), organic matter (5%), soil water (25%) and soil air (25%)
- Soil texture refers to the relative proportion of particles of various size in a given soil
- Soil texture refers to the relative percentage of sand, silt and clay in a soil
- The three basic processes of profile development of forest soils take place are Incorporation of humus (melanization), Leaching (podzolization) and Deoxidation by groundwater / gleization
- Humification is the process of transformation of raw organic matter into humus.
- Eluviation means washing out. It is the process of removal of constituents in suspension or solution by the percolating water from the upper to lower layers
- Calcification is the process of precipitation and accumulation of calcium carbonate ($CaCO_3$) in some part of the profile and the accumulation of $CaCO_3$ may result in the development of a calcic horizon
- Decalcification is the reverse of calcification that is the process of removal of $CaCO_3$ or calcium ions from the soil by leaching
- Podzolization is a process of soil formation resulting in the formation of Podzols and Podzolic soils
- Laterization is the process that removes silica, instead of sesquioxides from the upper layers and thereby leaving sesquioxides to concentrate in the solum
- The term *glei* is of Russian origin means blue, grey or green clay.
- Salinization is the process of accumulation of salts, such as sulphates and chlorides of calcium, magnesium, sodium and potassium, in soils in the form of a salty (salic) horizon
- Solonization or Alkalization is the process that involves the accumulation of sodium ions on the exchange complex of the clay, resulting in the formation of sodic soils (Solonetz).
- Solodization or dealkalization is the process refers to the removal of Na+ from the exchange sites.
- Pedoturbation is the process of mixing of the soil.

- The vertical section of the soil showing the various layers from the surface to the unaffected parent material is known as a soil profile
- A soil profile contains three main horizons. They are named as horizon A, horizon B and horizon C
- A hypothetical mineral soil profile will include O, A, B, C and R master horizons and all the possible sub-horizons
- `O' horizon is called organic horizon
- There are 12 orders in soils
- The soil order with the greatest coverage of the globe is Aridisols
- Entisols and Inceptisols provide much of the forestland of the temperate and boreal region.
- Alfisols are common under hardwood forests in areas with well-developed soil profiles.
- Biogeochemical cycle is the movement of chemical elements and involves biological organisms and their geological environment.
- The final product of composting is well rotted manure known as compost
- Fertilizers supply plant food and help to increase yield of crops through the improvement of soil fertility
- Nitrogen present in these fertilizers are in nitrate form, NO_3^- which are rapidly dissociated to release NO^{3-}ions and readily absorbed by the plants.
- The potassium content of potassic fertilizers is usually expressed as potassium oxide. K_2O, referred to as potash
- Gypsum salts are most widely used Mg fertilizers materials in dry fertilizer formulations.
- Critical nutrient concentration (CNC) is used in interpreting plant analysis results and diagnosing nutritional problems.
- Chlorophyll meter also called as SPAD (Soil and Plant Analysis Division, Japan) is a simple diagnostic tool to measure the chlorophyll content of leaves
- Mineralogy : It is that branch of Geology, which deals with the study of minerals.
- Paleontology: Branch of geology that deals with the mode of preservation of remains of plants and animals in the form of fossils, within rock beds and their proper utilization in elucidating the geological history of the earth
- The Sial is a heterogeneous mixture of rocks
- The Sima is a homogenous plastic or semi-plastic concentric layer that behaves like a solid
- Sial contains about 65-75% silica. Aluminium is the next important element in Sial, represented by the most common rocks like granite, and rhyolite
- Petrology is that branch of geology that deals with the study of rocks
- A rock may be defined as a mass of mineral water
- Igneous rocks formed by action of heat
- Example of extrusive rocks: Rhyolite, obsidian, trachyte, dolerite and basalt
- Example of Intrusive rocks: Granite, syenite, dolerite and gabbio
- Sedimentary rocks formed through the agency of water and called aqueous rocks
- The rocks that are mostly deposited in layers or strata is called stratified rocks

- Sedimentary rocks divided into six groups are as follows: Arenaceous, Argillaceous rocks, Calcareous rocks, Carbonaceous rocks Siliceous rocks and Precipitated salts
- Diamond is the hardest and Talc is the softest in the minerals
- Phenomenon of crystallization is polymorphism
- Substances of different but at the same time of allied chemical composition are capable of crystallizing in exactly related to Isomorphism
- Salinity is measured conductometrically and expressed as dSm^{-1}
- Salinty above 4 dSm^{-1} is harmful
- The soil organic matter content in most Indian soils is 1 to 2 per cent.
- Soil is a reservoir for supplying water and keeps salts in solution and used as plant nutrients.
- The solid phase provides mechanical support and nutrients to the plants.
- The liquid phase supply water and along with it dissolved nutrients to plant roots.
- Soil genesis – evolution of soil from its parent materials
- Soil classification – grouping of soils according to common properties
- Soil survey – the depiction of geographic distribution of soils and their relationship to landscapes.
- Soil Structure refers to the arrangement of various soil particles
- Soil texture is defined as the relative proportions of various sized particles in a given soil.
- The imaginary diameter of the soil particle is known as effective diameter or equivalent diameter
- Density is defined as the mass of a unit volume of soil solids
- In metric system, particle density is expressed as mega grams per cubic meter (Mg/m^3).
- The range for most mineral soils varies between 2.60 and 2.75 Mg/m^3
- Bulk density defined as the mass (weight) of unit volume of dry soil. This includes soil solids and pores (D_b)
- The volume of soil mass that is not occupied by soil particles is known as pore space
- Soil porosity is the percentage of pore space by volume in a soil. It can be calculated from the absolute and apparent specific gravity
- Colour is produced by the spectral reflectance of light by the soil
- Soil colour is measured by using Munsell colour chart and consists of 175 different coloured chips
- Hue is the dominant spectral colour
- Value refers the relative lightness of colour and is a function (approximately the square root) of the total amount of light
- Chroma is the relative purity or strength of the spectral colour
- The munsell notation is useful for international soil correlation
- Soil particles possess the power to adsorb moisture from atmosphere. The water adsorbed is known as hygroscopic water
- Hygroscopic coefficient is the maximum amount of hygroscopic water adsorbed by 100 g of dry soil at standard conditions of temperature and humidity

- The capillary of the soil to hold the maximum amount of total water is known as its water holding capacity
- The moisture content at which plants remain permanently wilted even when water is added to soil is called as Permanent Wilting Point (PWP)
- When the soil profile is wetted, the movement of more water flowing through the wetted soil is termed percolation
- The most common mathematical expression for the vertical water flow rate through soil is called Darcy's Law
- Darcy's law states that the rate of flow of liquid or flux through a porous medium is proportional to the Hg in the direction of flow of the liquid
- The main source of heat energy for soil is solar radiation
- pH is defined as the negative logarithm of hydrogen ion concentration or simply the log of the reciprocal of the hydrogen ion concentration
- The acidity or alkalinity of a solution can be expressed on the scale of acidity and alkalinity
- The scale of acidity or alkalinity is called pH scale. The unit of this scale is called pH value
- This scale runs from 0 to 14. The neutral point in this scale is at pH 7
- All the values above pH 7 represent alkalinity and below 7 denote acidity
- The degree of alkalinity increases as values go above pH 7 and the degree of acidity increases as the pH decreases below 7
- The electrical conductivity (E.C) measurement gives the total amount of soluble salts present in the soil and is expressed as millimhos / cm or dSm^{-1}
- The process of conversion of ammonia to nitrites (NO_2) and then to nitrate (NO_3) is known as nitrification
- The transformation of organic nitrogenous compounds into ammonia is called ammonification
- The process, which involves conversion of soil nitrate into gaseous nitrogen or nitrous oxide, is called Denitrification
- Aminization is the process of conversion of proteins to aminoacids
- The cation exchange capacity is expressed in terms of equivalents or more specifically, as milli equivalents per 100 gram and is written as meq /100g
- Indicators are chemicals used in volumetric analysis to indicate the end point of a reaction
- The 'formality', F, of a solution expresses the number of gram-formula weights of solute present in one litre of solution
- The 'Molarity' M, of a solution defines the number of gram-molecular weights (or moles) of a species in 1 litre of solution or the number of millimolecular weights in 1 ml of solution
- The 'normality' N, of a solution expresses the number milliequivalents of dissolved solute contained in 1 mil litre
- Milliequivalent represents 1/1000 of an equivalent weight
- A Molar solution of a substance contains one gram molecular weight or one mole of a substance in one litre of the solution. The strength of such a solution is represented by 1 M or simply M

- The molarity (M) of a solution is the number of gram molecular weights or moles of solute in one litre of the solution

MULTIPLE CHOICE QUESTION'S

1. Who is the Father of soil physics?
(a) Hillel (b) Richards
(c) Dokuchaiev (d) Thomas Way

2. What is the number of classification in size groups of soil particles by International Society of Soil Science (ISSS)?
(a) Three (b) Four
(c) Six (d) Seven

3. What is the size of fine clay particles in USDA classification of soil texture?
(a) <2mm (b) <0.2mm
(c) <0.02mm (d) <0.002mm

4. Particle smaller than ----- microns cannot be separated by sizing
(a) <2 (b) <20
(c) <50 (d) <100

5. In which of the soils, water infiltration is more rapid ?
(a) Sandy soils (b) Loamy soils
(c) Clay soils (d) All of these

6. The arrangement of primary and secondary soil particles is called as
(a) Soil structure
(b) Soil texture
(c) Soil consistency
(d) Soil plasticity

7. The relative proportion of soil particles is called as
(a) Soil structure
(b) Soil texture
(c) Soil consistency
(d) Soil plasticity

8. Wet bulk density g cm^{-3} =
(a) Weight of soil / Volume of soil
(b) Weight of dry soil / Volume of soil
(c) Weight of dry soil / Volume of soil particles
(d) Volume of soil / Weight of moist soil

9. Dry bulk density g cm^{-3} =
(a) Weight of soil / Volume of soil
(b) Weight of dry soil / Volume of soil
(c) Weight of dry soil / Volume of soil particles
(d) Volume of soil / Weight of moist soil

10. Particle density g cm^{-3} =
(a) Weight of soil / Volume of soil
(b) Weight of dry soil / Volume of soil
(c) Weight of dry soil / Volume of soil particles
(d) Volume of soil / Weight of moist soil

11. What is the bulk density of fine textured soils ?
(a) 1.1-1.3 g cm^{-3} (b) 2.65 g cm^{-3}
(c) 1.4-1.8 g cm^{-3} (d) 2.4-2.6 g cm^{-3}

12. What is the bulk density of coarse textured soils ?
(a) 2.4-2.6 g cm^{-3}
(b) 2.65 g cm^{-3}
(c) 1.1-1.3 g cm^{-3}
(d) 1.4-1.8 g cm^{-3}

13. Value of bulk density is
(a) Higher than particle density
(b) Lower than particle density
(c) Equal to particle density
(d) Unity

14. Element essential for the flocculation of soil particles is
(a) Sodium (b) Potassium
(c) Calcium (d) Magnesium

15. Element essential for the deflocculation of soil particles is
(a) Potassium (b) Calcium
(c) Magnesium (d) Sodium

16. More stable aggregates are formed by
(a) Humus substances
(b) Earthworms
(c) Plant roots decay
(d) All of these

17. The soil structural classification followed in India is
(a) ISSS classification
(b) USDA classification
(c) Atterberg classification
(d) All of these

18. Which soil is most desirable for agriculture ?
(a) Platy (b) Blocky
(c) Crumb (d) Granular

19. What are the total number of soil classes?
(a) Four (b) Five
(c) Six (d) Seven

20. what is the size of the non-capillary pore
(a) >6μ (b) >10μ
(c) > 15μ (d) >20μ

21. What is the size of micro or capillary pores ?
(a) >6μ (b) >10μ
(c) > 15μ (d) >20μ

22. Arrange the following soil texture in order of decreasing water retention capacity .
(a) Loam>Clay>Sandy
(b) Sandy>Loam>Clay
(c) Clay>Loam>sandy
(d) Sandy>Clay>Loam

23. The % of carbon dioxide in soil air ?
(a) 0.03 (b) 0.003
(c) 0.3 (d) 3

24. More than 90% of gaseous .exchange in soil takes place by
(a) Mass flow
(b) Diffusion
(c) Particle diffusion
(d) All of these

25. The dominant spectral colour in Munsell colour chart indicates
(a) Hue (b) Value
(c) Chroma (d) All of these

26. Black soils cover almost -------- %of the geographracial area of India.
(a) 9 (b) 20
(c) 21 (d) 22

27. Red soils cover almost ------ % of the geographical area of India.
(a) 9 (b) 20
(c) 21 (d) 22

28. Bulk density in soils can be estimated by
(a) Core method
(b) Excavation method
(c) Both (d) None of these

29. Gravitational water in soil move through the
(a) Microscopes (b) Macro pores
(c) Both a and b (d) None of these

30. Water held against the force of gravity in microspores is
(a) Heavy water
(b) Gravitational water
(c) Capillary water
(d) Hygroscopic water

31. Capillary rise of water is due to the force of
(a) Cohesion
(b) Adhesion
(c) Cohesion and adhesion
(d) Matric potential

32. Which type of water includes that which is available to plants?
(a) Hygroscopic water
(b) Gravitational water
(c) Capillary water
(d) All of these

33. Capillary water in the soil is that which is found in the
(a) Macropores (b) Micropores
(c) Root channels (d) Both a and b

34. The water held between -1/3and -1/5 bar is called
(a) Gravitational water
(b) Plant available water
(c) Wilting point
(d) Field capacity

35. A soil textural triangle measures
(a) Percentage of sand, silt, clay
(b) Minerals lives in a soil sample
(c) Percent slope of a land class
(d) Percentage of soil erosion on a given location

36. Soil water potential always has
(a) Positive sign (b) Negative sign
(c) No sign (d) None of these

37. Soil water suction always has
(a) Positive sign (b) Negative sign
(c) No sign (d) None of these

38. Since that deals with soil genesis classification is called as
(a) Petrology (b) Pedology
(c) Pedagogy (d) Petrography

39. Sandstone and limestone are
(a) Metamorphic rocks
(b) Sedimentary rocks
(c) Primary rocks
(d) Igneous rocks

40. Major component of igneous rocks are
(a) Quarts
(b) Amphiboles and pyroxenes
(c) Feldspars
(d) Micas

41. Of the following, which is an important chemical weathering process
(a) Carbonation (b) Oxidation
(c) Hydrolysis (d) Hydration

42. Weathered products of rocks and rock minerals are called as
(a) Pedon (b) Parent materials
(c) Solum (d) Poly pedons

43. The unconsolidated materials from which the solum develops
(a) Pedon
(b) Regolith
(c) Parent materials
(d) Both b and c

44. Materials transported and depoisited by water are called as
(a) Alluvium (b) Collivium
(c) Losses (d) Aeolian

45. Materials transported by the action of gravity are called as
(a) Alluvium (b) Lacustrine
(c) Glacial drift (d) Colluvium

46. Materials transported by water and settled out in lakes are called as
(a) Alluvium (b) Lacustrine
(c) Glacial drift (d) Colluvium

47. Silt material transported by wind are called as
(a) Delta (b) Meander
(c) Losses (d) Aeolian

48. Sand materials transported by water are called as
(a) Delta (b) Meander
(c) Loess (d) Aeolian

49. Azonal soils are
(a) Black soil (b) Red soil
(c) Lateriate soil (d) Alluvial soil

50. Process of eluviation of oxides of iron & aluminum, humus & carbonates from the upper horizons and their accumulation in subsurface horizons is called as
(a) Laterization (b) Podsolization
(c) Decalcification (d) Calcification

51. Process of removal of silica and accumulation of sesquioxides is called as
(a) Laterization (b) Podsolization
(c) Decalcification (d) Calcification

52. Cool and humid climatic conditions are favourable for
(a) Podsolization (b) Laterization
(c) Humification (d) Pedoturbation

53. Warm and humid climatic conditions are favourable for
(a) Podsolization (b) Laterization
(c) Humification (d) Pedoturbation

54. Bed rocks are obeviously absent in
(a) Red soils (b) Black soils
(c) Laterite soils (d) Alluvial soils

55. Master horizons are
(a) Oand C (b) O,A,E,B & C
(c) O,A,B and C (d) A,B and C

56. B horizon is a
(a) Illuvial horizon (b) Elluvial horizon
(c) Organic horizon (d) Bed rock

57. Elluvial horizon is
(a) A horizon (b) E horizon
(c) B horizon (d) O horizon

58. Soils under the soil order Alfish is/are
(a) Red soil (b) Laterite soil
(c) Black soil (d) Both a and b

59. Fertile soils of India are locally called as
(a) Urvara (b) Anurvara
(c) Usara (d) Kari

60. Barren soils of India are locally called as
(a) Usara (b) Anurvara
(c) Urvara (d) Both a and b

61. Total area of salt affected soils in india is
(a) 25mha (b) 28mha
(c) 29mha (d) 10mha

62. Total area of saline soils in india is
(a) 10mha (b) 7.2mha
(c) 2.8mha (d) 6.1mha

63. Total area of sodic or alkali soils in india is
(a) 10 mha (b) 7.2 mha
(c) 2.8 mha (d) 6.1 mha

64. What is the pHof saline-sodic soils?
(a) Less than 8.5 (b) More than 8.5
(c) 8.5-10.0 (d) 8.5

65. Who is the father of soil science?
(a) Dokuchaiev (b) Justin von Liebig
(c) J.W. Leather (d) Marbut

66. Density of the earth
(a) 4.0 g cm^{-3} (b) 5.5 g cm^{-3}
(c) 2.6-2.7 g cm^{-3} (d) 3.0-4.5 g cm^{-3}

67. Example(s) of light minerals is/are
(a) Quartz (b) Feldspars
(c) Muscovite (d) All of these

68. Example(s) of heavy minerals is/are
(a) Olivines (b) Heamatite
(c) Pyrite and Augite (d) All of these

69. Most important weathering for soil formation is/are
(a) Physical weathering
(b) Chemical weathering
(c) Biological weathering
(d) All of these

70. Active soil forming factors are
(a) Parent material
(b) Relief or Topography
(c) Time
(d) Climate and vegetation
(e) All the above

71. Passive soil forming factors are
(a) Parent material
(b) Relief or Topography
(c) Time
(d) All of these

72. The igneous rock basalt, in humid tropics may give rise to
(a) Black soil (b) Red soil
(c) Laterite soil (d) Alluvial soil

73. Soil that are formed from the same kind of parent materials but differing in their topographic positions within the landscape are called as
(a) Soil catena (b) Climosquence
(c) Toposquence (d) Clinosquence

74. Zonal soil forming process is
(a) Calcification (b) Salinization
(c) Solinization (d) Gleization

75. Saline soils are called as
(a) Solonetz (b) Solonchalks
(c) Soloth (d) Both b and c

76. Sodic soils are also called as
(a) Soloth (b) Solonetz
(c) Solonchalks (d) Both a and b

77. Moor soils are
(a) Normal soil (b) Transitional soil
(c) Abnormal soil (d) None of these

78. Tundra soils are
(a) Normal soil (b) Trantitional soil
(c) Abnormal soil (d) None of these

79. What is the total number of soil order?
(a) 10 (b) 11
(c) 12 (d) 13

80. Land capability class was given by
(a) Sys
(b) Klingebiel and Montgomery
(c) Sys and Verhegs
(d) Storie and Ricquier

81. Total number of categories in the land Capability classification is
(a) Three (b) Five
(c) Eight (d) Twelve

82. Colour of the land capability class-V on LCC Maps is
(a) Yellow (b) Pink
(c) Green (d) Red

83. The dominant soil order in India is
(a) Vertisols (b) Inceptisols
(c) Alfisols (d) Entisols

84. Soil erosion is mostly affected by
(a) Rainfall intensity
(b) Rainfall variability
(c) Rainfall erosivity
(d) Total annual rainfall

85. Clay minerals are called as .... minerals
(a) Primary minerals
(b) Secondary minerals
(c) Tertiary minerals
(d) Accessory minerals

86. The last layer in the soil profile is the
(a) A Horizon (b) B Horizon
(c) C Horizon (d) R Horizon

87. The soil orders which have highest sub (7) are
(a) Mollisols and Andosols
(b) Mollisols and Vertisols
(c) Vertisols and Andosols
(d) Inceptisols and Alfisols

88. The dominant soil order in India is
(a) Entisols (b) Alfisols
(c) Inceptisols (d) Vertisols

89. The soil horizon of maximum biological activity is
(a) A (b) B
(c) C (d) E

90. The C horizon is also called as
(a) Solum (b) Top soil
(c) Zone of loss (d) Parent material

91. Buffering capacity is high for
(a) Sand (b) Silt
(c) Clay (d) Humus

92. Red colour of the soil is due to the presence of
(a) Gibsite (b) Magnetite
(c) Hematite (d) Goethite

93. The capability of soil to produce a plant or plant parts is called as
(a) Soil fertility (b) Soil quality
(c) Soil productivity (d) All of these

94. The inherent capacity of soil to provide nutrients is called as
(a) Soil fertility (b) Soil quality
(c) Soil productivity (d) All of these

95. The total number of essential nutrient elements required for plant growth is
(a) 16 (b) 17
(c) 20 (d) 21

96. N, P and K are called as
(a) Macro nutrients
(b) Micro nutrients
(c) Secondary nutrients
(d) Beneficial nutrients

97. Ca, Mg and S are called as
(a) Macro nutrients
(b) Micro nutrients
(c) Secondary nutrients
(d) Beneficial nutrients

98. Iron, zinc, boron and molybdenum are called as
(a) Macro nutrients
(b) Micro nutrients
(c) Secondary nutrients
(d) Beneficial nutrients

99. Na, Ve, Ni, Si, Co, Se and I are the examples of
(a) Macro nutrients
(b) Micro nutrients
(c) Secondary nutrients
(d) Beneficial nutrients

100. The Law of Diminishing Return was given by
(a) Justin von Liebig (b) Spill man
(c) Mitcherlich (d) Schofield

101. Total number of classes in nutrient index is
(a) Three (b) Four
(c) Five (d) Six

102. What is the percentage of nitrogen in urea
(a) 48 (b) 46
(c) 25 (d) 42

103. The percentage of K_2O in muriate of potash is (MOP) IS
(a) 58.0 (b) 48.0
(c) 23.0 (d) 15.0

104. Which form of nitrogen is present in urea
(a) nitrate form (b) Ammonical form
(c) Amide form (d) Both a and b

105. The most deficient nutrient in Indian soil is
(a) Nitrogen (b) Zinc
(c) Copper (d) Boron

106. Soils of India are generally
(a) High in fertility
(b) Poor in fertility
(c) Moderate in fertility
(d) None of these

107. Erosion created by the activities of man and sometimes by animals is called as
(a) Natural geological erosion
(b) Geologic erosion
(c) Accelerated soil erosion
(d) All of these

108. Scattering of detachated soil particles by the impact of rain drops is called
(a) Splash erosion (b) Sheet erosion
(c) Gully erosion (d) Ravines

109. Removal by of a very thin layer of soil from the entire surface of Large area is called
(a) Splash erosion (b) Sheet erosion
(c) Rill erosion (d) Ravines

110. Sheet erosion is observed on
(a) Deeply sloping land
(b) Moderately sloping land
(c) Slopy land (d) Gently sloping land

111. In Universal soil equation, A = RKLSCP, K denotes
(a) Soil conservation practices
(b) Soil roughness
(c) Soil erodability (d) Rainfall factor

112. Size range of particles in saltation type of wind erosion is
(a) Less than 0.05mm
(b) 0.5-1.0mm
(c) 0.5-0.2mm (d) 0.05-0.5mm

113. Size range of particles in surface creep type of wind erosion is
(a) Less than 0.05mm
(b) 0.5-1.0mm
(c) 0.5-0.2mm (d) 0.05-0.5mm

114. Size range of particles in suspension type of wind erosion is
(a) Less than 0.05mm
(b) 0.5-1.0mm
(c) 0.5-0.2mm (d) 0.05-0.5mm

115. Wind erosion is calculated by
(a) Ramer's formula
(b) Chepil Equation
(c) Chepil and Woodruff equqtion
(d) All of these

116. Diameter of particles that is most suspectibles to wind erosion is
(a) 0.1mm (b) 0.2mm
(c) 0.02mm (d) 0.01mm

117. Dominant type of wind ersion in soil is
(a) Saltation (b) Surface creep
(c) uspension (d) All of these

118. Total number of classes in the group, land suitable for cultivation is
(a) Eight (b) Four
(c) Six (d) Two

119. Rills with more than ----------------cm depth are generally called as gullies
(a) 15 (b) 20
(c) 25 (d) 30

120. The most serious sheet erosion occurs in
(a) Red and Black soils
(b) Red and Alluvial soils
(c) Black and Alluvial soils
(d) Red and laterite soils

ANSWERSHEET
13. Soil Science

---

1. (a)	2. (b)	3. (d)	4. (b)	5. (a)	6. (a)	7. (b)	8. (a)	9. (b)	10. (c)
11. (a)	12. (d)	13. (b)	14. (c)	15. (d)	16. (c)	17. (b)	18. (c)	19. (a)	20. (c)
21. (a)	22. (c)	23. (c)	24. (b)	25. (a)	26. (d)	27. (b)	28. (c)	29. (b)	30. (c)
31. (c)	32. (c)	33. (b)	34. (b)	35. (a)	36. (b)	37. (a)	38. (b)	39. (b)	40. (c)
41. (c)	42. (b)	43. (c)	44. (a)	45. (d)	46. (b)	47. (c)	48. (d)	49. (d)	50. (b)
51. (a)	52. (a)	53. (b)	54. (d)	55. (c)	56. (a)	57. (b)	58. (d)	59. (a)	60. (d)
61. (b)	62. (b)	63. (c)	64. (a)	65. (a)	66. (b)	67. (d)	68. (d)	69. (b)	70. (d)
71. (d)	72. (b)	73. (c)	74. (a)	75. (b)	76. (b)	77. (c)	78. (a)	79. (c)	80. (b)
81. (a)	82. (c)	83. (b)	84. (a)	85. (b)	86. (d)	87. (a)	88. (c)	89. (a)	90. (d)
91. (d)	92. (c)	93. (c)	94. (a)	95. (b)	96. (a)	97. (c)	98. (b)	99. (d)	100. (c)
101. (a)	102. (b)	103. (a)	104. (c)	105. (a)	106. (a)	107. (c)	108. (a)	109. (b)	110. (d)
111. (c)	112. (d)	113. (c)	114. (a)	115. (c)	116. (a)	117. (a)	118. (b)	119. (d)	120. (a)

14

Forest Microbiology

KEY POINTS

- Microbiology is the study of microorganisms.
- The manner by which the disease develops is called Pathogenesis
- Single-celled fungi are called yeasts.
- Multi-celled fungi are called molds.
- Chemotherapy is the treatment of disease using chemicals administered to the diseased body.
- Plasmid is an extra chromosomal piece of genetic material which also independently replicates and thus remains established for many generations in the cytoplasm of the bacterial cell.
- Plasmid can be of two types: Conjugative and Non-conjugative.
- Transforming principle was identified as DNA by Avery, Machleod, and McCarty in 1944.
- Staphylococci are also known as pyogenic cocci (Pus-forming cocci).
- Acidophiles are the organisms whose optimum pH is relatively to highly acidic.
- Neutrophiles are the organisms whose optimum pH ranges about pH- 7, plus or minus approximately 1.5 pH units.
- Alkaphiles are the organisms whose optimum pH is relatively towards highly basic.
- Cold-adapted organisms are called Psychrophiles.
- Obligate psychrophiles are unable to grow above 20°C, whereas facultative psychrophiles are able to grow.
- Organisms whose optimum growth temperature is found between 20°C to 40 °C are termed mesophiles.
- Mesophilic organisms which can endure brief exposures to relatively high temperatures are termed thermoduric.
- High-temperature-adapted organisms are called thermophiles.
- Microaerophile are organisms that grow best when small amounts of oxygen are present.
- Facultative anaerobes tend to grow better and faster when O_2 is present.
- Capnophiles are organisms whose optimum growth requires relatively high concentrations of carbon- di-oxide

- Organisms that require high concentrations of dissolved salts to grow are termed halophiles.
- Algal taxon is polyphyletic.
- The word phycology is derived from the Greek *phykos,* meaning seaweed.
- Anaerobic able to grown in the absence of free or atmospheric oxygen.
- Antigen: Molecular structure that is capable of stimulating production of antibody.
- Autotroph: An organism that used inorganic compounds as nutrients and carbon dioxide as the sole source of carbon. (Contrast with heterotrophy)
- Auxotroph: Differing from the wild strain (prototroph) by an additional nutritional requirement.
- Catabolism: The biochemical processes involved in the breakdown or organic compounds, usually leading to the production of energy.
- Cell line: A cell culture that has been passed (Subcultured) *in vitro*.
- Thermophilic: Growing best at high temperatures (above 45 degrees C).
- Zoophilic: Fungi that grow preferentially on animals.
- Zoospore: A motile asexual spore
- Zygomycetes: Group of fungi with non septate hyphase and spores produced within a sporangium.
- Biological nitrogen fixation (BNF) is a unique prokaryotic phenomena estimated to be fixing 65% of the nitrogen (N) in agriculture.
- Nitrogen fixation by legumes ranges between 25-75 kg of nitrogen per hectare per year in a natural ecosystem and several hundred pounds in a cropping system.
- *Rhizobium* needs several gene products to perform symbiotic nitrogen fixation such as
  - 'nod genes' which synthesize nod factor that initiate nodule development.
  - 'nif genes' synthesize nitrogen-fixing apparatus.
  - 'fix genes' synthesize products required in respiration and metabolism of bacteroid cells.
- Acetylene reduction assay (ARA) and ^{15}N isotopic are the methods of estimation of nitrogenase activity in legume roots
- Actinomycetes *Frankia* forms the group of nitrogen-fixing bacteria that live in symbiosis with a large spectrum of dicot plants.
- The contribution to nitrogen fixation in the *Frankia*–alder symbiosis can be up to 300 kg/ ha/yr.
- The Actinomycete genus *Frankia* belongs to the recently emended family, Frankiaceae including gram positive bacteria, able to establish a nitrogen-fixing symbiosis with roots of angiospermous woody plants.
- Blue green algae or cyanobacteria are microscopic, prokaryotic cells with autotropic mode of nutrition, lack of flagella, exhibit presence of phycobilin pigments (C-phycocyanin and C-phycoerythrin), store food as myxophycean starch or cyanophycin.
- Heterocyst: Specialized rectangular or nearly round cells having thickened non cellulosic polysaccharide cell wall, homogenous transparent pale yellowish cytoplasm present at regular intervals (terminal or intercalary positions) in filamentous trichome forms.

- Hormogonia: Specialized motile cells formed by breakage of trichomes of filamentous cyanobacterial forms within sheath.
- Akinetes: Specialized large thick walled yellow to brown coloured cells of several shapers (Spherical, discoid, elongated) showing resistance to dessication and meant for perenation of heterocystous forms under unfavourable conditions.
- *Azotobacter* is very large sized (ranges from 2-10 x 1-2.5 mm) gram negative, pleopmorphic (different sizes and shapes) bacteria, produces copious amounts of polysaccharides and motile due to peritrichous flagella.
- Azotobacter sp. are sensitive to acidic pH, high salts and temperature above 35°C.
- Beijerinckia: It is a free living, aerobic chemoheterotrophic nitrogen fixer exhibits characteristic features.
- Derxia: Pleomorphic, motile gram negative bacilli occurring singly or in short chains, capable of nitrogen fixation under aerobic or microaerobic conditions.

MULTIPLE CHOICE QUESTION'S

1. The term 'rhizosphere' was coined by
(a) Beijernick (b) Hiltner
(c) Winogradsky (d) Wilfarth

2. The total number of 'nif' genes present in Klebsiella pneumoniae is
(a) 21 (b) 16
(c)28 (d) 35

3. The bacterial genes responsible for nodule formation in Rhizobium legume symbiosis are termed as
(a) 'Sym' genes (b) 'nif' genes
(c) 'nod' genes (d) Nodulins

4. The nitrogen fixing potential of a nodulating plant can be assessed by
(a) Nitrite reductase activity
(b) Acetylene reduction assay
(c) Nitrate reductase activity
(d) Estimation of ammonia formation

5. An example of phosphate solubilizing microorganisms is
(a) *Pseudomonsa striata*
(b) *Aspergillus niger*
(c) *Thiobacillus ferroxidans*
(d) *Nitrobacter sp.*

6. A group of microbes capable of utilizing CO_2 as a carbon source is termed as
(a) Autotrophs (b) Heterotrophs
(c) Organotrophs (d) Lithotrophs

7. The kind of microflora dominate in acidic environment is
(a) Bacteria (b) Actinomycetes
(c) Protozoa (d) Fungi

8. The endosymbiont responsible for nodulation in actinorhizal plants is an
(a) Actinomycetes (b) Fungi
(c) Bacteria (d)Algae

9. The plant genes responsible for nodule formation are termed as
(a)'nod' genes (b) Nodulins
(c) 'ntr' genes (d) 'fix' genes

10. VAM fungi develop characteristic structures called
(a) Arbuscules (b) Vesicles & arbuscules
(c) Mushrooms (d) Truffles

11. Parasponia is a non-leguminous plant nodulated by
(a) Azotobacter (b) Rhizobium
(c) Frankia (d) Azospirillum

12. The optimum temperature for the growth of mesophilic microbes ranges between
(a) 0° and 15°C (b) 30° and 40°C
(c) 40° and 60°C (d) 60° and 80°C

13. A group of microbes with symbiotic relationship between fungi and algae is termed as
(a) Actinorhizal plants
(b) Lichens
(c) Mycorrhizae
(d) Cyanobacteria

14. The site of nitrogen fixation in Frankia actinorhizal plants is
(a) Spores (b) Bacteriodes
(c) Vesicles (d) Conidia

15. An example of non-nodulating leguminous tree is
(a) Acacia nilotica
(b) Albizia lebbeck
(c) Hardwickia binata
(d) Acacia mellifera

16. Ectendomycorrhizas found in
(a) Amaranthaceae
(b) Ericaceae
(c) Monotrophaceae
(d) Orchidaceae

17. The quantum of nitrogen fixed by Leucaena leucocephala is
(a) 110 kg/ha/yr (b) 500 kg/ha/yr
(c) 50 kg/ha/yr (d) 300 kg/ha/yr

18. The microbes associated with seed surface as termed as
(a) Phyllosphere (b) Spermosphere
(c) Phylloplane (d) Rhizosphere

19. The organism associated with termite gut that is responsible for cellulose degradation is
(a) Bacteria (b) Fungi
(c) Algae (d) Protozoa

20. An ectomycorrhizal fungus forming symbiotic relationship relationship with most of the temperate trees is
(a) Glomus (b) Pisolithus
(c) Gigaspora (d) Acalospera

21. An example of an associate diazotroph is
(a) Azotobacter (b) Azopirillum
(c) Beijierinckia (d) Derxia

22. The substanes produced by microbes which can kill other microbes at low concentrations is known as
(a) Antibiotics (b) Antibody
(c) Immunogen (d) Immunoglobulins

23. The covalently closed circular DNA present in bacterium is
(a) Tranposons
(b) Insertional sequences
(c) Plasmids
(d) Cosmids

24. An example of cellulose and lignin degrading organism is
(a) Fusarium sp. (b) Pleurotus sp.
(c) Amanita sp. (d) Aspergillus sp.

25. The basic unit of lignin is
(a) Alpha 1-4 glucose
(b) Amylopectin
(c) Pectin
(d) Phenyl proponae

26. An example of plant growth promoting rhizobacteria is
(a) *Bacillus coagulans*
(b) *Pseudomonas fluorescens*
(c) *Rhodospirillum rubram*
(d) *Acetobacter diazotrophicus*

27. One of the mechanisms by which plant growth promoting rhizobacteria improve plant growth
(a) Siderophore
(b) Drought resistance
(c) Heavy metal tolerance
(d) Salt resistance

28. An example for biological insecticide is
(a) *Trichoderma viride*
(b) *Pseudomonas fluorescens*
(c) *Bacillus thuringiensis*
(d) *Nosem bombyci*

29. Rhizobium from leguminous root nodules was first isolated by
(a) Winogradsky
(b) Hellriegel
(c) Beijerinck
(d) Wilfarth

30. Mycorrhizae help in mobilizing
(a) Potassium (b) Phosphorus
(c) Nitrate (d) Boron

31. The enzyme involved in lignin degradation is
(a) Alpha 1-4 glycosidasa enzymes
(b) Beta 1-4 glycolytic enzymes
(c) Chitinase
(d) Laccase

32. The dominant insoluble form of phosphorus present in acidic soils is
(a) Tricalcium phosphate
(b) Iron phosphate
(c) Magnesium phosphate
(d) Dicalcium phosphate

33. An example of sulfur reducing bacterium is
(a) *Thiobacillus*
(b) *Desulfatomaculum*
(c) *Ferrobacillus*
(d) *Rhodo pseudomonas*

34. ATP generating system in clostridium
(a) EMP pathway
(b) TCA cycle
(c) Phosphoroclastic reaction
(d) Oxidative electron transport system

35. Total number of ATP molecules required for nitrogen fixation is
(a) 32 (b) 16
(c) 8 (d) 38

36. The nitrogenase of Azotobacter overcomes oxygen toxicity by
(a) Gum production
(b) Conformational production
(c) Leghaemoglobin
(d) Vesicle formation

37. The organism involved in iron transformation is
(a) Ferrobacillus (b) Thiobacillus
(c) Thiothrix (d) Chromatium

38. The process of conversion of nitrate into elemental nitrogen is termed as
(a) Nitrate reduction
(b) Denitrification
(c) Nitrification
(d) Ammonification

39. The acid preferring nitrogen fixing bacterium that dominates forest floor is
(a) Derxia (b) Thiobacillus
(c) Thiothrix (d) Chromaium

40. The rhizobial population of carrier based culture should be
(a) X x 10^8 cfus/g carrier material
(b) X x 10^{10} cfus/g carrier material
(c) X x 10^5 cfus/g carrier material
(b) X x 10^2 cfus/g carrier material

41. The expiry period of any carrier based bioinculants is
(a) 4 months (b) 6 months
(c) 8 months (d) 1 year

42. The tree species having association with both ecto and endomycorrhizae is
(a) Acacia nilotica
(b) Casuarina
(c) Albizia lebbeck
(d) Teak

43. The total number of tree families having association with Frankia
(a) 14 (b) 8
(c) 6 (d) 24

44. The rate of organic matter decomposition under tropical to subtropical condition over temperate is
(a) Slow (b) Slow and steady
(c) Moderate (d) Faster

45. One of the common mechanisms by which phosphate solubilizing microbes solubilize insoluble form of phosphorous
(a) Organic acid production
(b) Enzymatic hydrolysis
(c)H2S
(d) Soil respiration

46. Fermentation carried out in a limited volume of closed container is termed as
(a) Continous
(b) Batch
(c) Fed batch
(d)Air lift fermentation

47. The host plant that is normally used to multiply VAM culture
(a) Sorghum (b) Maize
(c) Chilli (d) Cenchrus grass

48. Azotobacter fixes nitrogen in
(a) Infertile soils
(b) Organic matter rich soil
(c) Alkali soils
(d) Acidic soils

49. The alkali preferring Azospirillum is
(a) *A. amazonense*
(b) *A. irakens*
(c) *A. halopraeferens*
(d) *A. brasilense*

50. The mycorrhizae is ________ microflora
(a) Ectorhizosphere
(b) Endorhizosphere
(c) Rhizosplane
(d) Rhizosphere

51. The Azospirillum was first isolated and developed by
(a) Beijerinck (b) Dobereiner
(c) Omeliansky (d) Frank

52. The native microflora are termed as
(a) Fermentative
(b) Autochthonous
(c) Spermosphere
(d) Rhizosphere

53. One of the quality control laboratories of National Biofertilizer Development Corporation is located at
(a) Chennai
(b) Coimbatore
(c) Bangalore
(d) Hyderabad

54. VAM spores are recovered from soil by
(a) Ultra centrification
(b) Wet sieving and decantation
(c) Serial dilution and planting technique
(d) Most probable number technique

55. The microbial process by which contaminants are removed from soil is termed as
(a) Phytoremediation
(b) Bioremediation
(c) Microbial remediation
(d) Decomposition

56. The carrier material normally used for multiplication of VAM is
(a) Lignite (b) Peat
(c) Vermiculite (d) Charcoal

57. The mycorrhizal infection is normally high in
(a) Phosphorous poor soil
(b) 'P' rich soil
(c)Molybdenum rich soils
(d)Calcium rich soils

58. A compound that regulates entry of O_2 into the site of nitrogen fixation in leguminous plants is
(a) Leghaemoglobin
(b) Nodulin
(c) Ferridoxin
(d) Flavodoxin

59. The maximum N fixing genera found in the sub-family of Leguminaceae
(a) Papilionoideae
(b) Mimosoideae
(c) Cesalpinoideae
(d) Rosaceae

60. The total microbial load of soil can be assessed by
(a) Soil phosphatase activity
(b) Urease acitivity
(c) Dehydrogenase
(d) Protease activity.

61. Diatoms are —
(a) Soil algae (b) Soil bacteria
(c) Parent rocks (d) Minerals

62. Blue-green algae grows best in flooded soils exposed to sun and occupied by—
(a) Sun hemp (b) rice crop
(c) jute (d) sorghum

63. Volume of soil under the influence of roots of growing plants is referred to as-
(a) Phyto-biosphere
(b) Phyto-biozone
(c) Rhizosphere (d) Phytozone

64. Symbiotic association of fungal hyphae with plant roots is referred to as—
(a) Mycorrhiza (b) Mutual infection
(c) Lichen (d) Co-parasitism

65. VAM-fungi have reported to increase the uptake of
(a) Phosphorus (b) Nitrogen
(c) Potash (d) None

66. Which of the followings acts as storage organ for VAM fungi?
(a) Arbuscule (b) Mycellium
(c) Hypha (d) Vesicle

67. Actinomycetes provide nutrients to the plant by
(a) Decomposing soil organic matter
(b) breaking down complex clays
(c) fixing from atmosphere secreting as metabolites

68. Soil organisms depending on dead tissue of both plants and animals as food source is called as
(a) Herbivores (b) Carnivores
(c) Detritovores (d) None of these

69. For thier carbon and energy needs, heterotrophs reply on
(a) Organic compounds
(b) Inorganic compounds
(c) Photosynthesis
(d) None of the above.

70. In soil, action of the microflora is mostly
(a) Physical
(b) Both physical and chemical
(c) Chemical
(d) None of the above

71. Which of the followings is used as food source by microphytic feeders?
(a) Microfauna (b) Macrofauna
(c) Microflora (d) Macroflora

72. Which of the following groups of soil organisms are termed as 'ultimate decomposers'?
(a) Microflora (b) Macroflora
(c) Microfauna (d) Mesofauna.

73. As energy source, photoautotrophs depend on
(a) Organic compounds
(b) Inorganic compounds
(c) Solar energy
(d) None of these

74. Chemoautotrophs dose not derive energy from
(a) CO_2 (b) N_2
(c) S (d) Fe

75. The size range of soil inhabiting protozoa is
(a) 0.1-0.5 μm (b) 1-2 μm
(c) 5-100 μm (d) 1000-5000 μm

76. The rhizosphere zone extends from root surface within
(a) 1-2 mm (b) 5-10 mm
(c) 15-20 mm (d) 50-100 mm

77. Which of the following groups of soil algae are prokaryotes?
(a) Green algae
(b) Blue-Green algae
(c) Yellow green algae
(d) Diatomes.

78. Fungi are
(a) Photoautotrophs
(b) Chemoautotrophs
(c) Heterotrophs
(d) None

79. Which of the following groups represent single celled fungi?
(a) Molds (b) Mushrooms
(c) Yeasts (d) None

80. Individual fungal filaments are called
(a) Hyphae
(b) Mycellium
(c) Pseudopodia
(d) Flagella

81. Penicillum, Mucor, Fusarium and aspergillus are genera of
(a) Yeasts
(b) Molds
(c) Mushroom fungi
(d) None

82. Mycorrhizae is a symbiotic association between
(a) Bacteria and plant roots
(b) Fungi and bacteria
(c) Two species of fungi
(d) Fungi and roots of higher plants

83. Mycotoxin is produced by
(a) Bacteria (b) Protozoa
(c) Fungi (d) Nematodes

84. Aflatoxin is a mycotoxin produced by
(a) *Bacillus polymyxa*
(b) *Aspergillus awamorii*
(c) *Pseudomonas striata*
(d) *Aspergillus flavus*

85. Which of the following statement is not true regarding arbuscules of VAM fungi?
(a) Transfer mineral nutrients from the fungi to host plant
(b) Transfer sugar from host plant to the fungi.
(c) Work as storage organ for the fungi
(d) A highly branched structure.

86. Which of the followings acts as storage organ for VAM fungi?
(a) Hypha (b) Arbuscule
(c) Mycellium (d) Vesicle

87. Which of the following groups of plants does not form mycorrhizae?
(a) Solanaceae (b) Poaceae
(c) Leguminosae (d) Cruciferae

88. Which of the following products are not origninated from actinomycetes?
(a) Actinomycin (b) Pencillum
(c) Neomycin (d) Streptomycin

89. Which of the following get oxidized during the first step of nitrification?
(a) Nitrate (b) Nitrite
(c) Ammonium (d) None

90. The nitrogen fixing organism living inside the leaves of the aquatic fern azolla is a
(a) Fungi
(b) Cyanobacteria
(c) Actinomycetes
(d) None

91. Siderophores show strongest affinity to bind
(a) Iron (b) Nitrogen
(c) Aluminum (d) Phosphorus

92. Who for the first time showed that legumes can obtain nitrogen from air?
(a) M.W. Beijerinck
(b) J.B. Bousssingault
(c) Robbert Koch
(d) Louis Pasteur

93. Bacteria from root nodules of legume was isolated for the first time by
(a). J.B. Boussingault
(b) M.W. Beijerinck
(c) Robert Koch
(d) Louis Pasteru

94. Who for the first time obtained pure culture of bacteria by serial dilution in liquid media?
(a) Louis Pasteur (b) John Needham
(c) Joseph Lister (d) M.W. Beijerinck

95. Cellulose degradation by anaerobic bacteria was first discovered by
(a) S.N. Wingradsky
(b) Omeliansky
(c) J.B. Boussingault
(d) Joseph Lister

96. Study of rhizophere was initiated by
(a) Hiltner
(b) Omeliansky
(c) Lipman and Brown
(d) S.N. Winogradsky

97. Penicillum was discovered by
(a) Robert Koch (b) Alexander Fleming
(c) Waksman (d) Winogradsky

98. Streptomycin was discovered by
(a) Robert Koch
(b) Winogradsky
(c) Alexander Fleming
(d) Waksman

99. Glomus and Acaulospora are examples of
(a) Ectomycorrhiza
(b) Actinomycetes
(c) Endomycorrhiza
(d) None of these

100. Rhizobium from root nodules of a non legume was first isolated by
(a) Quipsel (b) Nutman
(c) Bond (d) Trinick

101. Azobacterin and Phosphobacterin were first used in
(a) USA (b) U.K
(c) Australia (d)Russia

102. Bacteria thriving well within a temperature range of 15–45°C are called
(a) Thermophillic
(b) Psychrophillic
(c) Mesophillic
(d) None of the above.

103. The bacteria thiobacillus converts
(a) Ammonia to nitrite
(b) Inorganic sulphur to sulphate
(c) Ferrous iron to Ferric iron
(d) None of the above.

104. The most common genera of actinomycetes in soil is
(a) Nocardia (b) Micromonospora
(c) Actinomyces (d) Streptomyces

105. The smallest inhabitants of soil is
(a) Protozoa (b) Bacteria
(c) Actinomycetes (d) Bacteriophages

106. Which of the following does not represent anaerobic nitrogen fixing bacteria?
(a) Achromobacter
(b) Clostridium
(c) Chlorobium
(d) Chromatium

107. Acteylene Reduction Essay (ARE) is done to determine
(a) Nitrogenase activity
(b) Dehydrogenase activity
(c) Phosphatase activity
(d) None of the above.

108. The 15N technique for quantitative estimation of nitrogenase activity was developed by
(a) Hiltner
(b) Beijerinck
(c) Burris and Wilson
(d) None of the above

109. In cellulose molecule, individual glucose units are bound together by
(a) α-1-4 (b) β-1-4
(c) α-1-6 (d) β-1-6

110. Hemicelluloses represent various polymers
(a) Hexoses (b) Pentoses
(c) Uronic acid (d) All the above.

111. Pectin is an example of
(a) Cellulose (b) Lignin
(c) Hemicellulose (d) None of the above.

112. Which of the followings represents genera of lignin degrading fungi?
(a) Mycena and Marasmius
(b) Polyporus and pleurotus
(c) Citocybe and Collybia
(d) None

113. Fulvic acid is
(a) Acid soluble
(b) Alkali soluble
(c) Acid insoluble
(d) Acid and alkali soluble

114. Humin is
(a) Resistant to cold alkali
(b) Soluble in cold alkali
(c) Soluble in hot water
(d) Soluble in cold water

115. Which of the followings is a well known synthetic chelator?
(a) EBT (b) EDTA
(c) TTC (d) TPF

116. Iron is an important component of enzymes
(a) Nitrogenase
(b) Nitrate reductase
(c) Nitrite reductase
(d) All the above

117. Which of the followings is micro-aerophilic that fixes nitrogen freely or in associa-tion with grasses?
(a) Azotobacter (b) Azospirillum
(c)Rhizobium (d) Frankia

118. Which of the following bacteria genera take part in nitrification?
(a) Nitrosomanas and Nitrobacter
(b) Nitrosococcus and Nitrosopira
(c) Nitrosocystis and Nitrosogloea
(d) All the above.

119. Which of the followings plays a role in ripening of fruits?
(a) Auxin (b) Cytokinin
(c) Ethylene (d) None

120. The insecticide Aldrin is converted to Dieldrin by
(a) Trichoderma (b) Fusarium
(c) Penicillium (d) All of the above

121. The first organic fungicide is
(a) Zineb (b)Thiram
(c) Captan (d) None

122. Hartig net is developed by
(a) Endomycorrhizae
(b) Ectomycorrhizae
(c) Nodulating bacteria
(d) None of the above

123. The undefined constituents of root exudates required for growth/culture of ectomycorrhizae are called
(a) P-factor (b) G-factor
(c) M-factor (d) None

124. Which of the followings represents the liason tissue between the fungal sheath and the host cells?
(a) Cortex layer (b) Casparian strip
(c) Hartig net (d) None

125. Which of the following structures of VAMF connect the fungal ramifications inside roots with the mycelium of the fungus outside roots?
(a) Vesicles (b) Appressoria
(c) Arbuscules (d) None

126. The VAM fungi with spores bearing str-aight or angular stalks are grouped as
(a) Glomus (b) Gigaspora
(c) Acaulospora (d) Endogone

127. The VAM fungi with spores bearing bulbous stalks are grouped as
(a) Glomus (b) Sclerocystis
(c) Gigaspora (d) None

128. The VAM fungi with sessile spores are grouped as
(a) Sclerocystis (b) Acaulospora
(c) Gigaspora (d) Glomus

129. The VAM fungi with spores regularly arranged on a central core are grouped as
(a) Sclerocystis (b) Acaulospora
(c) Gigaspora (d) Glomus

130. Decomposition of larger polysaccharides are carried out by
(a) Intracellular enzymes
(b) Extracellular enzymes
(c) Both intracellular and extracellular enzymes
(d) None

131. Biological oxidation actually does not proceed by
(a) Addition of oxygen
(b) Removal of hydrogen
(c) Removal of electron
(d) All the above

132. The group of enzymes always produced in microbial cell irrespective of presence or absence of a substrate is called
(a) Inducible enzyme
(b) Intracellular enzyme
(c) Extracellular enzyme
(d) Constitutive enzyme

133. The term used to denote conversion of organic complexes of an element to inorganic state is
(a) Mineralization
(b) Immobilization
(c) Assimilation
(d) All the above.

134. The phenomenon of enhancement in native humus decomposition owing to external addition of organic matter is termed.
(a) Mineralization
(b) Assimilation
(c) Immobilization
(d) Priming

135. In woody perennials, the bulk of the organic matter lost is derived from
(a) Hemicellulose (b) Cellulose
(c) Lignin (d) All the above.

136. As the clay content of a soil increases, the decay rate
(a) Increases
(b) Decreases
(c) Remains unchanged
(d) Shows variable response

137. The most abundant organic compound in nature is
(a) Hemicellulose (b) Celulose
(c) Lignin (d) None

138. Beyond which of the following ratios between inorganic N and cellulose, the cellulose decomposition does not respond to further addition of inorganic N
(a) 1:10 (b) 1:20
(c) 1:30 (d) 1:35

139. Which of the following chemicals are formed during anaerobic decomposition of cellulose
(a) Ethanol
(b) Acetic and Formic acid
(c) Lactic and Butyric acid
(d) All the above.

140. Aerobic bacteria generally convert cellulose into
(a) CO_2 and cell substance
(b) H2 and ethanol
(c) Acetic and Formic acid
(d) Succinic and Butyric acid.

141. Hemicelluloses are polymers of
(a) Simple sugars
(b) Uronic acids
(c) Both simple sugars and uronic acids
(d) None

142. In the breakdown of xylan, the action of endo enzyme yields
(a) Cellobiose (b) Xylobiose
(c) Xylose (d) None

143. In the breakdown of xylan, the action of exo enzyme yields
(a) Cellobiose (b) Xylobiose
(c) Xylose (d) Glucose

144. The components of plant starches are
(a) Amylose (b)Amylopectin
(c) Both a and b (d) None

145. The anaerobic fermentation of starches yields
(a) Lactic acid (b) Acetic acid
(c) Butyric acid (d) All the above

146. The different enzymes catalyzing starch breakdown are
(a) α-amylase (b) β-amylase
(c) Glucoamylase (d) All the above

147. The main building block of pectin is
(a) Galactan
(b) Uronic acid
(c) Galactouronic acid
(d) None

148. Pectic substances are of types
(a) Protopectin and pectin
(b) Pectinic acid and pectic acid
(c) Both a and b
(d) None

149. Which of the following groups of bacterial genera represent pectin degrades
(a) Arthrobacter and Bacillus
(b)Clostridium and Flavobacterium
(c) Micrococcus and Pseudomonas
(d) All the above

150. The building block of chitin is
(a) Glucose
(b) N-aceetylglucosamine
(c) Fructose
(d) None

151. The enzymes responsible for breakdown of chitin are
(a) Chitinase (b) Chitobiase
(c) Both a and b (d)None

152. The anaerobic breakdown of protein rich materials is termed.
(a) Putrefaction
(b) Denitrification
(c) Depolymerization
(d) None

153. Which of the following groups represents final products of anaerobic protein breakdown?
(a) Ammonium, amines and CO_2
(b) Organic acids, mercaptans and hydrogen sulphide
(c) Both a and b
(d) None of the above

154. Which of the following fungal genera are protein decomposers?
(a) Alternaria and Aspergillus
(b) Mucor and Penicillium
(c) Rhizopus
(d) All the above.

155. Which of the following soil moisture conditions favour nitrogen mineralization the most?
(a) Continuous wetting
(b) Alternate drying and Wetting
(c) Continuous drying
(d) None

156. The rise in pH near the urea particles applied to soil is caused by production of
(a) Ammonia (b) Nitrate
(c) Nitrite (d) None

157. Which of the following bacterial genera produce enzyme urease?

(a) Bacillus and Micrococcus

(b) Pseudomonas and Klebsiella

(c) Cyanobacterium and Clostriium

(d) All the above.

158. 'Nitrogen factor' defines

(a) Number of units of inorganic nitrogen immobilized for each hundred units

(b) The amount of nitrogen need to be added to prevent a net immobilization from the environment

(c) Both a and b

(d) None

159. Which of the followings are the most dominant microorganisms in nitrification?

(a) Nitrosomonas (b) Nitrobacter

(c) Both a and b (d) Nitrosospira

160. In methemoglobinemia, the haemoglobin is converted to

(a) Methemoglobin

(b) Leghemoglobin

(c) Phytin

(d) None

161. Denitrification may also be termed as

(a) Nitrate assimilation

(b) Nitrate respiration

(c) Denitrification

(d) All the above

162. Which of the followings represents a dominant anaerobic N2 fixer?

(a) Azotobacter (b) Clostridium

(c) Beizerinckia (d) None

163. Which of the followings does not represent non leguminous N2 – fixing tree?

(a) Alnus (b) Casuarina

(c) Gmelina (d) Myrica

164. The rhizosphere effect can be quantitatively represented by

(a) C: N Ration (b) R:S Ratio

(c) C:P Ratio (d) None

165. The principle of gram staining is based on differences in

(a) Composition of bacterial cell wall

(b). Size of Bacteria

(c) Shape of Bacteria

(d) Colour of Bacteria

166. YEMA medium was introduced for rhizobium by

(a) Hiltner (b) Nobbe

(c) Fred (d) Joseph Lister

167. Which of the following genus under rhizobium contributes to stem nodulation

(a) Azorhizobium (b) Bradyrhizobium

(c) Rhizobium (d) None of these

168. Which of the following pigments is present in BGA?

(a) Carotene (b) Xanthophylls

(c) Phycocyanine (d) All the above

169. The minimum shelf of Rhizobium biofertilizer should be

(a) 4 Months (b) 6 Months

(c) 12 Months (d) None of these

170. The term autocolony stands for

(a) Any Microbial colony

(b) Microbial colony formed in the cells of parent colony

(c) Both a and b

(d) None

171. "Nif" gene is responsible for

(a) Nitrification

(b) Synthesis and regulation of Nitrogenase

(c) Dentrification

(d) None

172. 'Biofertiliser Newsletter' is published by

(a) IARI

(b) NifTAL

(c) National Biofertilizer Development centre

(d) None

173. Green plants are also called as
(a) Primary producers
(b) Secondary producers
(c) Consumers
(d) Decomposers

174. Of the following, which is soil micro fauna?
(a) Protozoa (b) Nematodes
(c) Earthworms (d) Diatoms

175. Penicillium is an example of
(a) Cellulose decomposer
(b) Protein Decomposer
(c) Lignin Decomposer
(d) Humus former

176. Cellulose is mainly decomposed by
(a) Penicillium (b) Aspergillus
(c) Mucar (d) All of these

177. Of the following, which is/are soil macro fauna?
(a) Earthworms (b) Moles
(c) Ants (d) All of these

178. Arrange the following microorganisms based on their population in soil.
(a) Fungi >Bacteria>Actinomycetes
(b) Bacteria>Fungi>Actinomycetes
(c) Bacteria>Actinomycetes>Fungi
(d) Fungi> Actinomycetes>Bacteria

179. Who is the 'Father of field experiment'?
(a) M.W. Beijerink
(b) J.B. Bousingault
(c) C. Theoder de Saucer
(d) Arnon and Stout

180. The bacterial population per gram of rhizosphere of soil is
(a) 6778×10^6 (b) 918×10^3
(c) 83×10^6 (d) 68×10^6

181. What is the temperature range of mesophiles?
(a) <10°C (b) 10 - 20°C
(c) 20 - 40°C (d) > 40°C

182. All the autotrophic organisms get carbon source from
(a) Organic matter
(b) Carbon dioxide
(c) Both a and b
(d) Minerals

183. All the heterotrophic organisms get carbon source from
(a) Organic matter
(b) Carbon dioxide
(c) Both a and b (d) Minerals

184. Phototrophic organisms derive their energy from
(a) Carbon dioxide (b) Organic Matter
(c) Sun light (d) Organic matter

185. What is the optimum temperature for the mesophylic organisms?
(a) 30 °C (b) 37°C
(c) 27°C (d) 17°C

186. The ideal pH range for the growth of fungi in soil is
(a) 6.5 and 8.0 (b) 5.5 and 7.0
(c) 6.0 and 7.0 (d) 4.5 and 6.5

187. The ideal pH range for the growth of bacteria in soil is
(a) 6.5 and 8.0 (b) 5.5 and 7.0
(c) 6.0 and 7.0 (d) 4.5 and 6.5

188. Soil protozoa are the predators of
(a) Bacteria (b) Actinomycetes
(c) Fungi (d) All of these

189. Lignin is mainly decomposed by
(a) Bacteria (b) Actinomycetes
(c) Fungi (d) All of these

190. Humus is primarily formed by
(a) Bacteria (b) Actinomycetes
(c) Fungi (d) Protozoa

191. VAM fungi are
(a) Obligate symbiont
(b) Facultative symbiont
(c) Free-Living
(d) Both a & b

192. Earthworms casts are rich in nutrients and
(a) Bacteria (b) Fungi
(c) Viruses (d) All of these

193. Earthworms are more abundant in
(a) Coarse textured soils
(b) Fine textured soils
(c) Medium textured soils
(d) Both a and b

194. Which is an example of green sulphur bacteria
(a) Chlorobium
(b) Rhodomicrobium
(c) Rhodosprillum
(d) All of these

195. Propagules of VAM fungi consist of
(a) Clamydospores
(b) Azygospores
(c) Soil – borne vesicle
(d) All of these

196. What is the optimum C/N ratio for organic matter
(a) <20:1 (b) 20:30-1
(c) >30:1 (d) None of these

197. Phytin is a type of organic
(a) Nitrogen (b) Phosphorous
(c) Sulphur (d) Molybedenum

198. Solidifying agent in a medium is called as
(a) Agar- agar (b) Starch
(c) Fats (d) Sugars

199. Biofertilizer inoculation technique was started during
(a) 1880 (b) 1885
(c) 1887 (d) 1895

200. 'Herting net' is produced by
(a) N- fixing rhizobium
(b) Ecototrophic mycorrhiza
(c) Both a and b
(d) None

201. Extra chromosomal genetic materials is
(a) Nucleoid (b) Nucleus
(c) Nucleolus (d) Plasmid

202. Structure involved in the transfer of genetic material during conjugation of bacteria
(a) Cilia
(b) Flagella
(c) Pilus
(d) Cytoplasmic membrane

203. Infecting a bacterium
(a) Bacteriophage
(b) Viroid
(c) Viroid
(d) Mycoplasma

204. Small, circular, self-replicating DNA found in *E. coli*
(a) Chromosomes (b) Plastids
(c) Plasmids (d) Ribosomes

205. A segment of DNA that is transcribed, but removed from the DNA sequences
(a) Introns (b) Exons
(c) Split genes (d) Jumping genes

206. Total genetic material in a cell constitutes
(a) Genome (b) Gene Bank
(c) Gene (d) Ganome DNA

207. Which of the following is not a prokaryote?
(a) Bacterium
(b) Amoeba
(c) Blue Green Algae
(d) Archaeobacteria

208. Thermophiles grow in conditions of high
(a) Temperature (b) Salt
(c) CO_2 (d) Food

209. Halophiles grow in conditions of high
(a) Salt (b) Temperature
(c) Acidity (d) Oxygen

210. Photoautotrophic metabolism involves

(1) Use of inorganic CO_2 as the only source of carbon

(2) Sun light as the source of energy

(3) Energy derived from the metabolism of organic components

(4) Organic carbon compounds as a source of energy

(5) Energy obtained from the oxidation of inorganic molecules

(a) 1,2 (b) 1

(c) 2,3,4 (d) 4,5

211. Bacteria multiplying in the dead sea are mostly likely to be

(a) Aerobic (b) Anaerobic

(c) Thermophiles (d)Holophiles

212. An organism that can grow either in the presence or in the absence of mol. O_2 is called a(n)

(a) Obligate aerobe

(b) Facultative anaerobe

(c) Obligate anaerobe

(d) Acidophiles

213. Bacteria which grow at a temperature around 10° are called as

(a) Phychrophiles

(b) Thermophiles

(c) Mesophiles

(d) Acidophiles

214. How many generation did the population go through, when a bacterial culture began with 2 cells and ended with 128 cells?

(a) 12 (b) 10

(c) 6 (d)5

215. Bacterial broth culture stops growing and enters a stationary phase, because it

(a) Depletes an essential nutrient

(b) Accumulates essential nutrients

(c) Becomes too crowded

(d) Produces plenty of oxygen

216. Which of the following processes does not result in recombination?

(a) Conjugation (b) Transformation

(c) Transduction (d)Translation

217. Which of the following can now be produced in microorganisms?

1. Insulin

2. Somatostatin

3. Vaccine for malaria

4. Blood clotting factors

(a) 1,2,4 (b)1,2,3,4

(c) 1,2 (d) 1,2

218. Five Kingdom scheme of classification was proposed by

(a) E.H. Haeckel

(b) Carolus Linnaeus

(c) Dobzhansky

(d) R.H. Whittaker

219. Yeast is a

(a) Bacteria (b) Virus

(c) Fungi (d)Algae

220. Scientist who for the first time demonstrated the principle of immunization

(a) Louis Pasteur (b)Edward Jenner

(c) John Tyndall (d) Robert Koch

221. Name of the scientist who observed micro-organisms for the first time

(a) Griffith

(b) A.V.Leeuwenhoek

(c) Lazaro Spallanzani

(d) Robert Hooke

222. Germ theory of disease was first proposed by

(a) Jacob Manle

(b) Father Sudrez

(c) Louis Pasteur

(d)Joseph Lister

223. Scientist who developed aseptic technique and isolated bacteria in pure culture

(a) Joseph Lister (b)Metchnikoff

(c) Robert Koch (d)Spencer

224. N_2 fixing bacteria in soil were discovered by
(a) S.N. Winogradsky
(b) John Needham
(c) Waddington (d) Walter Reed

225. In which of the following micro-organisms surface area/volume ratio is larges
(a) Fungi (b) Bacteria
(c) Yeast (d)Algae

226. Teichoi acids are present in
(a) Cell wall of G+ bacteria
(b) Cell wall of G+ bacteria
(c) Nucleoid of G+ bacteria
(d) Plasma membrane of G+ bacteria

227. The production of CAMP from ATP is catalysed by which of the following membrane bound enzyme
(a) Adenylate Cyclase
(b) Acetyl Coenzyme-A
(c) ATPase
(d) Catalases

228. L.T. Plasmid in *E.coli* produces
(a) Enterotoxin (b) Bacteriocin
(c) Tetanalysin (d) Tetanospasmin

229. Distinction between prokaryotic and eukaryotic cells was made on the basis of
(a) Structure of cell wall
(b) Presence or absence of Golgi bodies and Iysosomes
(c) Absence or presence of nuclear membrane
(d) Absence or presence of ribosomes

230. Organelles such as mitochondria and chloroplasts in eukaryotic cells evolved from prokaryotic organisms by the engulfment by primitive nucleated cells is known as
(a) Cell theory
(b) Protoplast theory
(c) Endosymbiotic theory
(d) Theory of abiogenesis

231. The correct way to represent the scientific name for a bacterium is
(a) *Escherichia Coli*
(b) *Escherichia coli*
(c) *Escherichia Coli*
(d) *Escherichia Coli*

232. Structure essential for living cells
(a) Ribosomes
(b) Capsule
(c) Cell wall
(d) Cytoplasmic membrane

233. The process of production of endospores from a vegetative cell is called as
(a) Germination
(b) Sporogenesis
(c) Encystment
(d) Conidia formation

234. Spherical bacteria arranged in a grape bunch manner
(a) Spirillum (b) Cocci
(c) Bacilli (d) Staphylococci

235. Resistance to antibiotics is carried on
(a) R factor plasmid
(b) Flagellum
(c) Slime layer
(d) Sex pilli

236. Which of the following has chain on conidia growing on aerial hypahe?
(a) Actinomyces
(b) Nocardia
(c) Streptomyces
(d) Nitrobacter

237. In some bacteria, capsule surrounding the cell wall is made up of
(a) Cellulose (b) Polysaccharides
(c) Polypeptide (d) Teichoic acids

238. Growth in bacteria reflects
(a) Growth in number
(b) Growth in weight
(c) Growth in volume
(d) Changes in shape and size

239. Which of the following bacteria oxidises ferrous compounds into ferric ones
(a) Leptothrix ochracae
(b) Pseudomonas
(c) Rhizobium
(d) Lactobacillus

240. Which of the following bacteria oxidizes molecular hydrogen into water in soil
(a) Penicillium notatum
(b) Rhizobium noduliformes
(c) Bacillus pentotrophus
(d) None of these.

ANSWERSHEET
14. Forest Microbiology

---

1. (b) 2. (b) 3. (a) 4. (b) 5. (a) 6. (a) 7. (d) 8. (a) 9. (a) 10. (b)
11. (b) 12. (b) 13. (b) 14. (c) 15. (c) 16. (c) 17. (a) 18. (b) 19. (a) 20. (b)
21. (b) 22. (a) 23. (c) 24. (b) 25. (d) 26. (b) 27. (a) 28. (c) 29. (c) 30. (b)
31. (d) 32. (b) 33. (b) 34. (c) 35. (b) 36. (b) 37. (a); 38. (b) 39. (d) 40. (a)
41. (b) 42. (b) 43. (b) 44. (d) 45. (a) 46. (b) 47. (b) 48. (b) 49. (c); 50. (b)
51. (b) 52. (b) 53. (c) 54. (b) 55. (c) 56. (c) 57. (b) 58. (a) 59. (a) 60. (c)
61. (a) 62. (b) 63. (c) 64. (a) 65. (a) 66. (d) 67. (c) 68. (c) 69. (a) 70. (c)
71. (c) 72. (a) 73. (c) 74. (a) 75. (c) 76. (a) 77. (b) 78. (d) 79. (c) 80. (a)
81. (b) 82. (d) 83. (c) 84. (d) 85. (c) 86. (d) 87. (d) 88. (b) 89. (c) 90. (b)
91. (a) 92. (b) 93. (b) 94. (c) 95. (b) 96. (a) 97. (b) 98. (d) 99. (c) 100. (d)
101. (d) 102. (c) 103. (b) 104. (d) 105. (d) 106. (a) 107. (a) 108. (c) 109. (b) 110. (d)
111. (c) 112. (d) 113. (d) 114. (a) 115. (b) 116. (d) 117. (b) 118. (d) 119. (c) 120. (d)
121. (b) 122. (b) 123. (c) 124. (c) 125. (b) 126. (a) 127. (c) 128. (b) 129. (a) 130. (b)
131. (a) 132. (d) 133. (a) 134. (d) 135. (b) 136. (b) 137. (b) 138. (d) 139. (d) 140. (a)
141. (c) 142. (b) 143. (c) 144. (c) 145. (d) 146. (d) 147. (c) 148. (c) 149. (d) 150. (b)
151. (c) 152. (a) 153. (c) 154. (d) 155. (b) 156. (a) 157. (d) 158. (c) 159. (c) 160. (a)
161. (b) 162. (b) 163. (c) 164. (b) 165. (a) 166. (c) 167. (a) 168. (d) 169. (b) 170. (b)
171. (b) 172. (c) 173. (a) 174. (a) 175. (a) 176. (d) 177. (d) 178. (c) 179. (b) 180. (a)
181. (c) 182. (b) 183. (a) 184. (c) 185. (b) 186. (d) 187. (a) 188. (b) 189. (a) 190. (a)
191. (a) 192. (a) 193. (b) 194. (a) 195. (d) 196. (a) 197. (b) 198. (a) 199. (d) 200. (b)
201. (d) 202. (c) 203. (a) 204. (c) 205. (a) 206. (a) 207. (b) 208. (a) 209. (a) 210. (a)
211. (d) 212. (b) 213. (a) 214. (c) 215. (d) 216. (d) 217. (a) 218. (d) 219. (c) 220. (a)
221. (b) 222. (a) 223. (a) 224. (a) 225. (b) 226. (a) 227. (a) 228. (a) 229. (c) 230. (c)
231. (c) 232 (d) 233. (b) 234. (d) 235. (a) 236. (c) 237. (b) 238. (a) 239. (a) 240. (c)

15

Forest Ecology and Environment

KEY POINTS

- The father of the world 'ecology' was Ernst Haeckel, a German zoologist
- It derived from the Greek words 'oikos' and 'logos' and spelt it Oecology. 'Oikos' means house and 'logos' study of.
- Ecology thus means the study of the house or the habitat of an organism, a living animal or plant.
- Autecology is the study of an individual organism, life history of the organism and it response to the environment
- Synecology is the study of properties of a community, i.e. a natural assemblage of different species of organisms
- Population is defined as a group of individuals of one species
- Density is anumber of individuals that occupy an unit area
- Natality is the rate at which new members are added to the population by reproduction
- Mortality is the rate at which members are lost on account of death
- Dispersal is the rate at which individuals immigrate into the population and emigrate out of the population. This attribute applies only to animals. Plants do not have this attribute as they are sessile organisms, not capable of movement
- Population growth rate is the combined effect of mortality, natality and dispersal on population
- Growth characteristic of populations is depicted by a J-shaped curve.
- Competition is the striving of two individuals for the same resource, which is not available in a quantity or manner so as to satisfy the needs of both
- Commensalism is the situation where one population benefits another, but is not itself significantly affected by the interaction
- Photocooperation is the two populations benefit each other, but do not play any significant role in each other's existence
- Mutualism is the association of two populations is vital for the existence of the either
- Charles Darwin, the father of the evolutionary theory, used the expression descent with modification to describe evolution
- The branch of ecology which pertains to study of intraspecific variation of plants in relation to environment is called genecology, a term originally used by Tueresson (1923).

- The local populations of a species, which are confined and adapted to particular and definable environmental conditions, are known as ecotypes
- A group of ecotypes constitute an ecospecies
- Race is defined as an intraspecific category characterized by conspicuous physiological, biological, geographical or ecological properties.
- Cline acoording to a particular direction measurable characters
- Ecocline is a cline apparently correlated with an observable ecological gradient.
- Ecotype is a particular range on an ecocline
- The term ecoclinal refers to continuous variation and ecotypic refers to discontinuous variation
- The process that brings about non-random (i.e. not associated with chance) and differential reproduction of genotypes which results in preserving favourable variants and eliminating less favourable variants, is known as natural selection.
- An environmental factor that forces natural selection is called a selection pressure.
- The expression, selection pressure is also used to denote the intensity of natural selection,
- Gene frequency is the proportion of any allele to the total of all alleles at the same locus (position of a gene on a chromosome) in the gene pool of the population and this situation of constant gene frequencies is known as Hardy-Weinberg equilibrium.
- The process of natural selection may be of three types stabilizing, directional and disruptive.
- Stabilizing selection takes place when physical environmental conditions remain stable
- Directional selection occurs in changing conditions of physical environment.
- Disruptive selection may be considered as a process opposite to that of stabilizing selection, as it tends to enhance the range of genotypic variation in a population
- Adaptation may be defined as a genotypic change in a population, which improves the chances of its survival and growth.
- Speciation is the physical environment of a particular ecosystem subject to rapid changes, new forms evolved by adaptation, which have often to be classified as new species Branched speciation is the process by which a single species branches down to form two or more new species.
- Graduation is the process by which a species changes gradually over time.
- Evolution is the adaptive modification of organisms through successive generations and it involves a change in gene frequencies with time.
- Alternating evolution of two species in face of mutual selection pressures is known as coevolution.
- Coevolution led to many organisms in an ecosystem becoming so dependent on each other that the survival of one depends on the existence of the other.
- Plant species possess physiological mechanisms that indicate particular seasons of the year, and thus plant species are known as biological clocks.
- A zone of transition, presenting a situation of special ecological interest between two different types of communities, is known as an ecotone.
- A life form is the sum of the adaptation of the plant to climate

- The percentage distribution of species among the various life forms of a flora is called the biological spectrum
- Succession is a natural process by which different groups or communities colonize the same area over a period of time in a definite sequence.
- The succession, which starts from a primitive substratum without previous living matter, is known as the primary succession.
- Secondary succession is the starting from previously built up substratum where living matter already exists.
- The existing community, its reaction with the environment, causes its own replacement; such a succession is known as autogenic succession.
- The replacement of the existing community takes place due to the influence of any external force, condition, etc., such course is known as the allogenic succession.
- The terminal community that can maintain itself more or less indefinitely in equilibrium with the prevailing environment, is known as the climax community and the stage is said to be the climax
- The amount of living material, present in a component population at any time, is known as the standing crop, which may be expressed in terms of numbers or weight per unit area
- Biomass is the standing crop expressed in terms of weight (i.e. organism mass) of the living matter present.
- Ecological pyramids are the graphic representations of the trophic structure and function at successive trophic levels of an ecosystem, which may be shown in terms of their number, biomass or energy content.
- The transfer of food energy from the source in plants through a series of organisms (arranged in a linear manner) with repeated eating and being eaten is known as a food chain in an ecosystem
- The interlocking pattern of a number of food chains forms a web-like arrangement known as a food-web
- The amount of organic matter accumulated in the living component of an ecosystem in unit time is referred to as the productivity of the ecosystem.
- Primary productivity is defined as the rate at which radiant energy of sun is stored by photosynthetic and chemosynthetic activity of producers (chiefly green plants) in the form of organic substances, used as food materials.
- The rates of energy storage at consumer levels are referred to as secondary productivity.
- Gross primary productivity is the total rate of photosynthesis, including the organic matter used up in respiration during the period of measurement and also called total photosynthesis or total assimilation.
- Net primary productivity is the rate of storage of organic matter in plant tissues in excess of that utilized in respiration by plants during the period of measurement and also called apparent photosynthesis or net assimilation
- Ecological niche of an organism includes the physical space occupied by it, its functional role in the community
- Organisms that occupy the same or similar ecological niches in different geographical regions are known as ecological equivalents.

- Biomes are the units of study with interaction between different communities of area
- The term antibiosis generally refers to the complete or partial inhibition or death of one organism by another through the production of some substance or environmental conditions as a result of metabolic pathways
- Intraspecific competition often called as scramble competition
- Interspecific competition is also called as contest competition

MULTIPLE CHOICE QUESTION'S

1. The Word 'Ecology' is derived from
   (a) Greek (b) Latin
   (c) French (d) None of these
2. The word 'oikos' in greek means
   (a) House (b) Land
   (c) Monkey (d) Area
3. The term Ecology was first coined by
   (a) Charles Darwin (b) Ernst Haeckel
   (c) Thomas Edison (d) Tansely
4. The branch of biology that deals with environment
   (a) Ecology (b) Ecosystem
   (c) Ethnology (d) Ecobotany
5. Basic unit of ecological hierarchy is
   (a) Organisms (b) Biomes
   (c) Microbes (d) Ecology
6. A distinct unit of life in nature
   (a) Organisms (b) Biomes
   (c) Microbes (d) Ecology
7. The term Niche was first coined by
   (a) Charles Darwin
   (b) Ernst Haeckel
   (c) Grinnel
   (d) Element
8. Forests and wildlife are
   (a) Renewable Resources
   (b) Non renewable resources
   (c) a and b
   (d) None of these
9. Biological species are
   (a) Renewable Resources
   (b) Non renewable resources
   (c) a and b
   (d) None of these
10. The residence time of water in the air is
    (a) 10 days (b) 15 days
    (c) 100 days (d) 150 days
11. The term 'Ecosystem' was introduced by
    (a) Ernst Haeckel
    (b) Arthur Tansley
    (c) Element (d) Whittakar
12. The structural and functional unit of biosphere is
    (a) Mass (b) Environment
    (c) Ecosystem (d) All the above
13. The relationship between biotic and abiotic environment
    (a) Biogenesis (b) Symbiosis
    (c) Holocoenosis (d) None
14. Key industry animals are
    (a) Primary consumers
    (b) Tertiary consumers
    (c) Secondary consumers
    (d) Primary producers
15. The term 'key industry animals' was coined by
    (a) Elton (b) Priestley
    (c) Tansley (d) Element

16. The rate of biomass production is called
(a) Productivity (b) Production
(c) Biodiversity (d) All the above

17. The rate of total production of organic matter by the producer per unit area and time is
(a) Primary Productivity
(b) Secondary productivity
(c) Net primary Productivity
(d) Gross primary productivity

18. The rate at which organic matter is stored after the basic metabolic processes is
(a) Primary Productivity
(b) Secondary Productivity
(c) Net primary Productivity
(d) Gross primary Productivity

19. The productivity that is left after meeting the cost of respiration and maintenance of producers is
(a) Primary Productivity
(b) Secondary productivity
(c) Net primary Productivity
(d) Gross primary productivity

20. The rate of increase in the biomass of consumers per unit area and time is
(a) Primary Productivity
(b) Secondary Productivity
(c) Net primary Productivity
(d) Gross primary Productivity

21. Ecological pyramids were developed by
(a) Charles Elton (b) Priestley
(c) Tansley (d) Element

22. The graphical representation of ecological parameters is called
(a) Ecological Pyramid
(b) Ecology
(c) Ecological hill
(d) Ecological Mountain

23. Ecological pyramids are also called as
(a) Eltonian Pyramids
(b) Charles Pyramids
(c) Hierarchical Pyramids
(d) All the above

24. ---------------------- Ecological pyramid is always straight
(a) Pyramid of biomass
(b) Pyramid of numbers
(c) Pyramid of energy
(d) All the above

25. Name the pioneer plants of xerarch
(a) Lichens (b) Grasses
(c) Toads (d) Frogs

26. A population genetically adapted to its particular environment is known as
(a) Ecotone (b) Ecotype
(c) Ecosystems (d) Ecology

27. The environment which has been modified by human activities is called
(a) Anthropogenic environment
(b) Natural environment
(c) Manmade environment
(d) a and c

28. Groups of species exploiting a common resources are called
(a) Builds
(b) Guilds
(c) Community
(d) Population

29. In our country the Van Mahotsav Day is observed on
(a) Second of October
(b) First week of June
(c) Twenty first of March
(d) First of July

30. The process of conversion of gas into a liquid is called
(a) Condensation (b) Sublimation
(c) Vaporisation (d) Solidification

31. The drying up of wet cloths represents the process of
(a) Evaporation (b) Boiling
(c) Condensation (d) Solidification

32. During change of state in a matter the temperature remains same due to
(a) Loss of heat (b) Less supply of heat
(c) Lattice energy (d) Latent heat

33. Which biome is characterised by very large, mostly evergreen trees and ultisols?
(a) Equatorial (b) Tropical forests
(c) a and b (d) None of these

34. Which Biome is characterised by rapid nutrient cycling and high biomass?
(a) Deserts
(b) Deciduous
(c) Tropical rain forests
(d) All the above

35. Biotic environment includes
(a) Producers (b) Consumers
(c) Decomposers (d) All the above

36. Decomposers include
(a) Bacteria (b) Fungi
(c) a and b (d) None

37. Abiotic environment does not include
(a) Air (b) Water
(c) Soil (d) Plants

38. Vermicomposting is done by
(a) Fungus (b) Bacteria
(c) Worms (d) Animals

39. The group of organisms which convert light into food are called
(a) Autotrophs (b) Heterotrophs
(c) Decomposers (d) Omnivores

40. Example of non-biodegradable?
(a) Vegetables (b) Fruits
(c) Earthworm (d) Aluminium foil

41. Air pollution is caused by
(a) Insecticides (b) Sewage
(c) Smoke (d) Loud speakers

42. An antibacterial chemical present in wood smoke is
(a) Formaldehyde
(b) Carbon
(c) Chlorine
(d) Carbohydrate

43. About half of earth's photosynthesis is carried out by
(a) Cyanobacteria
(b) Rain forest
(c) Protists
(d) Marine Phytoplankton

44. Ramsar convention is associated with
(a) Forest conservation
(b) Soil Conservation
(c) Wetland conservation
(d) Wildlife conservation

45. Plants are green because of the presence of a pigment called:
(a) Glucose (b) Nitrogen
(c) Chlorophyll (d) Oxygen

46. Air is composed of gases, water vapours and
(a) Dust particles (b) Rain fall
(c) Snowfall (d) Light

47. Medicine of quinine is provided by
(a) Eucalyptus (b) Aconite
(c) Cinchona (d) Pine

48. Primary source of energy in environment is
(a) Fire (b) Moon
(c) Sun (d) Stars

49. When trees are cut, amount of oxygen
(a) Decreases (b) Increases
(c) a and b (d) None of these

50. Process through which plants reproduce:
(a) Eating (b) Evaporation
(c) Pollination (d) Condensation

51. 71% of earth surface is covered with:
(a) Land (b) Air
(c) Nitrogen (d) Water

52. CNG stands for:
(a) Compressed Natural Gas
(b) Composed Natural Gas
(c) Compressed Nitrogen gas
(d) Composed Nitrogen gas

53. Green Petrol is
(a) Leaded petrol (b) Unleaded petrol
(c) Gasohol (d) Low sulphur Diesel

54. Bio-indicator of underground water is
(a) *Impatiens blasamina*
(b) *Prosopis cineraria*
(c) *Astragalus*
(d) *Haplopappus*

55. Air pollution is highest in
(a) Kolkata (b) Lucknow
(c) Panchmarhi (d) Bhopal

56. Which is a major pollutant causing acid rain?
(a) SO_2 (b) NO_2
(c) CO (d) O_3

57. The phenomenon of corrosion of marble due to acid rain:
(a) Marble fever (b) Marble rain
(c) Marble pain (d) Marble Cancer

58. An alternative source of energy to fuel energy:
(a) Tidal Energy
(b) Petroleum
(c) Nuclear Energy
(d) Compressed Natural Gas

59. The best way to dispose plant waste:
(a) Burning
(b) Composting
(c) Dumping in a water body
(d) Incineration

60. An eco-friendly slogan:
(a) 'Do or die'
(b) 'Birds of same feather fly together'
(c) Time and tide wait for none'
(d) 'Say no to crackers'

61. The term 'dead' with respect to a water body refers to:
(a) The inability of a water body to sustain aquatic life
(b) The ability of a water body to sustain aquatic life
(c) The inability of a water body to flow
(d) The heating or cooling of a water body

62. Example of water borne disease:
(a) Small Pox (b) Meningitis
(c) Diarrhea (d) Cholera

63. The water suitable for drinking:
(a) Pure water (b) Portable water
(c) Potable water (d) Pungent water

64. The best household method to obtain potable water:
(a) Filtration
(b) Boiling
(c) Chemical treatment
(d) Decantation

65. The Branch of Biology which is concerned with the inter-relationship between plants and animals is called
(a) Ecology (b) Botany
(c) Physiology (d) Ethonobotany

66. The largest unit of living organisms on earth is
(a) Biosphere (b) Ecotone
(c) Ecosystem (d) Biome

67. What are the components of an ecosystem are:
(a) Biotic (b) Abiotic
(c) a and b (d) None of these

68. The green plants are called:
(a) Producers (b) Consumers
(c) Decomposers (d) All the above

69. For domestic drinking water supply the total dissolved should not exceed
(a) 200 ppm (b) 500 ppm
(c) 700 ppm (d) 900 ppm

70. Which of the chemicals is responsible for the reduction of ozone content of the atmosphere?
(a) Carbon dioxide
(b) Sulphur dioxide
(c) Carbon monoxide
(d) Chlorofluoro carbon

71. Acid rains occur when atmosphere is heavily polluted with :
(a) CH_4 and O_3 (b) O_2 and NO_2
(c) SO_2 and NO_2 (d) CO_2 and SO_2

72. Spraying of DDT on crops causes pollution
(a) Soil (b) Water
(c) a and b (d) None of these

73. Soil erosion can be prevented by :
(a) Overgrazing
(b) Afforestation
(c) Removal of vegetation
(d) Deforestation

74. The main source of water pollution is :
(a) Smog (b) Sewage
(c) Industries (d) Detergents

75. Which gas is mainly responsible for green house effect?
(a) CO_2 (b) SO_2
(c) O_2 (d) N_2

76. Cutting of trees in a forest is called :
(a) Deforestation
(b) Afforestation
(c) Selection cutting
(d) All the above

77. The sound of which intensity is regarded as the safe noise level for a city?
(a) 45 dB (b) 75 dB
(c) 80 dB (d) 100 dB

78. Eco-labelling is governed by
(a) ISO 14001 (b) ISO 14010
(c) ISO 14020 (d) ISO 14002

79. Earth day is observed on:
(a) April 21 (b) April 22
(c) June 5 (d) September 16

80. State with the highest percentage of forests is:
(a) Mizoram
(b) Tamil Nadu
(c) Kerala
(d) Arunachala Pradesh

81. When was Project Tiger launched in India?
(a) 1973 (b) 1975
(c) 1981 (d) 1983

82. World Wildlife Fund was founded in :
(a) 1947 (b) 1961
(c) 1952 (d) 1988

83. World Environment Day is observed on :
(a) April 21 (b) April 22
(c) June 5 (d) September 16

84. The branch of biology dealing with interactions among organisms and between organisms and their environment is called
(a) Economy (b) Modelling
(c) Recycling (d) Ecology

85. All of the members of a particular species that live in one area are called a (an)
(a) Biome (b) Population
(c) Community (d) Ecosystem

86. Eco-Auditing focuses on
(a) Assessment of actual effects of exist activities
(b) Pollution effects
(c) Climatic effects
(d) water management

87. What is the original source of almost all the energy in most ecosystems?
(a) Carbohydrates (b) Water
(c) Carbon (d) Sunlight

88. International ozone day is celebrated on
(a) 5th June (b) 16th September
(c) 8th March (d) 21st March

89. An organism that uses energy to produce its own food supply from inorganic compounds is called a

(a) Autotroph (b) Heterotroph
(c) Chemotroph (d) Monotroph

90. Which organisms does NOT require sunlight to live?

(a) Chemosynthetic bacteria
(b) Algae
(c) Trees
(d) Photosynthetic bacteria

91. An organism that cannot make its own food is called a(an)

(a) Heterotroph (b) Chemotroph
(c) Autotroph (d) Producer.

92. Organisms that obtain nutrients by breaking down dead and decaying plants and animals are called

(a) Decomposers (b) Omnivores
(c) Autotrophs (d) Producers.

93. What is an organism that feeds only on plants called?

(a) Carnivore (b) Herbivore
(c) Omnivore (d) Detritivore

94. What is an ecological model of the relationships that form a network of complex interactions among organisms in a community from producers to decomposers?

(a) Food web (b) An ecosystem
(c) Food chain (d) A population

95. What animals eat both producers and consumers?

(a) Omnivores (b) Decomposers
(c) a and b (d) None of these

96. Only 10 percent of the energy stored in an organism can be passed on to the next trophic level. Of the remaining energy, some is used for the organism's life processes, and the rest is

(a) used in reproduction.
(b) stored as body tissue
(c) stored as fat.
(d) Eliminated as heat.

97. The repeated movement of water between Earth's surface and the atmosphere is called

(a) Water cycle (b) Nitrogen cycle
(c) Carbon cycle (d) Phosphorus cycle

98. Five R polices for waste minimization

(a) Reduction, Replacement, Recovery, Recycling and Reutilization
(b) Regain, Reuse, Resend, Re oxidation, Reduce
(c) Regain, Reuse, Resend, Reimbursement, Reduce
(d) Reduction, Reuse, Resend, Re oxidation, Reduce

99. The calorific value of natural gas varies from

(a) 10,000 -18,000 kCal/m³
(b) 12,000 -14,000 kCal/m³
(c) 13,000 -18,000 kCal/m³
(d) 16,000 -18,000 kCal/m³

100. Organisms need nutrients in order to

(a) Utilize hydrogen and oxygen.
(b) Carry out essential life functions.
(c) Recycle chemical compounds.
(d) Carry out nitrogen fixation.

101. The average year-after-year conditions of temperature and precipitation in a particular region is the region's

(a) Weather (b) Climate
(c) a and b (d) None of these

102. BOD of the effluents discharged on land for irrigation should not exceed

(a) 2[illegible] mg/l (b) 25 mg/l
(c) [illegible]0 mg/l (d) 35 mg/l

103. The greenhouse effect is

(a) The result of an excess of carbon dioxide in the atmosphere.
(b) A natural phenomenon that maintains earth's temperature range.
(c) The result of the differences in the angle of the sun's rays.
(d) An unnatural phenomenon that causes heat energy to be radiated back into the atmosphere.

104. Ecotone is
(a) Ecosystem
(b) A zone of an ecosystem
(c) An ecological study
(d) A zone between two ecosystem

105. Which is a biotic factor that affects the size of a population in a specific ecosystem?
(a) Average temperature of the ecosystem
(b) Type of soil in the ecosystem
(c) Number and kinds of predators in the ecosystem
(d) Concentration of oxygen in the ecosystem

106. Biogas is composed of
(a) Ethane (b) Methane
(c) CO_2 (d) All the above

107. An interaction in which one organism captures and feeds on another organism is called
(a) Predation. (b) Parasites
[illegible] Symbionts (d) Commensals

108. No tw[illegible] species can occupy the same niche in [illegible] the same habitat at the same time
(a) Because of the [illegible] interactions that shape the ecosyste[illegible]
(b) Unless the species req[illegible]uire different abiotic factors.
(c) Because of the competitive[illegible] exclusion principle.
(d) Unless the species require di[illegible]fferent biotic factors.

109. A symbiosis in which both species benefit is
(a) Commensalism (b) Mutualism
(c) Predation (d) Parasitism

[illegible] The series of predictable changes that [illegible]curs in a community over time is [illegible]

[illegible]ion growth.

[illegible]ccession.

(c) Climax community.
(d) Climate change.

111. Which two biomes have the least amount of precipitation?
(a) Tropical rain forest and temperate grassland
(b) Tropical savanna and tropical dry forest
(c) Tundra and desert
(d) Boreal forest and temperate woodland and shrubland

112. A biome is identified by its particular set of abiotic factors and its
(a) Average precipitation and temperature.
(b) Characteristic ecological community.
(c) Distance from the equator.
(d) Specific geographical location.

113. One of the main characteristics of a population is its
(a) Change over time
(b) Geographic distribution
(c) Dynamics
(d) Habitat

114. Which are two ways a population can decrease in size?
(a) immigration and emigration
(b) increased death rate and immigration
(c) decreased birthrate and emigration
(d) emigration and increased birthrate

115. When individuals in a population reproduce at a constant rate, it is called
(a) Logistic growth
(b) Growth density
(c) Exponential growth
(d) Multiple growth

116. As resources in a population become less available, population growth
(a) Declines rapidly
(b) Increases slowly
(c) Reaches carrying capacity
(d) Enters a phase of exponential growth

117. In a logistic growth curve, exponential growth is the phase in which the population

(a) Reaches carrying capacity

(b) Grows quickly and few animals are dying

(c) Growth begins to slow down

(d) Growth stops

118. A biotic or an abiotic resource in the environment that causes population size to decrease is a

(a) Carrying capacity

(b) Limiting nutrient

(c) Limiting factor

(d) Growth factor

119. Example of limiting factors

(a) Competition

(b) Predation

(c) Human disturbances

(d) All the above

120. If a population grows larger than the carrying capacity of the environment, the

(a) Death rate may rise

(b) Birthrate may rise

(c) Death rate must fall

(d) Birthrate must fall

121. Which of the following is a density-independent limiting factor?

(a) Earthquake (b) Disease

(c) Emigration (d) Parasitism

122. Example of density-dependent limiting factor

(a) Competition (b) Crowding

(c) Disease (d) All the above

123. Methanization is carried out by the process of

(a) Anaerobic fermentation

(b) Aerobic Fermentation

(c) Composting

(d) All the above

124. A disease resulting in the deaths of one-third of a dense population of bats in a cave would be a

(a) Density-dependent limiting factor.

(b) Result of exponential growth.

(c) Density-independent limiting factor.

(d) Nutrient-limiting factor.

125. What is the harm from the depletion of Earth's ozone layer

(a) The average temperature of earth's surface will increase gradually

(b) The oxygen content of the atmosphere will decrease

(c) Increased amount of Ultra violet radiation will reach earth's surface

(d) Sea levels will rise as the polar ice caps will gradually melt

126. Rapid EIA is called

(a) A study of three months

(b) A study of four months

(c) A study of six months

(d) A study of one year

127. Liquid ammonia can be used in refrigeration because of its

(a) High basicity

(b) High dipole moment

(c) High heat of vaporization

(d) Non-toxic nature

128. Which of the following pollutants causes ozone holes in the ozone layer?.

(a) CO_2 (b) SO_2

(c) CO (d) CFC

129. The most serious environmental effect posed by hazardous wastes is

(a) Air pollution.

(b) Contamination of groundwater.

(c) Increased use of land for landfills.

(d) Destruction of habitat.

130. The concentration of which gas is highest in our environment?

(a) Nitrogen (b) Oxygen

(c) Carbon (d) Hydrogen

131. The use of anthropoid hazard chemicals like DDT causes reduction in fish population because the chemical mimics the effect of

(a) Luteinizing hormone
(b) Testosterone
(c) Estrogen
(d) All the above

132. Biogas production is the outcome of

(a) Methanogensis
(b) Gasification
(c) a and b
(d) None

133. The increase in the concentration of CO_2 in our environment in last fifty years; since 1960 is about

(a) 20% (b) 10%
(c) 14% (d) 6%

134. The lowest layer of the atmosphere is called

(a) Troposphere (b) Thermosphere
(c) Stratophere (d) Mesosphere

135. A major in-stream use of water is for

(a) Producing hydroelectric power.
(b) Dissolving industrial wastes.
(c) Agricultural irrigation.
(d) Domestic use.

136. The example of Municipal and industrial discharge pipes

(a) Nonpoint sources of pollution.
(b) Violations of the clean water act.
(c) Point sources of pollution.
(d) Irrigation.

137. The presence of high coliform counts in water indicate

(a) Contamination by human wastes.
(b) Phosphorus contamination.
(c) Decreased biological oxygen demand.
(d) Hydrocarbon contamination

138. How the biological oxygen demand gets affected with the increased presence of organic matter in water?

(a) The oxygen demand increases
(b) The oxygen demand decreases
(c) The oxygen demand remains unchanged
(d) None of the above

139. Maximum density of water is at

(a) 4°C (b) 0°C
(c) 40°C (d) 100°C

140. The aggregation of all ecosystems on the earth is referred to as

(a) Atmosphere (b) Ecosphere
(c) Stratosphere (d) Ionosphere

141. The stage in which the biological processes is used to purify water in a wastewater treatment plants is called

(a) Primary sewage treatment
(b) Secondary sewage treatment
(c) Tertiary sewage treatment
(d) Tetra Sewage treatment

142. Groundwater mining in coastal areas can result into

(a) Increase in the salinity of groundwater.
(b) Decrease in the toxicity of groundwater.
(c) Decrease in the salinity of groundwater.
(d) Increase in the water table.

143. Thiobacillus and Beggiatoa play an important role in the

(a) Water cycle on earth
(b) Phosphorus cycle
(c) Sulphur cycle
(d) Breakdown of Sewage

144. The primary soil macronutrients are

(a) Potassium
(b) Phosphorus
(c) Nitrogen
(d) All the above

145. Bulking of sewage sludge is frequently associated with
(a) High C : P ratio
(b) Medium C : P ratio
(c) Low C : P ratio
(d) None

146. The process of mixing two or more gasses is called
(a) Condensation (b) Diffusion
(c) Sublimation (d) Solidification

147. A matter is defined as one which has
(a) No mass but occupies space
(b) Mass but occupies no space
(c) Mass and occupies space
(d) No mass and occupies no space

148. When water freezes into solid, its density
(a) Decreases (b) Increases
(c) Neutral (d) None of these

149. The water when converted into gaseous form is called as
(a) Water gas (b) Fog
(c) Snow (d) water vapour

150. When salt is added into water, its boiling point
(a) Decreases (b) Increases
(c) Neutral (d) None of these

151. The miscibility of two liquids is determined by:
(a) Molecular size: molecules of similar sizes tend to mix together
(b) Density: liquids of similar densities usually mix well together
(c) Polarity: polar and nonpolar liquids are not miscible while polar liquids can usually be mixed with each other
(d) All of the above

152. The particles vibration in gases takes place
(a) About their mean position
(b) About a vertical axis
(c) About a horizontal axis
(d) In any direction

153. The noise index Lgo refers to
(a) Background noise level
(b) Peak noise level
(c) Average noise level
(d) 90%

154. Which of the following property doesn't represent the fluid characteristic
(a) It can flow
(b) It has mass
(c) Can be perceived by our senses
(d) It has a definite shape

155. The optimum range of wind speeds for wind power generation is
(a) 2-4 m/sec (b) 4-12 m/sec
(c) 12-14 m/sec (d) 2-20 m/sec

156. Asbestos use is banned in many countries. it is known to cause
(a) Cardiac disease
(b) Cataract
(c) Lung Cancer
(d) Urinary disease

157. Sublimation is defined as:
(a) The process in which a solid transitions directly into the gas phase
(b) The process in which a gas transitions directly into the solid phase
(c) The process in which a gas transitions into the liquid phase
(d) The process in which a liquid transitions into the gas phase

158. Mobile phone frequencies are in the range of
(a) 1 kHz – 100 MHz
(b) 100 kHz – 100 MHz
(c) 100 kHz – 1000 MHz
(d) 1000 kHz – 10000 MHz

159. Ozone "hole" in the stratosphere was discovered over the continent of
(a) North America (b) Australia
(c) Green land (d) Antarctica

160. Population growth is
(a) Exponential (b) Linear
(c) Logistic (d) All of the above

161. Tertiary consumers are called
(a) Carnivores (b) Herbivores
(c) Omnivores (d) All of the above

162. Which one of the following is the cleanest source of energy?
(a) Hydropower
(b) Fossil fuel
(c) Nuclear power
(d) Wind energy

163. Which waves carry most energy ?
(a) UV light (b) Infra- red light
(c) Microwaves (d) Millimeter waves

164. Who developed the word hot spots?
(a) Odum (b) Norman Myers
(c) Clement (d) Brandis

165. The reason for signing 1987 Montreal Protocol was
(a) To stop global trade of products made from endangered animals
(b) To do away with the use of CFC's, which were found to be responsible for depletion of the ozone layer
(c) To prohibit and ban nuclear testing in tropical deserts and oceans
(d) To start using renewable sources of energy instead of fossil fuels to reduce the anthropogenic greenhouse effect

166. Which of the countries are the largest source of carbon dioxide emissions from fossil fuel burning?
(a) Countries of the former U.S.S.R.
(b) Developing nations
(c) Japan and China
(d) North America

167. Residence time of water molecule in the ocean is
(a) 3.5 years
(b) 35 Years
(c) 3.5 million years
(d) 350000 years

168. Group of individuals of the same species that share common attributes are called
(a) Community (b) Population
(c) Ecotype (d) Society

169. Which is the cleanest source of energy ?
(a) Hydropower (b) Fossil fuel
(c) Nuclear power (d) Wind energy

170. The atmospheric layer of ozone normally absorbs what proportion of UV radiation from sun?
(a) More than 99%
(b) Around a quarter
(c) Less than 10%
(d) About half

171. In order to address the problem of ozone depletion, the United Nations convened a meeting in 1987 in Canada. In that meeting, all the member nations agreed to reduce down the CFC production & use. What was the name of this agreement?
(a) Kyoto Agreement
(b) Montreal Protocol
(c) Chemical Stewardship Program
(d) Vancouver Convention

172. Acid rain is caused by
(a) CO and CO_2 (b) SO_2 and O_2
(c) SO_2 and NO_2 (d) NO_2 and O_2

173. What will be the outcome of eutrophication of surface waters?
(a) Overproduction of biomass
(b) Decrease in nitrogen concentration
(c) Decrease in phosphorus concentration
(d) Decrease in both nitrogen and phosphorus concentration

174. Example of non-biodegradable product
(a) Animal bones
(b) Pine leaves
(c) Wool
(d) Nylon

175. The chemical that is used to ripen mangoes is
(a) Calcium sulphide
(b) Calcium carbide
(c) Calcium carbonate
(d) Calcium chloride

176. Alpha diversity represents
(a) Species richness
(b) Species evenness
(c) Species richness and evenness
(d) Species richness and dominance

177. The combination of primary and secondary treatment reduces the original sewage BOD by
(a) 30-40% (b) 50-60%
(c) 60-70% (d) 80-90%

178. A natural phenomenon that becomes harmful due to pollution is
(a) global warming
(b) ecological balance
(c) greenhouse effect
(d) desertification

179. Earth emits
(a) Long and short radiation
(b) Ultra violet radiation
(c) Short wave radiation
(d) Long wave radiation

180. One of the best solutions to get rid of non-biodegradable wastes is
(a) Burning (b) Dumping
(c) Burying (d) Recycling

181. Animal dung is ____________ waste.
(a) Biodegradable (b) Non-biodegradable
(c) Hazardous (d) Toxic

182. Petroleum consists mainly of
(a) Hydrocarbons
(b) Sulphur
(c) Nitrogen
(d) All the above

183. Largest producer of Natural Gas in the World
(a) Russia (b) America
(c) India (d) Australia

184. Name one non-biodegradable waste which may pollute the earth to dangerous levels of toxicity, if not handled properly.
(a) DDT
(b) CFC
(c) Radioactive substances
(d) PAN

185. In a lake polluted with pesticides, which will contain the maximum amount of pesticides?
(a) Small fish
(b) Microscopic animals
(c) Big fish
(d) Water birds

186. Name the substance whose accumulation in pelicans of Lake Michigan led to the formation of thin shells of their eggs.
(a) CFC (b) PAN
(c) DDT (d) PAC

187. Name the process in which a harmful chemical enters the food chain and gets concentrated at each level in the food chain.
(a) Biomagnification
(b) Biological magnification
(c) Bio illumination
(d) a and b

188. Which one is non-renewable resources
(a) Coal and minerals
(b) Plants and coal
(c) Water and Natural gas
(d) Energy and water

189. Example of renewable resource?
(a) Soil (b) Water
(c) Flora (d) All the above

190. ______ of stratosphere provides protection to our life.
(a) Ozone (b) Nitrogen
(c) Carbon (d) Hydrogen

191. The life supporting gases such as O_2, CO_2 and N_2 are chiefly concentrated in the______.
(a) Troposphere (b) Stratosphere
(c) Mesosphere (d) Ionosphere

192. Which soil is the best for plant growth?
(a) Loamy soil (b) Sandy soil
(c) Sand soil (d) Clay soil

193. Both power and manure are provided by ______.
(a) Biogas plants (b) Compost
(c) a and b (d) None

194. In the atmosphere, the layer above the troposphere is ______.
(a) Stratosphere (b) Mesosphere
(c) Ionosphere (d) Thermosphere

195. ______ is the major raw material for biogas.
(a) Cow dung (b) Plants
(c) Earthworms (d) all the above

196. A biosphere reserve conserves and preserves______
(a) Wild animals
(b) Wild land
(c) Natural vegetation
(d) All the above

197. Atomic energy is obtained by using ores of______
(a) Copper (b) Uranium
(c) Neither (a) nor (b)
(d) Both (a) and (b)

198. The fission material for atomic explosion is
(a) Uranium – 235
(b) Plutonium – 239
(c) Radio Iodine
(d) a and b

199. The area reserved for the welfare of wildlife is called
(a) National park
(b) Botanical garden
(c) Sanctuary (d) Zoological park

200. The death of the last individual of a species is called
(a) Extinction (b) Endangered
(c) Vulnerable (d) Disappearance

201. Which is a fossil fuel?
(a) Natural gas (b) Petrol
(c) Coal (d) All the above

202. Biogas generation is mainly based on the principle of______
(a) Fermentation (b) Composting
(c) Gasification (d) Landfills

203. Red Data Book provides a list of ______
(a) Rare (b) Endangered
(c) Endemic species
(d) All the above

204. The CO_2 content is highest in
(a) Well manured soil
(b) Well aggregated soil
(c) Unmanured soil
(d) Barren soil

205. Who coined the term 'Chronobiology?
(a) Franz Halberg
(b) M.K. Chandrasekhar
(c) D.S. Farner
(d) E. Bunning

206. What is meaning of Chronobiology ?
(a) The study of the effects of time on living systems.
(b) Study of living systems
(c) Study of rocks
(d) Study of time

207. Energy flow in an ecosystem is
(a) Cyclic
(b) Unidirectional
(c) Multidirectional
(d) Sequential

208. The major types of flora and fauna found in aquatic ecosystems are determined primarily by

(a) Light (b) Temperature

(c) Soil (d) all the above

209. The most abundant alkaline component of the atmosphere is

(a) Carbon (b) Nitrogen

(c) Hydrogen (d) All the above

210. Oceans cover about ________ of the earth's surface.

(a) 70% (b) 30%

(c) 10% (d) 50%

211. The ocean zone that covers the continental shelf is the

(a) Estuary (b) Coastal zone

(c) Littoral zone (d) Benthic zone

212. The open sea contains ______% of the surface area of the oceans and ______% of its plant and animal life.

(a) 90 . . . 10 (b) 10 90

(c) 80 .... 20 (d) 20 80

213. The ecosystems with the world's highest net primary productivities per unit area are found in the

(a) Estuary. (b) Coastal zone.

(c) Littoral zone. (d) Benthic zone.

214. Which species is characteristic of tropical coastal wetlands?

(a) Cypress (b) Coconut

(c) Mangrove (d) Palm

215. The main pollutant in vehicle emission is

(a) CO_2 (b) SO_2

(c) NO (d) Ozone

216. Organisms of the intertidal zone

(a) Are constantly being swept away.

(b) Are exposed to constant salinity levels.

(c) Are exposed to both air and underwater conditions.

(d) Must avoid drowning at high tide.

217. Most of the wetlands that are lost are used for

(a) Mining

(b) Urban development

(c) Agriculture

(d) Forestry

218. "Mitigation banking" refers to

(a) The federal policy of zero net loss of wetlands.

(b) Creating or restoring as much wetlands as are destroyed.

(c) The world bank's support of wetlands protection policies.

(d) Debt-for-nature swaps.

219. What is CITES?

(a) A treaty banning barrier island development

(b) A treaty banning trade in endangered species

(c) A treaty banning dumping plastic in the ocean

(d) A treaty banning overfishing

220. What is the largest wetlands restoration project?

(a) The Okeefenokee

(b) The Everglades

(c) Dismal Swamp

(d) Chesapeake Bay

221. Desalination may be accomplished by

(a) Distillation.

(b) Osmosis.

(c) Salt-eating bacteria.

(d) Macrofiltration.

222. The first Environmental law in India was enacted in

(a) 1950 (b) 1960

(c) 1970 (d) 1980

223. Nitrates and phosphates are examples of

(a) Disease-causing agents.

(b) Oxygen-demanding wastes.

(c) Organic chemicals.

(d) Inorganic plant nutrients.

224. One class of pollutants that can cause excessive growth of algae is

(a) Radioactive substances.

(b) Oxygen-demanding wastes.

(c) Inorganic plant nutrients.

(d) Organic chemicals.

225. Thermal pollution is more prevalent near

(a) Hot water springs

(b) Coal based power plants

(c) Temperate zones

(d) Tropical zones

226. Which is a point source of water pollution?

(a) Offshore oil wells

(b) Livestock feedlots

(c) Urban lands

(d) Croplands

227. Which is a nonpoint source of water pollution?

(a) Sewage treatment plant

(b) Electric power plant

(c) Active and inactive coal mines

(d) Logged forest

228. The permissible Arsenic levels in drinking water is

(a) 20 µg/l (b) 30 µg/l

(c) 40 µg/l (d) 50 µg/l

229. Over half of the total mass of pollutants entering streams and lakes comes from

(a) Point pollution from sewage treatment plants.

(b) Point pollution from industrial plants.

(c) Non point pollution from roadside runoff

(d) Nonpoint pollution from agriculture

230. Due to the increasing energy use, the ozone layer over the Arctic is

(a) Decreasing (b) Increasing

(c) a and b (d) None of these

231. In cultural eutrophication, game fish die from

(a) Acid deposition.

(b) Suffocation from lack of oxygen.

(c) Toxic substances in the water.

(d) Salt.

232. When was the use of DDT banned for agricultural purposes in India ?

(a) 1985 (b) 1974

(c) 1971 (d) 1962

233. Sea water is -------------- nature

(a) Acidic (b) Alkaline

(c) Saline (d) All the above

234. In an aquatic ecosystem, the depth to which light penetrates is called

(a) Aphotic (b) Photic

(c) Euphotic (d) Nonphotic

235. High BOD values in a water sample reflects a

(a) High biodegradable organic content

(b) Low organic content

(c) High phosphorous content

(d) High nitrogen content

236. Zeitgeber means

(a) Time giving

(b) Photoperiod information

(c) Temperature information

(d) Rainfall information

237. Environmental Impact Assessment (EIA) seeks to achieve goals pertaining to

(a) Sustainable Development

(b) Prevention of pollution only

(c) Conservation of resources only

(d) Social benefits

238. The criteria indicators of water pollution are

(a) pH (b) Coliform

(c) COD (d) All the above

239. True xerophytes have
(a) Lower rate of transpiration than mesophytes
(b) Transpiration rate equal to mesophytes
(c) Transpiration rate higher than mesophytes
(d) No transpiration

240. Plants growing in extremely cold soils are
(a) Psychrophytes
(b) Halophytes
(c) Lithophytes
(d) Oxylophytes

241. Xerophytes have long roots
(a) Due to light
(b) To draw water from deep water beds
(c) To give mechanical support
(d) None

242. The orchid plants, which are found growing on the trees, are
(a) Parasites (b) Saprophytes
(c) Lithophytes (d) Epiphytes

243. In submerged hydrophytes functional stomata are found
(a) On lower surface of leaf
(b) On both the surface of leaf
(c) No where on the plant / reduced
(d) On upper surface of leaf

244. What is the relationship between water-fern Azolla and nitrogen-fixing cyanobacteria Anabaena called?
(a) Symbiotic (b) Predation
(c) Parasitic (d) Epiphytic

245. When animals feed on other dead animals, which had died naturally or had been killed by another animals, the relationship is termed as
(a) Predation
(b) Competition
(c) Symbiosis
(d) Scavenging

246. The first plants to reappear in a badly burned forest area will most probably be
(a) Liverworts (b) Grasses
(c) Mosses (d) Ferns

247. The transfer of energy from organism to organism in a natural community establishes
(a) Food habitat (b) Food chains
(c) Food nexus (d) all the above

248. Which one is a primary consumer?
(a) Carnivore (b) Scavenger
(c) Saprophyte (d) Herbivore

249. Energy and nutrients enter a community by way of the
(a) Producers (b) Consumers
(c) Detrivores (d) a and b

250. The food chain that starts from plants & goes from smaller to larger animals is
(a) Predator food chain
(b) Saprophytic food chain
(c) Detritus food chain
(d) Parasitic food chain

251. Driving force of ecosystem is
(a) Solar energy (b) Water
(c) Wind (d) All the above

252. The leguminous plants are important in agriculture because
(a) They need very little water to grow
(b) They are disease resistant
(c) They help in nitrogen economy of the plants
(d) Six crops of leguminous plants are harvested in a year

253. National Biodiversity Strategy and action plan (NBSAP) was established in
(a) 1986 (b) 1999
(c) 2002 (d) 2012

254. MIC & Chernobyl tragedies occurred at
(a) Bhopal 1984, Ukraine 1986
(b) Bhopal 1986, Ukraine 1984
(c) Bhopal 1988, Ukraine 1986
(d) Bhopal 1984, Ukraine 1988

255. Ecofriendly detergents do not contain
(a) PO_4 and bleaches (b) Alcohols
(c) Water softeners (d) Fluoresers

256. Nitrogen oxide can be absorbed by plants like
(a) Pinus (b) Quercus
(c) Junipers (d) All the above

257. Which satellite recorded the presence of an ozone hole?
(a) TIROS- N (b) GOES
(c) NIMBUS -7 (d) LANDSAT - 3

258. Natural sink of stratospheric ozone layer is
(a) SO_2
(b) HCF
(c) Sulphur flux of oceans
(d) Freons

259. Which plant brought from America for controlling water pollution has become a problem in India
(a) Opuntia (b) Lantana
(c) Kalmath weed (d) Eichhornia

260. The intensity of noise pollution is measured in
(a) Decibels (b) Mach
(c) Hertz (d) Noy

261. What is the intensity of sound in normal conversation
(a) 10 X 20 dB (b) 70 X 90 dB
(c) 0 X 60 dB (d) 30 X 60 dB

262. Range of human hearing is
(a) 20 X 1000 Hz
(b) 20 X 1500 Hz
(c) 20 X 10000 Hz
(d) 20 X 15000 Hz

263. The nutrient enrichment is known as
(a) Symbiosis
(b) Eutrophication
(c) Neutrification
(d) a and b

264. Eutrophication (mineral pollution) is maximum in
(a) Bottom layer of deep lakes
(b) Middle layer of deep lakes
(c) Upper layer of deep lakes
(d) All layers

265. A sewage is labeled as strong if its biological oxygen demand is around
(a) 300 mg/ litre
(b) 500 mg/ litre
(c) 600 mg/ litre
(d) 700 mg/ litre

266. Sound limit in industrial area is
(a) 30 dB (b) 40 dB
(c) 50 dB (d) 75 dB

267. Ambient noise standard in India for silence zone in day time is
(a) 30 dB (b) 40 dB
(c) 50 dB (d) 75 dB

268. Ambient noise standard in India for residential area at night time is
(a) 30 dB (b) 45 dB
(c) 50 dB (d) 75 dB

269. The standard permissible noise limits for refrigerator is
(a) 60 dB (b) 46 dB
(c) 68 dB (d) 75 dB

270. BOD is
(a) Amount of O_2 altered by organisms in water
(b) Amount of O_2 present in water at 37°C
(c) Amount of O_2 altered by microbes for decomposition of organic matter
(d) All the above

271. Safe limit of water fluoride level is
(a) 0.1 X 0.4 mg / litre
(b) 0.2 X 0.8 mg / litre
(c) 0.5 X 0.8 mg / litre
(d) 0.5 X 8.0 mg / litre

272. Mass killing of fishes occur in
(a) Eutrophic lakes
(b) Benthos lakes
(c) a and b
(d) None

273. Species ecology is
(a) Autecology
(b) Synecology
(c) Community ecology
(d) Palaeo ecology

274. Biosphere consists of
(a) Lithosphere (b) Atmosphere
(c) Hydrosphere (d) All the above

275. Biogeochemical cycles are also known as
(a) Sedimentary cycle
(b) Gaseous cycle
(c) Material cycling
(d)Cycle of water

276. Biosphere consists of interacting functional units called
(a) Ecosystems (b) Populations
(c) Communities (d) Species

277. Maximum oxygen comes from
(a) Phytoplanktons
(b) Grassland
(c) Forests
(d) Herbs and shrubs

278. The reservoir pool of carbon is
(a) Air and Water (b) Fossil fuels
(c) Rocks (d) All of the above

279. How many green house gases exist ?
(a) Two (b) Three
(c) Four (d) Six

280. Annual world precipitation is
(a) $7 X 10^{10}$ g (b) $7 X 10^{20}$ g
(c) $4.46 X 10^{20}$ g (d) $4.46 X 10^{20}$

281. Degradation of proteins plays a part in
(a) Calvin cycle (b) Water cycle
(c) Sulphur cycle (d) Nitrogen cycle

282. Biome is a
(a) Community of plants
(b) Community of animals
(c) Small sized ecosystems
(d) Large sized ecosystem

283. Ecological niche of an organism denotes
(a) Habitat of the organisms
(b) On whom the organisms feeds
(c) The status of an organism within a particular community
(d) All the above

284. The transition zone where two different types of communities meet is called
(a) Ecotone (b) Ecocline
(c) Ecotype (d) Ecosystem

285. Total organic matter present in an ecosystem is called
(a) Biomass (b) Biome
(c) Litter (d) Food

286. The most important organisms for an ecosystem are
(a) Green Plants (b) Herbivores
(c) Carnivores (d) Protozoa

287. A food chain is
(a) Group of organisms which eat the same type of food
(b) Animals eating the plants
(c) Series of plants/animals which are interrelated in the form of organism being eaten as food by the other
(d) None of these

288. A Food chain consists of
(a) Producers
(b) Consumers
(c) Decomposers
(d) a and b

289. A Food chain always starts with
(a) Photosynthesis
(b) Respiration
(c) Nitrogen fixation
(d) Decay

290. Relationship between different organisms can best studied by
(a) Eltonian pyramids
(b) Charles pyramids
(c) Hierarchical pyramids
(d) All the above

291. Controlling factor in ecosystem is
(a) Predation (b) Light
(c) Precipitation (d) Nutrients

292. Micro consumers are called
(a) Omnivores (b) Saprophyte
(c) a and b (d) None

293. In pond ecosystem, the shape of pyramid of number is
(a) Upright (b) Inverted
(c) Spindle like (d) Variable

294. In parasitic food chain, the pyramid of number is
(a) Upright (b) Inverted
(c) Spindle like (d) variable

295. In grassland ecosystem, the pyramid of biomass is
(a) Upright
(b) Inverted
(c) Spindle like
(d) Variable

296. Pyramid of energy in forest ecosystem is
(a) Always inverted
(b) Always upright
(c) Spindle shaped
(d) a and b

297. In pyramid of food, the producers occupy
(a) The base (b) Top
(c) Middle (d) All the above

298. Rate of storage of organic matter not used by heterotrophs is termed as
(a) Net Primary productivity
(b) Gross primary productivity
(c) Secondary productivity
(d) None of these

299. Graphic representation of biomass relationship between the producers and consumers in an ecosystem is called
(a) Ecological pyramid
(b) Ecological hill
(c) Ecological Mountain
(d) All the above

300. Organisms living at the bottom of a lake are
(a) Benthos (b) Biocoen
(c) Biocide (d) All the above

301. A recently discovered ecosystem is
(a) Tundra (b) Iceberg
(c) Vent (d) Crater

302. Factor related to structure of earth's surface is
(a) Topographic (b) Edaphic
(c) Physiographic (d) a and c

303. In upright pyramid of biomass, herbivores occupy the position
(a) One (b) Two
(c) Three (d) Four

304. Which is amphibious?
(a) Monogonum (b) Digonum
(c) Trigonum (d) Polygonum

305. Which forests have maximum diversity?
(a) Tropical forests
(b) Temperate forests
(c) Alpine forest
(d) Thorn forest

306. A treeless biome is
(a) Tundra (b) Grassland
(c) Desert (d) All the above

307. Lentic ecosystem occurs in
(a) Standing water
(b) Running water
(c) River water
(d) All the above

308. Ecosystem of a pond is
(a) Lentic (b) Lotic
(c) a and b (d) None of these

309. Alpine tundra is found in
(a) Siberia (b) Greenland
(c) Himalayas (d) a and b

310. Type of plants having adaptations to check transpiration is
(a) Xerophytes (b) Lithophytes
(c) Halophytes (d) Epiphytes

311. Vivipariy and Pneumatophores are features of
(a) Hydrophytes (b) Mesophytes
(c) Xerophytes (d) Halophytes

312. Which part of the world has high density of organisms?
(a) Savannahs (b) Grassland
(c) Deciduous forest
(d) Tropical rain forest

313. The most favoured habitat of a plant is called
(a) Ecological habitat
(b) Microhabitat
(c) Ecotype (d) Ecoline

314. Alpine plants show
(a) Halophytism
(b) Mesophytism
(c) Luxurian growth
(d) Xerophytism

315. Temperate forests occur in India in
(a) Indo-Gangetic plants
(b) Himalayas
(c) Eastern India
(d) Southern Peninsula

316. The forest colours in autumn is
(a) Temperate evergreen forest
(b) Temperate deciduous forest
(c) Tropical evergreen forest
(d) Tropical deciduous forest

317. Ephemerals are a type of xerophytes
(a) Drought escaping
(b) Drought Resistant
(c) Drought enduring
(d) None of these

318. Succulents occur in
(a) Deserts
(b) Tundra
(c) Temperate deciduous forest
(d) Tropical rain forest

319. The proximate causes of behavior are interactions with the environment, but behavior is ultimately shaped by
(a) Evolution
(b) Mutation
(c) Natural selection
(d) Genetic drift

320. Which of these ecosystems has the highest primary productivity per square meter?
(a) Thorn forest
(b) Temperate forest
(c) Deciduous forest
(d) Tropical rain forest

321. The total biomass of photosynthetic autotrophs present in an ecosystem is known as the
(a) Standing state
(b) Standing crop.
(c) Standing vegetation
(d) All the above

322. Starting on bare rock, what is the usual ecological succession of organisms?
(a) Lichens --> grasses --> shrubs --> trees
(b) Trees --> grasses --> shrubs --> lichens
(c) Grasses --> shrubs --> trees --> lichens
(d) None

323. The stable stage that is established in an area as a result of the process of ecological succession is known as the
(a) Population growth
(b) Ecological succession
(c) Climate change
(d) Climax community

324. The place an organism lives is called its ________________

(a) Habit (b) Habitat

(c) Niche (d) Community

325. This is a symbiotic relationship between organisms where both are helped (+,+) This relationship is called __________.

(a) Mutualism (b) Parasitism

(c) Commensalism (d) Parasite

326. This is a symbiotic relationship between organisms where one organism is helped, while the other is harmed. (+,-) This relationship is called __________.

(a) Mutualism (b) Parasitism

(c) Commensalism (d) Parasite

327. This is a symbiotic relationship between organisms where one organism is helped, while the other it is associated with is basically unaffected. (+, 0) This relationship is called

(a) Mutualism (b) Parasitism

(c) Commensalism (d) Parasite

328. The feeding role of an organism is called a(n) _____

(a) Biche (b) Niche

(c) Creche (d) None of these

329. As a population reaches its carrying capacity, there is an increase in competition for

(a) Food (b) Shelter

(c) Mates (d) All the above

330. An ecosystem consists of

(a) A community of organisms

(b) Energy

(c) The soil, water and weather

(d) All the above

331. Organisms that manufacture organic nutrients for an ecosystem are called

(a) Primary producers

(b) Primary consumers

(c) Secondary producers

(d) Secondary consumers

332. In a food web, which type of organism receives energy from every other type?

(a) Herbivore (b) Producers

(c) Carnivore (d) None

333. Animals that feed on plants are at least in the

(a) Second trophic level

(b) Third trophic level

(c) Fourth trophic level

(d) Fifth trophic level

334. Because energy diminishes at each successive trophic level, few ecosystems can contain more than

(a) Two trophic levels

(b) Three trophic levels

(c) Four trophic levels

(d) Five trophic levels

335. Precipitation and evaporation are important components of the

(a) Nitrogen cycle (b) Carbon cycle

(c) Potassium cycle (d) Water cycle

336. The paths of water, carbon, nitrogen, and phosphorus pass from the non-living environment to living organisms and back to the non-living environment in closed circles called

(a) Biogeochemical cycles

(b) Chemical cycle

(c) Atmosphere cycle

(d) None

337. The process by which species evolve in response to other living members of their ecosystem is called

(a) Competition (b) Coevolution

(c) Coincidence (d) Predation

338. The diversity that enables life on earth to adapt and survive environmental changes is called

(a) Functional diversity

(b) Species diversity

(c) Ecosystem diversity

(d) Genetic diversity

339. The variety of processes, including matter cycling and energy flow within ecosystems, that result from species interacting with one another in food webs is called

(a) Functional diversity
(b) Species diversity
(c) Ecosystem diversity
(d) Genetic diversity

340. A change in the genetic characteristics of a population from one generation to another is called

(a) Emigration (b) Mutation
(c) Genetic drift (d) Evolution

341. Some individuals in a population have genetic traits that enhance their ability to survive and produce offspring and process is called

(a) Natural selection
(b) Emigration
(c) Mutation
(d) Genetic drift

342. Biodiversity is believed to be the result of the interaction between

(a) Geographic isolation and reproductive isolation
(b) Speciation and extinction
(c) Mass extinction and background extinction
(d) Speciation and genetic engineering

343. The number of different species an ecosystem contains is its

(a) Speciation (b) Species evenness
(c) Species niche (d) Species richness

344. A species in an ecosystem that plays a central role in the health of that ecosystem and whose removal may cause the collapse of the ecosystem, is called a(n)

(a) Indicator species
(b) Keystone species
(c) Specialist species
(d) Foundation species

345. A species in an ecosystem that shapes communities by creating and enhancing habitats in ways that benefit other species is called

(a) Indicator species
(b) Keystone species
(c) Specialist species
(d) Foundation species

346. In which eco systems, producers are of large size

(a) Grassland ecosystem
(b) Tropical deciduous
(c) Temperate
(d) Desert

347. The test has self purification capacity of water body

(a) BOD test (b) DOD test
(c) COD test (d) All the above

348. Sources of soil pollutants are

(a) Herbicides (b) Pesticides
(c) Weedicides (d) All the above

349. Most Poisonous pollutant in water is

(a) Arsenic (b) Nickel
(c) Cobalt (d) All the above

350. Pollutant free alternative to petrol for automobiles

(a) Propane (b) Methane
(c) Ethane (d) All the above

351. Upright pyramid of biomass in forest ecosystem is

(a) Producer-herbivores-carnivores
(b) Carnivores -herbivores- Producer
(c) a and b
(d) None

352. Thickness of ozone layer is measured in

(a) Pascal (b) Dabson units
(c) cm /sec (d) a and c

353. The gas responsible for Bhopal tragedy

(a) Methyl isocyanide
(b) Methyl isocyanate
(c) BHC
(d) All the above

354. The gas does not pollute air
(a) Nitrogen Dioxide (b) SO_2
(c) CO (d) NO

355. Term ecology was proposed by
(a) Odum (b) Reiter
(c) Grinnel (d) Lack

356. Who coined the term symbiosis
(a) De Bary (b) Clements
(c) Tansley (d) McDougall

357. Plants which are attached to the stones and rocks are
(a) Halophytes (b) Psychrophytes
(c) Lithophytes (d) Oxylophytes

358. The residence time of N_20 in atmosphere is approximately
(a) 15 years (b) 150 years
(c) 15 weeks (d) 15 days

359. Plants growing in soils having sand and gravel are
(a) Psammophytes (b) Halophytes
(c) Psychrophytes (d) Oxylophytes

360. Plant growing well in full sunlight are called
(a) Halophytes (b) Psychrophytes
(c) Lithophytes (d) Oxylophytes

361. Plants growing in saline soil water are called
(a) Halophytes (b) Psychrophytes
(c) Lithophytes (d) Oxylophytes

362. The Ministry of Environment was set up in
(a) 1952 (b) 1980
(c) 1986 (d) 1988

363. The National Institute of Oceanography is presently situated at
(a) Kerala (b) Calicut
(c) Cochin (d) Goa

364. The process used for removal of water hardness is
(a) Zeolite process (b) Heber's process
(c) Ostwald process (d) None of these

365. The first biosphere reserve in India was
(a) Nilgiri (b) Gulf of Manna
(c) Sundarbans (d) Nanda Devi

366. Permanent hardness of water is caused by
(a) Carbonates and Chlorides
(b) Bicarbonates and Chlorides
(c) Carbonates and Sulphates
(d) Chlorides & Sulphates

367. In which year Biodiversity act was proposed
(a) 2000 (b) 2002
(c) 2012 (d) 1998

368. Kyoto protocol came into force on
(a) 16 February 2004
(b) 14 February 2006
(c) 14 February 2004
(d) 16 February 2006

369. The compound used for artificial rain making /cloud seeding
(a) Agl (b) Ni
(c) Fe (d) Hg

370. Number of species in single community
(a) Alpha diversity
(b) Beta diversity
(c) Gamma diversity
(d) All the above

371. The degree to which species composition changes along an environmental gradient
(a) Alpha diversity
(b) Beta diversity
(c) Gamma diversity
(d) All the above

372. Which diversity applied to larger geographical scales?.
(a) Alpha diversity
(b) Beta diversity
(c) Gamma diversity
(d) All the above

373. What is frequency?
(a) Total number of quaqrats in which species has occurred / Total number of quadrats studied
(b) Total number of individual of the species / Total number of quadrats studied
(c) Total number of individual of the species / Total number of quadrats in which the species has occurred
(d) None

374. What is density ?
(a) Total number of quaqrats in which species has occurred / Total number of quadrats studied
(b) Total number of individual of the species / Total number of quadrats studied
(c) Total number of individual of the species / Total number of quadrats in which the species has occurred
(d) None

375. What is abundance ?
(a) Total number of quaqrats in which species has occurred / Total number of quadrats studied
(b) Total number of individual of the species / Total number of quadrats studied
(c) Total number of individual of the species / Total number of quadrats in which the species has occurred
(d) None

376. Study of energy transformation in living systems called
(a) Bioenergetics
(b) Biotic potential
(c) Bio concentration
(d) Energy flow

377. In a symbiosis which one organisms is inhibited by other but the latter remains unaffected is called
(a) Commensalism (b) Amensalism
(c) Neutralism (d) Mutualism

378. The science of relationship of life to weather is called
(a) Biometeorology
(b) Biotic potential
(c) Biogeosphere
(d) Biological amplification

379. A plant which grows on ice or snow is called
(a) Cryophyte (b) Hydrophyte
(c) Halophyte (d) Hygrophyte

380. Science of communication and control mechanisms in system is called
(a) Cybernetics (b) Sciophyte
(c) Toxicology (d) Genecology

381. The relation between number of species and number of individuals is called
(a) Diversity (b) Frequency
(c) Diversity index (d) Abundance

382. The process of successful establishment of a species in a new area
(a) Ecesis (b) Ecad
(c) a and b (d) None

383. Maintenance of equilibrium between living components of an ecosystem, so that it remains stable system is called
(a) Ecological balance
(b) Ecological succession
(c) Ecological equivalents
(d) Ecotone

384. One or more ecotype in a single coenospecies is called
(a) Ecospecies
(b) Ecospecies
(c) Endemic species
(d) Indicator species

385. Science of improving the human race by improving environment
(a) Euthenics
(b) Ecology
(c) Ecospecies
(d) Eutrophic

386. Science which deals with the genetic of population in relation to the environment is called
(a) Genecology (b) Toxicology
(c) Autecology (d) Synecology

387. Study of changes in population densities in an area
(a) Population dynamics
(b) Population
(c) a and b
(d) None of these

388. Species diversity is lowest in this ecosystems
(a) Tundra
(b) Temperate grasslands
(c) Desert
(d) Tropical forest

389. Best soil for growth of plants in summer
(a) Clay (b) Sand
(c) Red (d) Loamy

390. Ecological succession on sand is
(a) Halosere (b) Xerosere
(c) Hydrosere (d) Psammomere

391. Succession showing changes in communities at a place is called
(a) Biotic succession
(b) Climatic succession
(c) Geographical succession
(d) Edaphic succession

392. Secondary succession is not completed if the barren area is invaded by
(a) Sphagnum (b) Lantana
(c) Eichhornia (d) All of the above

393. Xerosere is
(a) Xerophyte
(b) Halophyte
(c) Ecological succession on rock
(d) None of these

394. Maximum diversity is found in
(a) Pioneer community
(b) Seral community
(c) Climax community
(d) Moss stage

395. Minimum diversity is found in
(a) Pioneer community
(b) Seral community
(c) Climax community
(d) Moss stage

396. Long term stability of a biotic community is governed by its
(a) More diversity
(b) Population
(c) Pioneer community
(d) Environment

397. Which seral stage in a succession takes maximum time?
(a) Pioneer stage (b) Climax stage
(c) Seral stage (d) None of these

398. Regeneration of community is due to
(a) Primary succession
(b) Secondary succession
(c) Climax community
(d) a and b

399. Last succession in a sere is called
(a) Autogenic succession
(b) Allogenic succession
(c) Biotic climax
(d) Climatic climax

400. Most wide spread succession is
(a) Autotrophic succession
(b) Autogenic succession
(c) Allogenic succession
(d) None of these

401. Largest vegetation group is known as
(a) Association (b) Consociation
(c) Formation (d) Ecosystems

ANSWERSHEET
15. Forest Ecology and Environment

1. (a)	2. (a)	3. (b)	4. (a)	5. (a)	6. (a)	7. (c)	8. (a)	9. (b)	10. (a)
11. (b)	12. (c)	13. (c)	14. (a)	15. (a)	16. (a)	17. (d)	18. (c)	19. (c)	20. (b)
21. (a)	22. (a)	23. (a)	24. (b)	25. (a)	26. (b)	27. (d)	28. (b)	29. (d)	30. (a)
31. (a)	32. (d)	33. (c)	34. (c)	35. (d)	36. (c)	37. (d)	38. (c)	39. (a)	40. (d)
41. (c)	42. (a)	43. (a)	44. (c)	45. (c)	46. (a)	47. (c)	48. (c)	49. (a)	50. (c)
51. (d)	52. (a)	53. (b)	54. (b)	55. (a)	56. (a)	57. (d)	58. (a)	59. (b)	60. (d)
61. (a)	62. (d)	63. (c)	64. (b)	65. (a)	66. (a)	67. (c)	68. (a)	69. (b)	70. (d)
71. (c)	72. (c)	73. (b)	74. (b)	75. (a)	76. (a)	77. (a)	78. (a)	79. (b)	80. (a)
81. (a)	82. (b)	83. (c)	84. (d)	85. (b)	86. (a)	87. (d)	88. (b)	89. (a)	90. (a)
91. (a)	92. (a)	93. (b)	94. (a)	95. (a)	96. (d)	97. (a)	98. (a)	99. (b)	100. (b)
101. (a)	102. (c)	103. (b)	104. (d)	105. (c)	106. (d)	107. (a)	108. (c)	109. (b)	110. (b)
111. (c)	112. (b)	113. (b)	114. (c)	115. (c)	116. (c)	117. (b)	118. (c)	119. (d)	120. (a)
121. (a)	122. (d)	123. (a)	124. (a)	125. (c)	126. (a)	127. (b)	128. (d)	129. (b)	130. (a)
131. (a)	132. (c)	133. (c)	134. (a)	135. (c)	136. (a)	137. (a)	138. (a)	139. (a)	140. (b)
141. (b)	142. (a)	143. (d)	144. (d)	145. (a)	146. (b)	147. (c)	148. (a)	149. (d)	150. (b)
151. (c)	152. (d)	153. (b)	154. (d)	155. (b)	156. (c)	157. (a)	158. (b)	159. (d)	160. (a)
161. (a)	162. (d)	163. (a)	164. (b)	165. (b)	166. (d)	167. (b)	168. (c)	169. (d)	170. (a)
171. (b)	172. (c)	173. (a)	174. (d)	175. (b)	176. (a)	177. (c)	178. (c)	179. (d)	180. (d)
181. (a)	182. (a)	183. (a)	184. (c)	185. (d)	186. (c)	187. (d)	188. (a)	189. (d)	190. (a)
191. (a)	192. (a)	193. (a)	194. (a)	195. (a)	196. (d)	197. (b)	198. (d)	199. (a)	200. (a)
201. (d)	202. (a)	203. (d)	204. (a)	205. (a)	206. (a)	207. (b)	208. (a)	209. (b)	210. (a)
211. (b)	212. (a)	213. (b)	214. (c)	215. (a)	216. (c)	217. (c)	218. (b)	219. (b)	220. (b)
221. (a)	222. (a)	223. (d)	224. (c)	225. (b)	226. (a)	227. (d)	228. (a)	229. (d)	230. (a)
231. (b)	232 (a)	233. (c)	234. (b)	235. (a)	236. (b)	237. (a)	238. (d)	239. (a)	240. (a)
241. (b)	242. (d)	243. (c)	244. (a)	245. (d)	246. (d)	247. (b)	248. (d)	249. (a)	250. (a)
251. (a)	252. (c)	253. (b)	254. (a)	255. (a)	256. (d)	257. (c)	258. (d)	259. (d)	260. (a)
261. (d)	262. (d)	263. (b)	264. (a)	265. (a)	266. (d)	267. (c)	268. (b)	269. (b)	270. (c)
271. (c)	272. (a)	273. (a)	274. (d)	275. (c)	276. (a)	277. (a)	278. (d)	279. (d)	280. (d)
281. (d)	282. (d)	283. (c)	284. (a)	285. (a)	286. (a)	287. (c)	288. (d)	289. (a)	290. (a)
291. (a)	292. (b)	293. (a)	294. (b)	295. (a)	296. (b)	297. (a)	298. (a)	299. (a)	300. (a)

301. **(d)** 302. **(a)** 303. **(b)** 304. **(d)** 305. **(a)** 306. **(a)** 307. **(a)** 308. **(a)** 309. **(c)** 310. **(a)**
311. **(d)** 312. **(a)** 313. **(b)** 314. **(d)** 315. **(b)** 316. **(b)** 317. **(a)** 318. **(a)** 319. **(a)** 320. **(d)**
321. **(b)** 322. **(a)** 323. **(d)** 324. **(b)** 325. **(a)** 326. **(b)** 327. **(c)** 328. **(b)** 329. **(d)** 330. **(d)**
331. **(a)** 332. **(c)** 333. **(a)** 334. **(d)** 335. **(d)** 336. **(a)** 337. **(b)** 338. **(d)** 339. **(a)** 340. **(d)**
341. **(a)** 342. **(b)** 343. **(d)** 344. **(b)** 345. **(d)** 346. **(a)** 347. **(a)** 348. **(d)** 349. **(a)** 350. **(a)**
351. **(a)** 352. **(b)** 353. **(b)** 354. **(a)** 355. **(b)** 356. **(a)** 357. **(c)** 358. **(b)** 359. **(a)** 360. **(a)**
361. **(a)** 362. **(b)** 363. **(d)** 364. **(a)** 365. **(a)** 366. **(d)** 367. **(b)** 368. **(a)** 369. **(a)** 370. **(a)**
371 **(b)** 372. **(c)** 373. **(a)** 374. **(b)** 375. **(c)** 376. **(a)** 377. **(b)** 378. **(a)** 379. **(a)** 380. **(a)**
381. **(c)** 382. **(a)** 383. **(a)** 384. **(b)** 385. **(a)** 386. **(a)** 387. **(a)** 388. **(a)** 389. **(a)** 390. **(d)**
391. **(a)** 392. **(d)** 393. **(c)** 394 **(c)** 395. **(a)** 396. **(a)** 397. **(a)** 398. **(c)** 399. **(d)** 400. **(c)**
401. **(b)**

16

Wildlife

KEY POINTS

- Wings - Carry the large fans of feather that are used in flight, but folded away when at rest
- Bill – It is adapted for all feeding (line hand)
- Skeleton - It has a thinner structure and is lightly made to aid flight.
- Reflexes - Simplest – Fixed responses to stimulus
- Physic reflexes – small movements
- Tonie reflexes : long lasting adjustment
- Kineses - An animal respond to a change or alteration in the stimulus intensity by changing its land of activity
- Taxes - Movement which is oriented in relation to the direction of a particular stimulus
- Instinct - Many mysterious actions of animals
- Biological rhythms - All living organisms have the capacity to measure the passage of time
- Cireadian rhythms - Cyetical variation in the intensity of a physiological process on a pattern of behavior with a periodicity of about 24 hours – behavioural activities – in a day
- A total count or census involves estimation of the population in the entire area of the reserve or protected area
- Road side index for population trends - This is simple technique for monitoring the population trends of animals such as deers, antelopes, goats and sheep
- Waterhole survey - This method is based on the assumption that in dry areas all animals come to drink water at waterholes which are usually limited in number
- Calls for census - Calls of birds and sounds emitted by other animals are used for estimation of their population
- Pug mark count method - This is a widely accepted method for the census of large carnivores in India
- Single team block census - This technique is based on the sample count of animals using actual sightings
- Habitat is the place where an organism or community lives; it is a spatial entity on the surface of the earth

- A wildlife habitat has four basic components viz., food, cover, water and space
- Qualifiers: Any event or activity which affects the quality of a habitat, by modifying it, is known as a qualifier
- Territory, home-range and cruising radius: The environment of an animal comprises of biotic and abiotic components; the biotic components are vegetal and faunal, with which an animal interacts
- Cruising radius - Some workers consider the extent of an animals movement as the outer boundary of a circle; half of this is treated as diameter to express the cruising radius
- An edge is the place of contact between plant communities or successional stages or vegetative conditions
- Snag means a standing dead tree, which is devoid of leaves and branches
- Snags can be classified as hard or soft; hard snags are usually marketable and they are free from decay, whereas soft snags show stages of deterioration
- Juxtaposition is the measure of the proximity of different habitat units
- Habitat analysis means a report of the characteristics of a given habitat
- National parks have a sound legal status; they are created exclusively for the conservation of wildlife in their natural environment
- Sanctuaries also have sound legal status with a strong, but not exclusive, wildlife oriented management
- The status of a sanctuary can always be elevated to that of a national park
- Sanctuaries and national parks usually suffer from biotic interference in the form of village settlements, collection of forest produce by local inhabitants, cattle grazing, lopping and human movement on foot, vehicles or boats
- Project tiger launched in 1973-74, is one of our most successful conservation projects
- The Government of Gujarat, launched the Gir lion project in 1972
- Crocodile breeding project - This project was undertaken from the 1st of April 1975 for all the three crocodilian species found in our country, based on the advice of an FAO expert.
- The Wildlife Protection Act of 1972 has listed all the crocodilian species under schedule I, thereby affording total protection to these animals
- The Wildlife Protection Act of 1972
- The zoo is a very complex organization which has to fulfill many requirements, some of which may be conflicting also
- Barriers help to keep the animals in and the public out. An animal may feel secure in a well designed cage, especially when it finds the human beings at a close distance
- Birds (class Aves) are winged, bipedal, endothermic (warm-blooded), egg-laying, vertebrate animals
- A list of the taxonomic orders in the subclass Neornithes, or modern birds
- Struthioniformes — Eg. Ostriches , emus, kiwis, and allies
- Tinamiformes — Eg. tinamous
- Birds live and breed in most terrestrial habitats and on all seven continents, reaching their southern extreme in the Snow Petrel's breeding colonies up to 440 kilometres (270 mi) inland in Antarctica

- Feathers are a feature characteristic of birds
- They facilitate flight, provide insulation that aids in thermo regulation and are used in display, camouflage, and signaling
- Plumage is regularly moulted; the standard plumage of a bird that has moulted after breeding is known as the "non-breeding" plumage
- The scales of birds are composed of the same keratin as beaks, claws, and spurs
- They are found mainly on the toes and metatarsus, but may be found further up on the ankle in some birds
- Most bird scales do not overlap significantly, except in the cases of kingfishers and woodpeckers
- The scales of birds are thought to be homologous to those of reptiles and mammals
- Many bird species migrate to take advantage of global differences of seasonal temperatures, therefore optimizing availability of food sources and breeding habitat
- Elephant preservation Act 1898
- Rhinoceros Preservation Act 1909
- Bombay wild animals and wild birds preservation act 1951
- Schedule I : contains the list of mammals, birds, amphibians and reptiles
- Schedule II : includes the animals classed as special game
- Schedule III : consists of the big game animals
- Schedule IV : consists of small game animals
- Schedule V : consists of vermin
- Type A – large defended area, various requirements as food, mating, rearing young from this area and some small mammals, arboreal lizards and insectivorous birds.
- Type B- typical of some birds, used for almost all breeding activities but food gathering somewhat restricted.
- Type C- sea gulls and colonial nesting birds, smallest territory with defended area
- Type D- lakes, only for mating and pairing. In small gallinaceous birds
- Type E- very small, consists of roosting positions maintained by certain birds and bats

MULITPLE CHOICE QUESTION'S

1. The Latin name for black buck is
(a) *Antilope cervicarpa*
(b) *Moschiala memima*
(c) *Rusa unicolor*
(d) *Axis axis*

2. The floating national park is situated in the state of
(a) Kerala
(b) Assam
(c) Manipur
(d) Tamil Nadu

3. Wildlife protection act was enacted during
(a) 1975 (b) 1980
(c) 1972 (d) 1995

4. The project tiger was launched in the year
(a) 1972 (b) 1995
(c) 1973 (d) 1980

5. Corbet National Park is located in
(a) Madhya Pradesh
(b) Uttranchal
(c) Uttar Pradesh
(d) Himachal Pradesh

6. The Latin name for lion tailed macaque is
(a) *Macaca silenus*
(b) *Macaca aratoides*
(c) *Macaca nemtoetrina*
(d) *Macaca assamensis*

7. Wildlife protection act of 1972 was amended during the year
(a) 1988 (b) 1990
(c) 1991 (d) 1980

8. The Keoladeo Naitonal Park is situated in the state of
(a) Rajasthan (b) Orissa
(c) Assam (d) Uttar Pradesh

9. Ponking call is attributable to
(a) Sambar (b) Chittal
(c) Wolf (d) Tiger

10. In Wildlife Protection Act 1972, Nilgiri Tahr comes under
(a) Schedule I Part II
(b) Schedule I Part I
(c) Schedule II Part II
(d) Schedule III

11. The census method followed for Asiatic lion is
(a) Pugmark (b) Call
(c) Waterhole (d) Transect

12. Wild Asiatic Water Buffalo is distributed in the state of
(a) Assam (b) Himachal Pradesh
(c) Karnataka (d) Kerala

13. Kanha National Park is established in the year
(a) 1948 (b) 1950
(c) 1955 (d) 1961

14. Kanha National Park is located in the state
(a) Karnataka (b) Uttar Pradesh
(c) Madhya Pradesh
(d) Assam

15. Which of the following is the largest bird in India?
(a) Hornbill
(b) Great Indian Bustard
(c) Sarus crane (d) Monal Pheasant

16. Andaman teal (*Anas gibberifrons algularis*) is a
(a) Mammal (b) Reptile
(c) Amphibian (d) Bird

17. India has seen the extinction of one of the finest cats namely
(a) *Panthera Pardus*
(b) *Axis tigris*
(c) *Actnonyx jubatus venaticus*
(d) *Neofelis nebulosa*

18. *Chelonia mydas* is commonly called as
(a) Olive Ridley turtle
(b) Green turtle
(c) Travancore tortoise
(d) Starred tortoise

19. *Ophiophagus hannah* is commonly called as
(a) Indian cobra (b) King cobra
(c) Rat snake (d) Python

20. Common langur *(Presbytis entellus)* is the animal coming under
(a) Schedule I (b) Schedule II
(c) Schedule III (d) Schedule IV

21. Which one of the following animals is not a vermin?
(a) Common crow (b) Fruit bat
(c) Field rat (d) Wild cat

22. Who wrote the autobiographical book, My Life : My Trees ?
(a) Joy Adamson
(b) Richard St. Barbe Baker
(c) Konard Lorenz
(d) K.M. Tewari

23. In which biogeographical realm does most part of the Asian continent fall ?
(a) Nearctic realm
(b) Neotropical realm
(c) Palaearctic realm
(d) Oriental realm

24. Which tree yields the heaviest wood in the World ?
(a) Balsa *(Ochroma Pyramidale)*
(b) Black Iron Wood *(Olea laurifolia)*
(c) Durian *(Durio Zibethinus)*
(d) Mango *(Mangifera indica)*

25. Which is the only animal in the world to have four horns ?
(a) Elk *(Alces alces)*
(b) Chausingha *(Tetracerus Quadricornis)*
(c) Black Buck *(Antilope Cervicapra)*
(d) Bharal *(Pseudonis Nayaur)*

26. Name the only country in the world where both the Lion *(Panthera Leo)* and the tiger (*Panthera tigris*) are found in the wild ?
(a) Uganda (b) India
(c) Zaire (d) Pakistan

27. Which bird constructs the largest unit nest structure in the world ?
(a) Red-billed Quelea *(Quelea quelea)*
(b) Sociable Weaver *(Philetairus Socius)*
(c) Mountain Drongo *(Chaetorhynchus Papuensis)*
(d) Ostrich *(Struthio Camelus)*

28. Which is the longest reptile in the world?
(a) King Cobra *(Ophiophagus hannah)*
(b) Reticulated Python *(Python reticulatus)*
(c) Emerald Tree Boa *(Corallus Caninus)*
(d) Black Mamba *(Dendroapsis polylepsis)*

29. Which is the world's largest private international nature conservation organization?
(a) Bombay Natural History Society
(b) Wold Wide Fund for nature
(c) Audubon Society
(d) Ranthambhore Foundation

30. The largest in terms of area (673 sq km), this bird sanctuary is also a Ramsar site
(a) Neelapattu (b) Kolleru
(c) Rollapadu (d) Rudrasagar Lake

31. The first man made bird sanctuary is
(a) Kaladeo Ghana (b) Vedanthangal
(c) Ranganthitoo (d) Bharatpur

32. India has -----------number of tiger reserves
(a) 50 (b) 23
(c) 14 (d) 20

33. What is Ictheologists?
(a) Study of fish
(b) Study of Birds
(c) Study of Animals
(d) Study of Mammals

34. India`s oldest national park is
(a) Kaziranga (b) Corbett
(c) Kanha (d) Kaladeo Ghana

35. To the indiscriminate slaughter to which wild animals were subjected to prior to the 1920s, the idea of the .....was born.
(a) Game sanctuary
(b) National park
(c) Forest reserve
(d) Wild life Sanctuary

36. The Rajaji Sanctuary is located in the state of
(a) Tamil Nadu
(b) Karnataka
(c) Andhra Pradesh
(d) Uttarakhand

37. The Vettankudi Bird Sanctuary in Tamil Nadu is situated in the district of
(a) Tirunelveli
(b) Chengalpattu
(c) Ramanathapuram
(d) Thanjavur

38. Which one of the following is incorrectly matched?
(a) Asiantic Lion – *Panthera leopersica*
(b) Leopard – *Panthera bengalensis*
(c) Tiger – *Panthera tigris*
(d) Snow leopard – *Panthera uncia*

39. The Indian Board for Wildlife was first constituted in the year
(a) 1972 (b) 1982
(c) 1958 (d) 1952

40. The scientific name of Nilgiri tahr is
(a) *Asinus hemionus*
(b) *Felis sylvestris*
(c) *Canis lupus*
(d) *Hemitrogus hylocrius*

41. One horned rhinoceros is found in
(a) Kanha National Park
(b) Periyar Tiger Reserve
(c) Kaziranga National Park
(d) Corbet National Park

42. How many Biosphere reserve exist in India?
(a) 18 (b) 23
(c) 15 (d) 17

43. Which of the following areas have the highest number of endemic species of Amphibians?
(a) Eastern Ghats
(b) Terais (c) Western Ghats
(d) Central Indian Highlands

44. The state animal of Tamil Nadu is
(a) Sambar (b) Nilgiri thar
(c) Tiger (d) Lion-tailed Macaque

45. How many project tiger areas are there in India now ?
(a) 18 (b) 44
(c) 19 (d) 30

46. Special game is included in which of the following schedules of the Wildlife Protection Act of 1972?
(a) II (b) III
(c) IV (d) V

47. The simplest behavior of animal is
(a) Kineses (b) Taxes
(c) Reflexes (d) None of these

48. Kalakad Mundanthurai Tiger Reserve is located in
(a) Andhra Pradesh
(b) Kerala
(c) Karnataka
(d) Tamil Nadu

49. Nilgiris Biosphere reserve is found in
(a) Tamil Nadu (b) Kerala
(c) Karnataka (d) All of the above

50. The Tamil Nadu Animals and Birds Sacrifices Prohibition Act was enacted in the year
(a) 1950 (b) 1960
(c) 1970 (d) 1980

51. The Staffing of wild animals is
(a) Animal article
(b) Animal trophy
(c) Taxidermy
(d) None of these

52. The animal coming under Schedule I part II of the Wildlife Protection Act, 1972 is
(a) Agra Monitor Lizard
(b) Black Buck
(c) Caracal
(d) Dugong

53. *Ptyas mucosus* is commonly called as
(a) Indian cobra (b) King cobra
(c) Rat snake (d) Python

54. Indian cobra is scientifically called as
(a) *Ophiophagus hannah*
(b) *Naja naja*
(c) *Python molurus*
(d) *Ptyas mucosus*

55. Digitigrade locomotion is used by
(a) Dogs (b) Cats
(c) Wolves (d) All of the above

56. Unguligrade locomotion is found in
(a) Cow (b) Deer
(c) Horse (d) All of the above

57. The dropping of carnivores are called as
(a) Dung (b) Pellet
(c) Bolus (d) Scat

58. Olive Ridley turtle is scientifically called as
(a) *Chelonia mydas*
(b) *Lepidochelys divacea*
(c) *Geochelone travancoria*
(d) *Geochelone elegans*

59. Green turtle is scientifically called as
(a) *Chelonia mydas*
(b) *Geochelone elegans*
(c) *Geochelone travancoria*
(d) *Lepidochelys divacea*

60. *Geochelone elegans* is called as
(a) Starred tortoise
(b) Travancore tortoise
(c) Green turtle
(d) Olive Ridley turtle

61. Animal adaption to climbing is
(a) Scansorial (b) Fossorial
(c) Cursorial (d) Volant

62. Animal adaption to flight is
(a) Scansorial (b) Fossorial
(c) Cursorial (d) Volant

63. The animal coming under perissodactyla includes
(a) Rhinoceros (b) Horse
(c) Zebra (d) All of the above

64. The study of the combination of animal behavior and ecology is
(a) Bioecology (b) Ethology
(c) Autecology (d) Synecology

65. The competitive exclusion principle in animal is called as
(a) Liebig's law minimum
(b) Shelford's law of tolerance
(c) Gause's principle
(d) None of these

66. ---------------- National Park in Kerala celebrates its 25th year of declaration as a national park in 2010?
(a) Silent Valley National Park
(b) Periyar National park
(c) Eravikulam National park
(d) Pambadum Shola National park

67. Which animals are captured in large numbers for pet trade?
(a) Parrots (b) Love birds
(c) Corals (d) All of the above

68. What is wild life symbol of WWF?
(a) Red Panda (b) Tiger
(c) Elephant (d) Leopard

69. When does the wild life week is celebrated in India?
(a) First week of June
(b) First week of October
(c) First week of March
(d) First week of August

70. Name the only Indian state where all three species of Indian crocodiles are found wild
(a) Uttar Pradesh
(b) Bihar
(c) Orissa
(d) Andhra Pradesh

71. Cheetah is found in
(a) Gir forest (b) West Bengal
(c) Deccan Plains (d) Extinct

72. The first marine national park is in
(a) Tamil Nadu (b) Gujarat
(c) Bihar (d) West Bengal

73. Rhino usually produces one calf after a gestation period of ....months
(a) 12 months (b) 15 months
(c) 18 months (d) 24 months

74. Asian elephants have a life span of --------------- years.
(a) 20-30 years (b) 30-40 years
(c) 50-60 years (d) 60-70 years

ANSWERSHEET
16. Wildlife

1. (a)	2. (b)	3. (c)	4. (c)	5. (b)	6. (a)	7. (c)	8. (a)	9. (a)	10. (b)
11. (c)	12. (a)	13. (c)	14. (c)	15. (c)	16. (d)	17. (c)	18. (b)	19. (b)	20. (b)
21. (d)	22. (b)	23. (c)	24. (b)	25. (b)	26. (b)	27. (b)	28. (b)	29. (b)	30. (b)
31. (a)	32. (b)	33. (a)	34. (b)	35. (a)	36. (d)	37. (c)	38. (b)	39. (d)	40. (d)
41. (c)	42. (d)	43. (c)	44. (b)	45. (b)	46. (a)	47. (c)	48. (d)	49. (d)	50. (a)
51. (c)	52. (a)	53. (c)	54. (b)	55. (d)	56. (d)	57. (d)	58. (b)	59. (a)	60. (a)
61. (a)	62. (d)	63. (d)	64. (b)	65. (c)	66. (a)	67. (d)	68. (a)	69. (b)	70. (c)
71. (d)	72. (b)	73. (c)	74. (d)						

17

Forest Entomology

KEY POINTS

- The study of insects are called entomology
- Antennae in insects are also called feelers
- The antenna is set in a socket of the cranium called antennal socket
- The basal segment of antennae is called scape.
- A mass of sensory cells called Johnston's organ is present in the pedicel
- The second antennal segment is called pedic el, which immediately follow the scape
- Antènna is useful to detect chemicals including food and pheromones.
- Pheromones Achemicals secreted into air by opposite sex

Types of Antenna

- Setaceous (Bristle like): Size of the segments decrease from base to apex. e.g. Dragonfly, Damselfly.
- Filiform (Thread like): Segments are usually cylindrical. Thickness of segments remains same throughout e.g. Grasshopper.
- Moniliform (Beaded): Segments are either globular or spherical with prominent constriction in between e.g. Termite.
- Serrate (Saw like): Segments have short triangular projections on one side. e.g. Longicorn beetle
- Unipectinate (Comb like):Segments with long slender processes on one side e.g. Sawfly
- Bipectinate (Double comb like):Segments with long slender lateral processes on both the sides e.g. Silkworm moth
- Clavate (Clubbed):Antenna enlarges gradually towards the tip. e.g. Blister beetle
- Capitate : (Knobbed):Terminal segments become enlarged suddenly e.g. butterfly
- Lamellate (Plate like): Antennal tip is expanded laterally on one side to form flat plates e.g. lamellicorn beetle
- Aristate: The terminal segment is enlarged. It bears a conspicuous dorsal bristle called arista e.g. House fly
- Stylate: Terminal segment bears a style like process e.g. Horse fly, Robber fly.
- Plumose (Feathery):Segments with long whorls of hairs e.g. male mosquito

- Pilose (Hairy): Antenna is less feathery with few hairs at the junction of flagellomeres. e.g. Female mosquito.
- Geniculate (Elbowed): Scape is long remaining segments are small and are arranged at an angle to the first resembling an elbow joint. e.g. Ant, weevil and honey bee.

Types of Insect Heads

- Hypognathous (Hypo-below; gnathous-jaw): This type is called Orthopteroid type. The long axis of the head is vertical. e.g. grasshopper
- Prognathous : (Pro-infront; ganthous-jaw): This type is also called Coleopteroid type. The long axis of the head is horizontal. e.g. ground beetle.
- Opisthognathous : (Opistho-behind; gnathous-jaw): This is also called Hemipteroid type or opisthorhynchous. Head is deflexed. e.g. stink bug

Types of Insect Mouthparts

- Biting and chewing type of mouth parts seen in Cockroach
- Piercing and sucking or Hemipterous or Bug type e.g. Plant bugs
- Piercing and sucking or Dipterous or Mosquito type :e.g. Female mosquito
- Chewing and lapping type : e.g. honey bee.
- The glossa terminates into a small circular spoon shaped lobe called spoon or bouton or flabellum, which is useful to lick the nectar.
- Rasping and sucking : e.g. Thrips
- Mandibulosuctorial type : e.g. grub of antlion
- Sponging type : e.g. House fly
- Siphoning type : e.g. Moths and butterflies

Types of Insect Legs

- Ambulatorial (Ambulate - to walk; walking leg): e.g. Fore leg and middle leg of grasshopper.
- Cursorial (Cursorial = adapted for running: Running leg): e.g. All the three pairs of legs of cockroach.
- Saltatorial (Saltatorial = Leaping; Jumping Leg): e.g. Hind leg of grasshopper.
- Scansorial (Scansorial = Climbing; climbing or clinging leg): e.g. All the three pairs of legs of head louse.
- Fossorial (Fossorial = Digging; Burrowing leg): e.g. Fore legs of mole cricket.
- Raptorial (Raptorial = predatory ; Grasping leg): e.g. Forelegs of preying mantis.
- Natatorial (Natatorial = pertaining to swimming; Swimming leg): e.g. Hing legs of water bug and water beetle.
- Sticking Leg : e.g. All three pairs of legs of house fly.
- Basket Like Leg : e.g. Legs of dragonfly and damselfly.
- Clasping Leg: e.g. Forelegs of male water beetle.
- Foragial Leg (Forage = to collect food material) : e.g. Legs of honey bee

- Wing is a flattened double - layered expansion of body wall with a dorsal and ventral lamina having the same structure as the integument
- The arrangement of veins on the wings is called venation.
- In insects like dragonfly and damselfly, there is an opaque spot near the coastal margin of the wing called pterostigma
- The anterior area of the wing supported by veins is usually called remigium.
- The flexible posterior area is termed vannus.
- The proximal part of vannus is called jugum, when well developed is separated by a jugal fold. The area containing wing articulation sclerites, pteralia is called axilla

Types of Insect Wings

- Tegmina : (Singular : Tegmen):. e.g. Forewings of cockroach and grasshopper.
- Elytra : (Singular : Elytron):. e.g. Fore wings of beetles and weevils.
- Hemelytra : (Singular : Hemelytron)
- Halteres: (Singular : Haltere): In true flies the hind wings are modified into small knobbed vibrating organs called haltere. e.g. true flies, mosquito, male scale insect.
- Fringed wings: e.g. Thrips.
- Scaly wings: e.g. Moths.
- Membranous wings:. e g. Dragonfly.
- Hamulate: e.g. bees.
- Amplexiform :. e.g. butterflies.
- The first stage of development in all insects is egg.
- The outer protective shell of the egg is called chorion.
- Pupa is the resting and inactive stage in all holometabolous insects. During this stage, the insect is incapable of feeding and is quiescent
- Pupa of mosquito is called tumbler
- Metamorphosis is the change in growth and development of an insect undergoes during its life cycle from birth to maturity
- Insect is a six legged arthropod
- Skeletonizing: Insect feeds on the soft material between the veins and leaves the veins as a "skeleton" of the leaf
- Teak skeletoniser: *Eutectona machaeralis*
- Window feeding: Insect feeds on only one surface of leaf that allows the light to penetrate through the remaining leaf layer.
- Pollen basket is also called corbicula. The pollen basket enables the bee to carry a larger load of pollen and propolis from the field to the hive.
- Pollen packer is also called pollen press. It is a row of stout bristles at the distal end of tibia. Auricle is a small plate fringed with hairs at the. Pollen packer is useful to load pollen in corbicula.
- The pollen comb is used to collect pollen from middle legs and from posterior part of the body
- Obesity of abdomen of queen termite is called physogastry
- In ants, bees and wasps, the first abdominal segment is fused with the metathorax and is called propodeum

- The egg laying organ found in female insect is called ovipositor
- The ovipositor of house fly is called pseudoovipositor or ovitubus or oviscapt.
- Larva of butterfly is called caterpillar
- The young ones in Cockroach are called nymphs
- The young ones in Dragonfly, damselfly and may fly are aquatic and are called naiads
- Mouth leads into pharynx which leads immediately into a narrow tube called oesophagus.
- The distal end of the oesophagus enlarges into a large sac like structure called crop which is useful to store the food prior to digestion.
- Associated with the alimentary tract at the junction of midgut and hindgut are many yellowish, hair like structures called Malpighian tubules which eliminate nitrogenous waste from blood
- Male gonopore is surrounded by chitinous plates called gonapophyses which aid in copulation
- Opening into the genital pouch are a pair of extensively branched, fibrous glands called colleteral glands
- Powder post beetle damage: Insect bores into the dead trees and wood products producing very fine dustlike wood powder
- Honey bees belong to the family Apidae, suborder Apocrita, of the order Hymenoptera. Five species of honey bees are found in India.

Giant or Rock bee	: *Apis dorsata*
Indian bee	: *Apis cerana indica*
Italian bee	: *Apis mellifera ligustica*
Little bee or Dwarf bee	: *Apis florea*
Dammer bee or mosquito bee	: *Melipona iridipennis (or) Trigona iridipennis*

Important Pest in Tree Species

TEAK - Tectona grandis

Borers	: *Cossus cadambae, Sahyadrassus malabaricus, Dihammus cervinus, Indarbela quadrinotata, Indarbela tetraonis*
Teak defoliator	: *Hyblaea puera, Spodoptera litura*
Teak skeletoniser	: *Eutectona machaeralis*
Root feeder	: *Clinteria klugi*

Sal

Borer	: *Haploceramby x spinicornis, Batocera rufomaculata*
Leaf feeders	: *Ingura subapicalis, Hyposidra successaria*
Shoot feeder	: *Parmmene theristis*

Eucalyptus

Shoot feeders	: *Celosterna scabrator, Bactocera rufomaculata*

Casuarina

Borers : *Indarbela quadrinotata, Zeuzera coffeae*
Sap feeder : *Naiacoccus serpentinus*

Ailanthus

Borers : *Sthenias grisator*
Leaf feeders : *Eligma narcissus, Attacus atlas*
Web worm : *Atteva fabriciella*
Leaf feeders : *Eupterote fabia*

Sissoo

Borer : *Acamaeodera kerremansi*
Leaf feeder : *Plecoptera reflexa*

Neem

Leaf Feeders : *Cryptocephalus ovulum, Laspeyresis aurantiana*
Sap Feeders : *Helopeltis antonii*

Sandal

Borer : *Sinoxylon atratum, Zeuzera coffeae*
Vector of Spike disease : *Jassus indicus*
Sap feeder : *Sarima nigrohypeata*

Bamboo Spp.

Borer : *Dinoderus brevis, Dinoderus minutus*
Shoot feeder : *Estigmena chinensis*
Leaf feeder : *Cryptocephalus sexsignatus*
Sap feeder : *Trichoregama insulers*

MULTIPLE CHOICE QUESTION'S

1. The econfriendly insecticide is
(a) Malathion (b) Thiodan
(c) Nimbicidine (d) Permethrin

2. Pheromone trap attracts
(a) Female bugs (b) Caterpillars
(c) Female moths (d) Male moths

3. Which of the following is not a bio-pesticide?
(a) Bioneem (b) Biolep
(c) Dipel (d) Carbaryl

4. Which one of the following is a carbamate pesticide?
(a) Aldicarb (b) Heptachlor
(c) Malathion (d) Endosulfan

5. Dimethoate is
(a) Contact insecticide
(b) Systemic insecticide
(c) Fumigant
(d) Stomach poison

6. Erisilkworm is reared on
(a) Mulbery (b) Castor
(c) Coconut (d) Banana

7. The worker bee is
(a) Drone (b) Sterile male
(c) Sterile female (d) None of these

8. One of the major insect enemy of honey bee is
(a) *Vespa cincta*
(b) *Galleria mellonella*
(c) *Acherontia styx*
(d) *Formicaria rufa*

9. The major predator of lac insect is
(a) *Mantis religiosa*
(b) *Eublemma amabilis*
(c) *Coccinella septempunctata*
(d) *Chrysoperla carnea*

10. The most widely used bacterium in insect control is
(a) *Fusarium* sp.
(b) *Bacillus papillae*
(c) *Bacillus cereus*
(d) *Bacillus thuringiensis*

11. Use of resistant varieties in the IPM of insect is an example of
(a) Legal control
(b) Cultural control
(c) Chemical control
(d) Biological control

12. Which one of the following is the example of insect hormone?
(a) Lindane (b) Ecdysone
(c) Chiti (d) Allethrin

13. Certain secretions attract the opposite sex of the same species which are termed as
(a) Hormones (b) Kairomones
(c) Synomones (d) Pheromones

14. The dwelling place of termites is also known as
(a) Canal (b) Tunnel
(c) Mound (d) Cave

15. Chlorpyriphos belongs to --------------- group of insecticide
(a) Organochloride (b) Organocarbanates
(c) Cyclodiere (d) Organophosphate

16. Following is the acaricide widely recommended for mite management
(a) Kelthane (b) Rogor
(c) DDVP (d) Monocrotophos

17. Following is the pesticide which is of plant origin
(a) Spinoside (b) Bulldoc
(c) NSKE (d) Allethrin

18. Following granular insecticide is having long residual effect
(a) Phorate (b) Carbofuran
(c) Aldicarb (d) Quinalphos

19. Following insecticide belongs to organophosphate group
(a) Sulphur (b) Carbofuran
(c) Malathion (d) Allethirn

20. The insecticide which is used in mosquito mats is
(a) Endosulphan (b) Diazinon
(c) Malathion (d) Allethrin

21. Following anticoagulant is used for rat control
(a) Zinc-phosphide
(b) Al2Po4
(c) Hydrogen Cyanide
(d) Warfarin

22. Aphids and white flies are effectively controlled by
(a) Fumigants
(b) Stomach poisons
(c) Contact insecticides
(d) Systemic insecticide

23. IPM relies heavily on
(a) Economic threshold level
(b) Total loss of crops
(c) Resistant pest
(d) Susceptible pest

24. Project Directorate of Biological Control is located at
(a) Bangalore (b) New Delhi
(c) Faridabad (d) Jaipur

25. The following is the most notorious polyphagous pest
(a) *Chilo incertulas*
(b) *Mythimna separate*
(c) *Helicoverpa armigera*
(d) *Gangara thyrsis*

26. Anamorphosis is encountered the following insect order
(a) Protura
(b) Dictyoptera
(c) Ephemereptera
(d) Coleoptera

27. Sting in worker honey bee is a modified part of
(a) Antenna (b) Ovipositor
(c) Styli (d) Cerci

28. Following is an example for secondarily wingless insect
(a) Blister beetle (b) Head louse
(c) Collembolan (d) Silver fish

29. ____________ is an important major pest species in the complex of pests attacking a crop
(a) Sporadic pests (b) Regular pest
(c) Key pest (d) Endemic pest

30. Flooding of the fields will help in control of
(a) White grubs (b) Termites
(c) Cut worms (d) All of these

31. Destructive Insects Pest Act (DIPA) was passed on
(a) 1914 (b) 1920
(c) 1937 (d) 1968

32. Directorate of Plant Protection Quarantine and Storage (DPPQS) was established in 1946 with head quarters
(a) Ghaziabad (b) Faisabad
(c) Faridabad (d) Faislabad

33. Biocontrol agent for lantana weed *(Lantana camera)* is
(a) *Ophiomyia lantanae*
(b) *Neochetina eichorniae*
(c) *Crytobagus singularis*
(d) *Zygogramma bicolorata*

34. Bacillus thuringiensis was discovered by
(a) S.Ishiwata (b) Berliner
(c) Flacher (d) Frisch

35. Commercially granulosis virus (GV) used for the control of
(a) *Helicoverpa armigera*
(b) *Spodoptera litura*
(c) *Chilo infuscatellus*
(d) None

36. Which of the following repellent is used against termites?
(a) Oil of citronella
(b) Deet
(c) Pentachloro phenol
(d) Bordeaux mixture

37. The term "third generation pesticides" is given for
(a) Juvenile hormone
(b) Moulting hormone
(c) Brain hormone
(d) Anti juvenile hormone

38. The non-preference plant resistance to insects is also known as
(a) Antibiosis (b) Antixenosis
(c) Tolerance (d) None of these

39. Father of host plant resistance is
(a) Kogan & Ortman
(b) R.H. Painter
(c) Zeidler
(d) Paul Muller

40. Which of the following pairs is not correct?

Insecticide	Trade name
(a) Endosulfan	Thiodan
(b) Chlorpyrifos	Dursban

(c) Dimethoate Rogar
(d) Phorate Nuvacron

41. Which one of the following pairs is not correctly matched?

Insecticide	Trade name
(a) Phosphomidon	Dimecron
(b) Aldicarb	Temik
(c) Carbofuran	Furadan
(d) Carbaryl	Celphos

42. Bacillus thuringiensis is available in market as
(a) Vertilec
(b) Dipel, Delfin
(c) Larvocel, Boverin
(d) Biomax

43. Atropine is a / an
(a) Repellent (b) Antifeedant
(c) Antidote (d) Pheromone

44. Central Insecticides Laboratory (CIL) is located at
(a) Gurgaon (b) Faridabad
(c) Ghaziabad (d) Kanpur

45. The host of *Bombax mori* is -----------
(a) Castor (b) Terminalia
(c) Tamarid (d) Mulberry

46. Scientific name of the Dammeer bee is :
(a) *Apis dorsata*
(b) *Melipona irridipennis*
(c) *Apis mellifera*
(d) *Apis indica*

47. The process of leaving off the colony by queen is known as
(a) Absconding (b) Swarming
(c) Supersedure (d) Queen excluder

48. The first abdominal segment when fused with metathorax in bees is known as:
(a) Propodeum (b) Mandibles
(c) Basitarsus (d) Mesopodeum

49. Safest chemical for honey bee is
(a) Malathion (b) Methyl parathion
(c) Endosulfan (d) None of these

50. Bee venom is used for ________ and released from acid glands
(a) Curing pains at joints
(b) Repellent against greater wax moth
(c) Pollinization
(d) None of these

51. In honey bees, the queen substance is secreted by
(a) Mandibular glands
(b) Acid glands
(c) Pharyngeal glands
(d) Wax glands

52. Rangeeni strain is related to
(a) Bee (b) Silk worm
(c) Lac insect (d) None of these

53. In India more honey is obtained from which bee?
(a) Little bee (b) Indian bee
(c) Italian bee (d) Rock bee

54. Which pesticide is suitable for controlling the white grub?
(a) Malathion (b) Chlorpyriphos
(c) Phorate (d) Aldicarb

55. Insects that attack young twigs, stems or buds are ___.
(a) Bark borers (b) Defoliators
(c) Sap suckers (d) Tip feeders

56. Caterpillars are an example of ___ who may destroy all or part of the leaf.
(a) Bark borers (b) Defoliators
(c) Sap suckers (d) Tip feeders

57. The study of forest insects is called _______
(a) Forest insectology
(b) Forest entomology
(c) Forest pestology
(d) Forest ornithology

58. Aphids damage forest trees by ________.
(a) Chewing (b) Sucking
(c) Defoliating (d) All of the above

59. Most insect damage is done while the insects are in the _____ stage.

(a) Egg (b) Larva

(c) Pupa (d) Adult

60. Soft cambium and newly formed wood is damaged by these enemies which leave tunnels as a sign of damage.

(a) Bark borers (b) Defoliators

(c) Sap suckers (d) Tip feeders

61. Aphids and scales are examples of _____ with penetrating mouth parts.

(a) Bark borers (b) Defoliators

(c) Sap suckers (d) Tip feeders

62. Insects that produce abnormal growth on twigs, limbs, trunks or leaves are ___.

(a) Gall forming insects

(b) Root feeders

(c) Seed eaters

(d) Wood borers

63. Which one of the following insect pests lays down a silky trail on foliage or bark?

(a) Aphid (b) Gypsy moth

(c) Spider mite (d) Tent caterpillar

64. Grubs and wire worms are examples of ___.

(a) Defoliators (b) Gall makers

(c) Root feeders (d) Wood borers

65. The term biological control refers to ___.

(a) Pest control chemicals containing mostly carbon

(b) Pest control using natural agents

(c) Pest control chemicals that are created in a lab by biologists

(d) Synthetic pest chemicals

66. Caterpillars are an example of _____ which may destroy all or part of the leaf.

(a) Bark borers

(b) Defoliators

(c) Sap suckers

(d) Tip feeders

67. Which of the following is a biological pest management strategy?

(a) Coat uninfest wood with paint

(b) Harvest mature trees or stands

(c) Release parasites or predators

(d) Use pesticides

68. Convulsions, secretions from nose and ears, loss or consciousness and inability to breathe are symptoms of ________.

(a) Heat stroke

(b) Hypothermia

(c) Severe pesticide poisoning

(d) Smoke inhalation

69. The control of a damaging insect population by its natural enemies is an example of ________.

(a) Mechanical control

(b) Chemical control

(c) Management control

(d) Natural control

70. The most destructive group of forest insects are the ____________.

(a) Borers

(b) Leaf feeders

(c) Plant sucking insects and mites

(d) Gall insects and mites

71. The severity of damage to forest trees from leaf feeders depends on ________.

(a) The season when the foliage damage takes place

(b) The extent of the damage

(c) The frequency of the damage

(d) All of the above

72. Severe defoliation that occurs ________ is most harmful to trees.

(a) late in the season

(b) In September and October

(c) In July and August

(d) As the spring foliage is unfolding

73. The gypsy moth over winters in the _____ stage.

(a) Adult (b) Pupae

(c) Egg (d) Larva

74. The damage done by the gypsy moth is when it is in the _______ stage.
(a) Adult (b) Pupae
(c) Egg (d) Larva

75. The male gypsy moth locates the female because she _______________.
(a) Secretes a powerful scent called a pheromone
(b) Is bright orange in color
(c) Is about 5 inches long and quite colorful
(d) Emits a shrill mating call

76. Chemicals used to control only forest insects are called _____.
(a) Forest pesticides
(b) Forest herbicides
(c) Forest insecticides
(d) Forest fungicides

77. A notorious weed, *Lantana camara* is controlled by
(a) Lantana bug (b) Shoot fly
(c) Cochnial insect (d) None

78. Directorate of plant protection, quarantine and storage started in
(a) 1940 (b) 1946
(c) 1950 (d) 1552

79. Fossil records indicate that insect evolved during :
(a) Cambrian period
(b) Devonian period
(c) Silurian period
(d) Precambrian period

80. Economic threshold level is always
(a) Lower than economic injury level (EIL)
(b) Equal to EIL
(c) More than EIL
(d) Reached when severe damage is done

81. Entomological Society of India started in
(a) 1934 (b) 1936
(c) 1938 (d) 1940

82. Factor on which penetration of insecticide through insect cuticle does not depend :
(a) Thickness of the cuticle
(b) Chemical nature of the cuticle
(c) Components of the cuticle
(d) Nature of the carriers and solvents

83. Feeding in insects can be checked by the use of :
(a) Hormone
(b) Pheromone
(c) Feeding deterrent
(d) Antibiosis

84. In India, lac is mainly produced in :
(a) Tamil Nadu (b) Bihar
(c) Kerala (d) Rajasthan

85. India is exporting shellac to :
(a) China (b) Japan
(c) USA (d) Pakistan

86. India's place in the production of mulberry silk in the world is :
(a) First (b) Second
(c) Third (d) Fourth

87. Insecticide having antagonistic effect with NPV :
(a) DDT (b) Aldrin
(c) Endosulfan (d) Malathion

88. Insecticides approved for aerial spray :
(a) Lindane 20 EC
(b) Malathion 50 EC
(c) Phos amidon 85 WSC
(d) All the above

89. Most of the soil fumigants are applied with:
(a) Applicator (b) Injector
(c) Generator (d) Dispenser

90. Light traps are used for the collection of:
(a) Positively Phototropic
(b) Negatively Phototropic
(c) Geotropic
(d) Hydrotropic

91. Malathion is a ---------- insecticide
(a) Slow fumigant
(b) Contact plus systemic
(c) Contact plus stomach
(d) Systemic

92. Most of the insects possess :
(a) One pair of wings
(b) Two pair of wings
(c) Three pair of wings
(d) None of these

93. Number of insects known :
(a) About 50000
(b) About 1 million
(c) About 2 million
(d) About 3 million

94. Pheromones are also known as :
(a) Antibiotics (b) Ectohomones
(c) Hormones (d) Attractants

95. Pheromones are synthesized by :
(a) Thoracic glands
(b) Glandular epidermal
(c) Neuro-secretory cells
(d) All the above

96. Phorate is ------------- insecticide
(a) Systemic
(b) Contact plus systemic
(c) Contact
(d) Contact plus stomach

97. Termite control can be also done by :
(a) Tilling (b) Clean cultivation
(c) Irrigation (d) Trap cropping

98. The author of "Indian insect life"
(a) Hussain (b) Fletcher
(c) Lefroy (d) Mani

99. The central insecticide Act came into force from :
(a) 1950 (b) 1966
(c) 1971 (d) 1975

100. The damaging stage of cut worm is :
(a) Caterpillar (b) Nymph
(c) Adult (d) All of these

101. The density of pest population at which control measures should be applied :
(a) GEL (b) PBL
(c) EIL (d) ETL

102. The first Government entomologist of India :
(a) Niceville (b) Stebbing
(c) Pruthi
(d) Ramakrishna Ayyar

103. The first discovered protective insecticide was :
(a) B.H.C (b) D.D.T
(c) Diazinon (d) None of these

104. The first Indian imperial Entomologist
(a) Hem singh pruthi
(b) Afzal Hussain
(c) Ram Chandra Rao
(d) C.S. Mishra

105. The grasshopper possess
(a) Natatortial legs
(b) Foragial legs
(c) Saltatorial legs
(d) Ambulatorial legs

106. The grinding of food in grasshopper is done by :
(a) Mandibles (b) Crop
(c) Gizzard (d) Midgut

107. The Indian lac Research Institute is located at :
(a) Patna (b) Cuttack
(c) Gaya (d) Ranchi

108. The latest philosophy in insect pest is control pest control is :
(a) Eradication (b) Pest management
(c) Control (d) Integrated control

109. The main characteristic feature of an insect is :
(a) Presence of wings
(b) Three pairs of legs
(c) Three body regions
(d) Pair of antennae

110. The host tree for the cultivation of lac insect is :
(a) Pipal (b) Palas
(c) Khejri (d) Babul

111. The main host tree of eri-silk worm is :
(a) Mulberry (b) Sal
(c) Castor (d) Ber

112. The nervous system of insect is derive from:
(a) Ectoderm (b) Mesoderm
(c) Endoderm
(d) Endoderm and mesoderm

113. The paralysis of honey bees is caused by:
(a) Bacteria (b) Virus
(c) Fungus (d) Mycoplasma

114. The pupation of fruit flies takes place in
(a) Soil (b) Rind of fruit
(c) Stubbles in the field
(d) Pulp of fruit

115. The prothoracicotropic hormone is secreted by
(a) Prothoracic glands
(b) Neurosecretory cells
(c) Corpora allata
(d) Corpora cardiac

116. The queen in the honey colony feeds on :
(a) Royal jelly (b) Nector
(c) Queen substance (d) Honey

117. The ratio of Zinc phosphide in poison bait for rats is :
(a) 1% (b) 2%
(c) 3% (d) 4%

118. The queen termite produces --------------- egg per second
(a) 40 (b) 50
(c) 60 (d) 70

119. The reproductive system of insect is derived from :
(a) Ectoderm (b) Endoderm
(c) Ectoderm and mesoderm
(d) Ectoderm and Endoderm

120. Use of resistant varieties constitutes methods of :
(a) Genetic control
(b) Self control
(c) Biotic control
(d) Natural control

121. Vertical resistance is also called as :
(a) Oligogenic (b) Specific
(c) Monogenic (d) All the above

122. Where the Directorate of plant protection, Quarnatine and storage is situated :
(a) New Delhi (b) Faridabad
(c) Hissar (d) Gurgaon

123. Which animal acts as natural tiller of soil?
(a) Termite (b) Earthworm
(c) Snake (d) Millipede

124. Which insecticides belongs to synthetic pyrethroid group?
(a) Pyrethrum (b) Cypermethrin
(c) Rotenone (d) Sabadilla

125. Which insecticides has contact and fumigant action group?
(a) Phosphamidon
(b) Diazinon
(c) Lindane
(d) Endosulfan

126. Which insecticide has systemic action?
(a) Endosulfan (b) Chlorpyriphos
(c) Methyldemeton (d) Diazinon

127. Which is the best suiting species for bee culture?
(a) *Apis dorsata*
(b) *Apis florae*
(c) *Apis cerana indica*
(d) *Milipona* sp.

128. Which are the following are systemeic fungicides?
(a) Vitravax (b) Benlate
(c) Brestonal (d) All the above

129. Which of the following is called "water scorpion"?
(a) Belostoma (b) Nepa
(c) Corixa (d) Notonecta

130. Which of the following is larval parasitoid?
(a) Xanthopimpla (b) Tetrastichus
(c) Apanteles (d) None of the above

131. Which of the following is pupal patrasitoid?
(a) Elasmus (b) Tetrasticus
(c) Xanthopimpla (d) All of the above

132. Which of the following is systemic insecticide?
(a) Parathion (b) Malathion
(c) Cabaryl (d) Carbofuran

133. Which of the following organism is not storage pest?
(a) Rodents (b) Nematodes
(c) Bacteria (d) Fungi

134. Which one is used for inoculation of host tree for lac cultivation?
(a) Seed lac (b) Brood lac
(c) Phunki lac (d) Eri lac

135. Which one of the following insects has not much changed from its fossil records?
(a) Butterfly (b) Praying mantis
(c) Cockroach (d) Grasshopper

136. Which state is famous for sericulture?
(a) Kerala (b) Karnataka
(c) Andhra Pradesh
(d) Uttar Pradesh

137. Which type of special reproduction is found in aphid?
(a) Polyembryony (b) Paedogenesis
(c) Parthenogenesis (d) Viviparity

138. Which is often referred to as the 'Father of animal classification'?
(a) Aristotle (b) Carlous Linnnaeus
(c) Wallace (d) Hugo de Varies

139. Honey bee queen produces------------ thousand eggs per day
(a) 0.5 – 1.5 (b) 2.0 – 3.0
(c) 4.0 – 4.5 (d) 5.0 – 5.5

140. Inorganic salts are toxic to insects by
(a) Contact (b) Ingestion
(c) a and b (d) inhalation

141. Insects feeding on plants of several genera within a family are called
(a) Phytophagous
(b) Polyphagous
(c) Oilgophagous
(d) Monophagous

142. Integrated Pest Management (IPM) aims at
(a) Keeping pest population below injurious levels
(b) Increasing natural enemies of the pest
(c) Strengthening the host
(d) Billing the pest

143. It has been estimated that insects had existed on this earth for at least ------------- million years
(a) 100 (b) 150
(c) 200 (d) 250

144. Lac research institute is situated at
(a) Bangalore (b) Behrampur
(c) Lucknow (d) Namkum

145. Name the insecticide which is harmless to the honey bees
(a) B.H.C (b) Carbaryl
(c) Endosulfan (d) Permethrin

146. Odontotermes obesus is the scientific name of
(a) Cutworm (b) Termite
(c) Shoot borer (d) Top Borer

147. Originally the phytopahgous insects were
(a) Monophagous (b) Oligophagous
(c) Polyphagous (d) All the above

148. Rodenticide are used to kill
(a) Rats & mites (b) Insects
(c) Weed (d) None of these

149. Coine the term 'Pest management"
(a) Geier
(b) Bartlett
(c) Dr. Perry Adkisson
(d) Dr.Ray F. Smith

150. Which insecticide is more persistent in the environment ?
(a) Parathion (b) Aldrin
(c) DDT (d) BHC

151. Marine borer: (pick the odd man out)
(a) Limnoria (b) Teredo
(c) Banksia (d) Cerambicids

152. Sal borer is
(a) *Hoplocerambyx spinicornis*
(b) *Sthenias grisator*
(c) *Hyblea peura* (d) *Atteiva fabriciella*

153. Sisham defoliator is
(a) *Hyblea peura*
(b) *Plecoptera reflexa*
(c) *Calosomma beesoni*
(d) *Hypsipyla robusta*

154. *Hyblaea puera* is a defoliator on
(a) Teak (b) Deodar
(c) Sal (d) Toon

155. What is another term for complete metamorphosis?
(a) Ametabolous
(b) Hemimetabolous
(c) Holometabolous
(d) Paurometabolous

156. Which insect develops using complete metamorphosis?
(a) Caddisfly (b) Katydid
(c) Mayfly (d) Water strider

157. Attack by scales and aphids would result in:
(a) Death by girdling
(b) Defoliation
(c) Forked and crooked trees
(d) Stunted growth

ANSWERSHEET
17. Forest Entomology

---

1. (c)	2. (d)	3. (d)	4. (a)	5. (b)	6. (b)	7. (c)	8. (b)	9. (b)	10. (d)
11. (b)	12. (b)	13. (d)	14. (c)	15. (d)	16. (a)	17. (c)	18. (c)	19. (c)	20. (d)
21. (d)	22. (d)	23. (a)	24. (a)	25. (c)	26. (a)	27. (b)	28. (b)	29. (c)	30. (d)
31. (a)	32. (c)	33. (a)	34. (a)	35. (c)	36. (c)	37. (a)	38. (b)	39. (b)	40. (d)
41. (d)	42. (b)	43. (c)	44. (b)	45. (d)	46. (b)	47. (b)	48. (a)	49. (c)	50. (a)
51. (a)	52. (c)	53. (d)	54. (c)	55. (d)	56. (b)	57. (b)	58. (b)	59. (b)	60. (a)
61. (c)	62. (a)	63. (d)	64. (c)	65. (b)	66. (b)	67. (c)	68. (c)	69. (d)	70. (a)
71. (d)	72. (d)	73. (c)	74. (d)	75. (a)	76. (c)	77. (a)	78. (a)	79. (b)	80. (a)
81. (c)	82. (b)	83. (c)	84. (b)	85. (a)	86. (c)	87. (d)	88. (d)	89. (b)	90. (a)
91. (c)	92. (b)	93. (b)	94. (b)	95. (b)	96. (b)	97. (c)	98. (c)	99. (c)	100. (a)
101. (d)	102. (a)	103. (b)	104. (a)	105. (c)	106. (a)	107. (d)	108. (b)	109. (b)	110. (b)

111. **(c)**	112. **(a)**	113. **(b)**	114. **(a)**	115. **(b)**	116. **(a)**	117. **(b)**	118. **(d)**	119. **(b)**	120. **(a)**
121. **(d)**	122. **(b)**	123. **(b)**	124. **(b)**	125. **(b)**	126. **(c)**	127. **(c)**	128. **(d)**	129. **(b)**	130. **(c)**
131. **(d)**	132. **(d)**	133. **(b)**	134. **(b)**	135. **(c)**	136. **(b)**	137. **(c)**	138. **(b)**	139. **(b)**	140. **(b)**
141. **(c)**	142. **(a)**	143. **(c)**	144. **(d)**	145. **(c)**	146. **(b)**	147. **(c)**	148. **(a)**	149. **(a)**	150. **(c)**
151. **(d)**	152. **(a)**	153. **(b)**	154. **(a)**	155. **(c)**	156. **(a)**	157. **(d)**			

18

Forest Pathology

KEY POINTS

- Fungi are microscopic, non-chlorophyllous and filamentous organisms
- The vegetative body of the fungi consists of elongated filaments, which are individually called as hypha.
- The entire vegetative body or group of hyphae is called mycelium
- Spore is a Minute propagative units of the fungus
- Hyphae is made up of tubular wall filled with protoplasm and may be continuous or separated by cross wall called septum
- Absence of septa is referred as aseptate or coenocytic.
- Spores: (Gr. Sporos – seed, spore) Minute propagative units of the fungus
- Zygospore: It is a thick walled spore which results from the fusion of two gametangia in Zygomycotina . Eg: *Mucor, Rhizopus*
- Oospores : (Gr. Oon = egg + Sporos = seed, spore). It is a thick walled spore developed from the union of male and female gametangia in Mastogomycotina Eg: *Pythium, Phytophthora, Sclerospora, Plasmopara*
- Ascospores : It is a spore which results from meiosis, borne in the ascus.
- Ascus: It is a sac like structure containing a defnite number of ascospores (typically eight), which are formed as a result of karyogamy and meiosis. Eg: *Endothia, Claviceps, Venturia, Erysiphae*
- Ascocarp : The sexual fruiting body containing asci is termed ascocarp. Ascocorp may be formed singly or in groups.They may be superficial, erumpent or deeply embedded inside the substratum.
- Smut refers to a soft, charcoal like substance or a sooty powder
- The pathogens that belong to the order Ustilaginales causes smut disease

  Ovariculus smut: *Ustilago avenae, Ustilago tritici*

  Culmiculus smut :*Ustilago scitaminae*

  Loose smut: *Ustilago tritici*

  Grain smut/ Covered smut: *Sphacelotheca sorghi*
- Rusts is caused by the fungi belonging to the order Uredinales

  Pungam Rust: Causal organism: *Ravenelia hobsoni*

  Eucalyptus Rust: Casual organism is *Puccinia psidi*

Teak Rust: Causal organism is *Olivea tectonae (= Uredo tectonae)*

Albizia-- *Ravenelia japonica* and *Uredo albizziae*

Mulberry-*Aecidium mori* and *Cerotelium fici*

Chir Pine-Swertia felt rust- *Cronartium himalayense*

Chir Pine Needle rust- *Coleosporium campanulae*

Blue pine- Stem blister rust- *Cronartium ribicola*

Blue pine- Needle rust- *Melampsora oblonga*

Poplar- *Melampsora rostrupii*

- Powdery Mildew: The term is restricted to diseases caused by the members of the family Erysiphaceae.
- Powdery mildews are recognised by the presence of fungal mycelium and conidiophores on the host.

Teak : *Uncinula tectonae*

Eucalyptus : *Oidium eucaluti*

Poplar : *Uncinula salicis*

- Downy mildew: Development of white / grey growth of the fungus on the leaf and stem owing to the production of sporangiophores of the casual fungi
- Fungal leaf spots:

Neem : *Cercospora meliae, Septoria meliae, Phyllosticta azadirachis*

Teak : *Cercospora tectonae, Phyllosticta tectonae*

Albizia : *Cercospora albiziae, Helminthosporium albiziicolum, Phyllosticta albizzinae*

Mulberry : *Cercospora moricola*

Eucalyptus : *Cercospora eucalypti*

Pungam : *Cercospora pongamiae*

Jack : *Septoria artocarpi, Phyllosticta artocarpina*

Rubber : *Helminthosporium heveae*

Sissoo : *Phyllosticta sissoo*

Semul : *Phyllosticta bambacis*

- Leaf blights: Blight implies sudden and extensive damage to leaves resulting in necrosis of tissue including veins
- Brown Leaf blight: *Alternaria alternata*- Acacia, Jatropha, Neem Pungam, Eucalyptus
- Grey leaf blight

Pestatiopsis heveae - Rubber

Pestatiopsis funereal - Chir Pine

Pestatiopsis banksiana - Silver oak

Pestatiopsis elastica - Jack

Pestatiopsis theae - tea

Pestatiopsis indica - Sapota, guava

Pestatiopsis palmarum - Coconut

- Anthracnose – means 'like coal'. The term is now used to describe disease caused by the members of the family Melanconiaceae

Colletotrichum sissoo - Sissoo
Colletotrichum heveae - Rubber
Colletotrichum gloeosporioides - Cashew, Neem. Mango
Gloeosporium artocarpi - Jack

- Damping off causal organism is *Pythium aphanidermatum, P. vexans, P. ultimum*
- Root rot causal organism is *Macrophomina phaseolina*
- Fusarial wilt in sissoo : *Fusarium solani*
- Bacteria are non- filamentous, unicellular organisms that lack chlorophyll and reproduce by binary fission

  Bacterial leaf blight : Causal organism- *Xanthomonas spp*
  Citrus canker : Causal organism: *Xanthomonas campestris* pv. *citri*
  Teak leaf blight : *Xanthomonas melhusi*
  Eucalyptus leaf blight : *Xanthomonas eucalypti*
  Mulberry leaf spot : *Pseudomonas mori*
  Cashew leaf spot : *Pseudomonas mangiferae- indicae*
  Neem leaf spot : *Pseudomonas azadirachtae*
- Viruses are submicroscopic infective entities that multiply intercellularly and are potentially pathogenic
- Viruses are ultramicrospic; neither living nor non-living, which replicate with any of the nucleic acid either RNA or DNA which are covered by a proteinaceous covering called nucleocapsid
  - ♦ Mosaic – Alternate dark and light green areas on leaves which are irregular e.g. TMV, jatropha, neem, eucalyptus
  - ♦ Vein banding – a broader band of chloritic tissues along the veins
  - ♦ Vein clearing – Chlorosis of tissue in or immediately adjacent to vein. Eg. Mulberry vein netting
  - ♦ Yellow mosaic – Yellow pigments predominant. Black gram yellow mosaic virus, Wheat yellow mosaic, Rice transitory yellowing.
  - ♦ Leaf curl - The leaves curl downward with a thickened margins. Mulberry, Eucalyptus
  - ♦ Leaf crinckle- The leaves become rough, thickened with undulated lamina showing minute swellings. Eg. Eucalyptus
  - ♦ Leaf roll- The leaves become thick, rough and roll upwards with a swollen margins.
  - ♦ Witches broom – Production of excess branches giving an appearance of broom. Eg. Eucalyptus, neem.
  - ♦ Dwarf / stunt – plant growth is very much affected.
  - ♦ Phytoplasma diseases are confined to sieve elements of phloem and seldom found in Xylem
  - ♦ Witches broom: Excess production of branches giving the appearance of a leaf broom Eg. Eucalyptus

- Spike: The infected trees exhibit shortened internodal length leading to crowding of the leaves at the top
- Phanerogamic parasites are flowering plants that parasitise on other plants and depend upon the host for food and water.
- Partial root parasite: Eg. *Striga*
- Complete or total root parasite: Eg. *Orabanche cernua* (Broom rape)
- Partial stem parasites: Mistle toes (Loranthus) : Eg. *Dendrophthoe falcata.*
- Total stem parasites: Dodder: Eg. Cuscuta *reflexa*
- Per cent disease Incidence = $\frac{\text{Number of infected plants}}{\text{Total number of plants observed}} \times 100$
- Percent Disease Index (PDI)= $\frac{\text{Sum of individual ratings}}{\text{Total number of leaves}} \times \frac{100}{\text{Maximum disease observed grade}}$
- The word 'fungicide' originated from two latin words, viz., 'fungus' and 'caedo'. The word 'caedo' means 'to kill'
- The chemicals do not kill the fungal pathogens and arrest the growth of the fungus temporarily. These chemicals called fungistat
- The chemicals may inhibit the spore production without affecting the growth of vegetative hyphae are called Antisporulant
- Necrosis: Death of plant tissue sometimes in spots
- Chlorosis: Yellowing of plant time due to limitation on chlorophyll synthesis
- Fungicide which is capable of eradicating a fungus after it has caused infection and there by curing the plant is called chemotherapeutant. eg. Carboxin, Oxycarboxin antibiotics like Aureofungin.
- Eradicant are those which remove pathogenic fungi from aninfection court (area of the host around a propagating unit of a fungus in which infection could possibly occur). eg. Organic mercurials, lime, sulphur, dodine etc
- Antibiotic is defined as a chemical substance produced by one micro-organism which is low concentration can inhibit or even kill other micro-organism
- Streptomycin is an antibacterial, antibiotic produced by streptomyces griseus
- Aureofungin is a hepataene antibiotic produced in sub-merged culture of *Streptoverticillium cinnamomeum* var. terricola.
- Griseofulvin: This antifungal antibiotic was first discovered to be produced by *Penicillium griseofulvum*
- The fungus, *Tichoderma vitide* is a biocontrol agent, mainly used for the control of root rot diseases
- *Pseudomonas fluorescens* is bacterium effectively used in controlling sheath blight and wilt diseases.
- *Pseudomonas fluorescens* is a gram negative, rod shaped nonspore forming bacteria which may be mono or lopotrichous or non motile.
- Bordeaux paste is a wound dresser and used to protect the wounded portions, cut ends of trees etc., against the infection by fungal pathogens.
- Bordeaux paste is prepared by mixing 1 kg of copper sulphate and 1 kg of lime in 10 litres of water (1:1:10)

- Burgundy mixture is prepared by mixing 1 kg of copper sulphate and 1 kg of sodium carbonate in 100 litres of water and also called as 'Soda Bordeaux'
- Burgundy mixture used in copper-sensitive crops.
- Cheshunt compound is compound usually prepared by mixing 2 parts of copper sulphate and 11 parts of ammonium carbonate
- Lime Sulphur is a product obtained by the combination of lime with sulphur and chemically it is a mixture of calcium thiosulphate and calcium polysulphides.

Forest Trees Diseases

Teak (*Tectona grandis*)

Powdery mildew : *Uncinula tectonae*
Rust : *Olivea tectonae (Uredo tectonae)*
Leaf spot/blight : *Phomopsis tectonae*
Leaf blight : *Alternaria alternata*
Butt rot/ Root rot : *Polyporus zonalis*
Ganoderma Root disease : *Ganoderma lucidum*

Jatropha (*Jatropha curcus*)

Damping off : *Fusarium solani*
Root rot : *Macrophomina phaseolina*
Powdery Mildew : *Oidium* sp.
Leaf blight : *Alternaria alternata*
Leaf blight : *Phoma* sp.
Leaf spot : *Cercopsora jatropha- curcus*
Anthracnose : *Colletotrichum gloeosporioides*
Mosaic : Virus

Neem (*Azadirachta indica*)

Nursery Root rot/ Web blight : *Rhizoctonia solani*
Leaf spot : *Cercospora meliae*
Leaf blight : *Alternaria alternata*
Root rot : *Ganoderma lucidum*
Flowering parasite/Giant Mistle toe- *Dendrophthoe falcata*

Eucalytus (*Eucalyptus tereticonis*)

Leaf spot/Die back : *Colletotrichum eucalytii*
Leaf spot : *Cylindrocaldium quinqueseptatum*
Leaf blight : *Alternaria alternate*
Stem canker : *Diaporthe cubensis*
Rust : *Puccinia psidii*

Casuarina (*Casuarina equisetifolia*)

Leaf spot/ Needle blight : *Phoma casuarinae*

Root rot : *Ganoderma lucidum*
Flowering parasite : *Cuscuta campestris*
Flowering Parasite/Giant Mistle toe : *Dendrophthoe falcata.*

Bamboo (*Bambusa bambos*)

Leaf spot : *Bipolaris bambusae and Bipolaris maydis*
Leaf blight : *Helminthosporium turcicum.*
Leaf blight : *Alternaria alternata*
Anthracnose : *Colletotrichum gloeosporioides*
Rust : *Puccinia bambusicola*
Culm rot : *Fusarium moniliforme, F.equiseti, F.solani*
Black mildew : *Meliola bambusicola*
Sooty mould : *Capnodium braziliense*
Stripe mosaic : Bamboo Mosaic virus
Streak mosaic : wheat streak mosaic virus

Simaruba (*Simarouba glauca*)

Leaf spot : *Cercospora* sp
Leaf blight/spot : *Helminthosporium* sp.
Root rot : *Ganoderma lucidum*
Sooty mould : *Capnodium braziliense*

Pungam (*Pongamia pinnata*)

Leaf spot : *Cercospora pongamiae*
Leaf blight : *Alternaria alternata*
Leaf blight : *Phoma indica*
Sooty mould : *Capnodium braziliense*
Flowering Parasite : *Dendrophthoe falcata.*

Sissoo (*Dalbergia sissoo*)

Leaf spot : *Cercospora sissoo*
Leaf blight : *Alternaria alternata*
Leaf blight : *Phoma indica*
Anthracnose : *Colletotrichum sissoo*
Powdery mildew : *Phyllactinia sissoo*
Rust : *Uredo sissoo*
Root rot : *Ganoderma lucidum*

Silver oak *(Grevillea robusta)*

Grey leaf spot : *Botryodiplodia theobromae*
Leaf spot : *Cercospora agharker*
Flowering Parasite : *Dendrophthoe falcata.*

Ailanthus *(Ailanthus excelsa)*

Leaf spot : *Cercopsora glandulosa*
Leaf blight : *Alternaria* sp.

Semul (*Bombax ceiba*)

Leaf spot : *Cercospora bambicina*
Leaf spot : *Myrothecium roridum*
Sooty mould : *Capnodium braziliense*

Sandal (*Santalam album*)

Spike : Phytoplasma
Flowering Parasite : Dendrophthoe falcata.

Albizia (*Albizia saman)*

Leaf spot : *Cercospora albiziae*
Leaf spot : *Helminthosporium albizicolum*
Root rot : *Ganoderma lucidum*
Flowering Parasite : *Dendrophthoe falcate*

Gmelina (*Gmelina arborea*)

Leaf spot : *Cercospora ranjita*
Powdery mildew : *Phyllactinia corylea*

Peltophorum *(Peltophorum ferugineum)*

Leaf Blight : *Alternaria alternata*
Root rot : *Ganoderma lucidum*

Mahua *(Mathuca latifolia)*

Grey Leaf spot : *Pestatiopsis mathucae*
Leaf Blight : *Phoma* sp.
Sooty Mould : *Capnodium indicum*
Red rust : *Cephaleuros parasiticus*

Bauhinia *(Bauhinia purpurea)*

Leaf spot : *Cercospora bauhiniae, Phyllosticta bauhiniae*

Leucaena *(Leucaena leucophloea)*

Leaf Spot : *Cercospora cruenta*
Leaf blight : *Alternaria alternata*
Sooty Mould : *Capnodium indicum*
Root rot : *Ganoderma lucidum*

Hardwickia (*Hardwickia blnata*)

Leaf spot : *Phoma* spp.

MULTIPLE CHOICE QUESTION'S

1. Sandal spike disease of sandal *(Santalum album)* is caused by
   (a) Bacteria (b) Fungi
   (c) MLO (d) Virus
2. Fungi imperfecti includes in
   (a) Deuteromycotina
   (b) Basidiomycotina
   (c) Ascomycotina
   (d) Oomycetes
3. Rust includes in
   (a) Deuteromycotina
   (b) Basidiomycotina
   (c) Ascomycotina
   (d) Oomycetes
4. Smut includes in
   (a) Deuteromycotina
   (b) Basidiomycotina
   (c) Ascomycotina (d) Oomycetes
5. Powdery mildew include in
   (a) Deuteromycotina
   (b) Basidiomycotina
   (c) Ascomycotina (d) Oomycetes
6. Downy mildew include in
   (a) Oomycetes (b) Chytridiomycetes
   (c) Zygomycetes
   (d) Hypochytridiomycetes
7. Father of Indian Mycology
   (a) EJ Butler (b) KC Mehta
   (c) Mundakur (d) RS Singh
8. Micrografting is used to produce plants free from
   (a) Virus (b) RLO's
   (c) MLO's (d) Bacteria
9. Bacterial diseases are controlled by use of chemicals
   (a) Kelthane (b) Fungicide
   (c) Antibiotics (d) Viricides
10. Application of potash increases
    (a) Resistance to water logging
    (b) Frost resistance in plants
    (c) Disease resistance in plants
    (d) None of these
11. Decomposition of organic matter in submerged soil is carried out by
    (a) Bacteria (b) Actinomycetes
    (c) Fungi (d) Earthworm
12. Select the pair which is not correctly matched
    (a) Longidorus – Needle nematode
    (b) Pratylenchus – Lesion nematode
    (c) Ditylenchus – Bulb nematode
    (d) Tylenchulus – Cyst nematode
13. The root – knot nematode generally have interaction with
    (a) *Pseudomonas solanacearum*
    (b) *Xanthomonas citri*
    (c) *Bacillus subtilis*
    (d) *Streptomyces scabies*
14. Select the systemic fungicide was first time discovered by Von Schmeling and Kulka?
    (a) Pyrimidines (b) Triazoles
    (c) Metalaxyl (d) All of these
15. Major genes contributes disease resistance in the form of
    (a) Vertical resistance
    (b) Horizontal resistance
    (c) Both (a) & (b) (d) None
16. Most widely used fungicide for smut fungi is
    (a) Vitavax (b) Plantvax
    (c) Dithane M-45 (d) Dithane Z-78
17. Race specific resistance called
    (a) Vertical resistance
    (b) Horizontal resistance
    (c) Adult plant resistance
    (d) Apparent resistance

18. When plant showed the partial resistance against all the races of pathogen then it is a type of?
(a) Horizontal resistance
(b) Vertical resistance
(c) Induced resistance
(d) Non-host resistance

19. Nuclear polyhedrosis virus (NPV) is the most effective for control of
(a) *Chilo partellus*
(b) *Pectionophora gossypiella*
(c) *Helicoverpa armigera*
(d) *Diacrasia oblique*

20. Solar treatment is used for
(a) Loose smut (b) Stem rust
(c) Powery mildew (d) Blast

21. A fungus disease which may cause death or damage to newly emerging seedlings is called ______.
(a) Ground rot (b) Damping off
(c) Mulching (d) Powdery mildew

22. The study of diseases of forest trees is called:
(a) Forest dendrology
(b) Forest entomology
(c) Forest pathology
(d) Forest technology

23. For which of the following pests would a fungicide be used?
(a) Aphids (b) Damping-off
(c) Scale (d) Sunscald

24. Pathogens are:
(a) Disease causing organisms
(b) Wood boring insects
(c) Used in biological control
(d) a and c

25. Other than wildfires, which of the following forest enemies are the most damaging?
(a) Diseases and insects
(b) Domestic livestock
(c) Environment
(d) Wildlife

26. 2,4-D is a(n):
(a) Fungicide (b) Herbicide
(c) Insecticide (d) Pesticide

27. Fungicides are chemicals used to control ___.
(a) Rust diseases
(b) Fungus diseases
(c) Powdery mildew
(d) All of the above

28. The disease fire blight is caused by ___.
(a) Fungus
(b) Virus
(c) Bacteria
(d) All of the above

29. Salt damage to a tree is identified by______.
(a) Bent or broken stems
(b) Foliage necrosis
(c) Tip curl
(d) Twisting of needles or leaves

30. A symptom of winter injury in conifers may be _____.
(a) Blackened and dead roots
(b) Foliage turns red and dies
(c) Small spots on the needles
(d) Stem cankers

31. Which of the following is symptomatic of needle cast diseases?
(a) Brown needles, die back and top kill
(b) Needles bent at the end of the fascicle
(c) Orange discoloration at the base of the needles
(d) Spots on the needles, death of the affected needles

32. For which of the following would a fungicide be used?
(a) Aphids
(b) Damping-off
(c) Scale
(d) Sunscald

33. The introduction of a pathogen that attacks certain insects is called a ______.
(a) Biological control
(b) Chemical control
(c) Management control
(d) Mechanical control

34. Forest patholcgy is study of _________.
(a) Forest ecosystems
(b) Forest insects
(c) Tree diseases
(d) Tree identification

35. Abnormal tissue growth on trees that is cussed by insects or mites is called ____
(a) Conk (b) Fungus
(c) Gall (d) Pitch

36. A canker would best be described as a:
(a) Dead tree top (b) Globule swelling
(c) Sunken area (d) Tumor

37. The best way to control forest insects or diseases is to _________.
(a) Spray as soon as a disease or insect is noticed
(b) Use integrated pest management
(c) Harvest all trees near the infestation immediately
(d) All of the above

38. The vector for spike disease in sandal is
(a) Jassid (b) Psyllid
(c) Green plant hopper
(d) White fly

39. Root rot in sal is caused by
(a) Pseudomonas (b) Polyporus
(c) Ganoderma (d) Pythium

40. *Corticium salmonicolor* is a disease which attacks
(a) Casuarina (b) Acacia
(c) Eucalyptus (d) Tamarind

41. An organism on or within which another organism feeds and develops is called
(a) Parasites (b) Epiphytes
(c) Host (d) None of the above

42. Stemwilt in Casuarina is caused by
(a) Pseudomonas (b) Trichosporium
(c) Corticium (d) Ganoderma

43. Fungal roots are otherwise called as
(a) Frankia (b) Mycorrhiza
(c) Rhizobium (d) Azospirillum

44. Which of the following is considered a parasitic cause of disease?
(a) Drought (b) Fungi
(c) Ice storms (d) Salt

45. __________, before 1910 initiated exhaustive study of Indian fungi and disease caused by them.
(a) E.J. Butler (b) J.C. Mosemam
(c) R.E. Nelson (d) A. Livey

46. A bacterium is devoid of:
(a) Chlorophyll
(b) Well defined nucleus
(c) Cell wall
(d) Both a and b

47. A virus particle with both nucleic acid and protein coat is known as:
(a) Virion (b) Viriod
(c) Capsid (d) All above

48. Agar-agar is produced from:
(a) Nostoc (b) Macrocrocystis
(c) Gelidium (d) oscillatoria

49. Rust is a:
(a) Obligate parasite
(b) Saprophyte parasite
(c) Facultative parasite
(d) predator

50. Alexander fleming (1929) discovered:
(a) Streptomycetease
(b) Penicillin
(c) Tyrocidine
(d) Giseofulvin

51. Association of tinsel flagellum is characteristics of:
(a) Lichen (b) Mycorrhiza
(c) Ascolichen (d) Basidiolichen

52. B.B. Mundkur wrote:
(a) Fungi and disease in plants
(b) Fungi and plant diseases
(c) The fungi
(d) All of the above

53. Bacilli are:
(a) Spiral shaped
(b) Spherical shaped
(c) Rod shaped
(d) None of the above

54. Bacterial cell surrounded by flagella is called:
(a) Monotrichous
(b) Lophotrichous
(c) Amph itricous
(d) Peritrichous

55. Plant pathology is ____________ word.
(a) English (b) Latin
(c) Greek (d) Mandarin

56. Bordeaux mixture was discovered by P.A. Millardet of france during the year 1882 following his chance observationof farmer's practice for protection against:
(a) *Plasmopara viticola* on grapevine
(b) *Uncinula nector* on grapevine
(c) *Podosphaera leucotricha* on apple
(d) *Venturia inequalis* on apple

57. Carborundum powder is:
(a) Silicon carbide
(b) Aluminium oxide
(c) Diatomaceous earth
(d) Calcium carbonate

58. Cocci are:
(a) Spiral shaped
(b) Spherical shaped
(c) Rod shaped
(d) None of the above

59. Crystalization of virus infected plant described as:
(a) Mayer (b) Stanely
(c) Iwanowski (d) Bejerinck

60. Cysts are formed by:
(a) Meloidogyne (b) Aphelenchoides
(c) Heterodera (d) None of the above

61. Death of tissue in virus infected plant described as:
(a) Mottling (b) Mosaic
(c) Necrosis (d) Curling

62. Dithane M-45 is a:
(a) Bactericide (b) Fungus
(c) Fungicide (d) Nematicide

63. First plant parasitic bacteria was reported by:
(a) T.J. Burrill (b) Needham
(c) Louis pasteur (d) Leeuwenhoeck

64. First plant parasitic nematode was discovered by:
(a) Meloidogyne (b) Heterodera
(c) Anguina (d) Globodera

65. First plant parasitic nematode was reported by:
(a) Needham (b) Spallanzani
(c) Cobb (d) Mayer

66. Fungi which can grow on living host plant are called:
(a) Obligate saprophyte
(b) Obligate parasite
(c) Facultative parasite
(d) Saprophytes

67. Fusarium belong to order:
(a) Melaconiales
(b) Moniliales
(c) Sporobolomycetales
(d) Cryptococcales

68. Gelatin plate technique was discovered by:
(a) Burrill (b) Fanny hesse
(c) Robert Koch (d) Louis Pasteur

69. How much quantity of nematicide is used to control the nematodes in one hectare are?
(a) 10-15 kg (b) 20-25 kg
(c) 50-60 kg (d) 100-200 kg

70. In fermentation industry, Pasteurization is done at temperature:
(a) 90.7°C (b) 100°C
(c) 62.8°C (d) 80.0°C

71. Indian journal of Mycology and Plant Pathology is published from:
(a) IARI, New Delhi
(b) Botany Department, Lucknow University
(c) R.C.A., Udaipur
(d) Botany Department, Madras University

72. Indian Phytopathological Society was established by:
(a) J.F. Dastur (b) E.J. Butler
(c) B.B. Mundkar (d) K.C. Mehta

73. Lichen is an association between:
(a) Algae and Fungus
(b) Bacterium and Fungus
(c) Fungus and Virus
(d) Fungus and Nematode

74. Most effective fungicide against powdery mildew disease:
(a) Dithane (b) Captan
(c) Karathane (d) Bavistin

75. Most of the plant parasitic nematodes belong to order:
(a) Tylenchida (b) Dorylaimida
(c) Aphelenchida (d) Rhabditida

76. Mycology is the study of:
(a) Bacteria (b) Algae
(c) Virus (d) Fungi

77. Mycoplasma is sensitive to:
(a) Penicillin (b) Steptomycin
(c) Bavistin (d) Tetracycline

78. Nematodes are also known as:
(a) Tapeworms
(b) Trematodes
(c) Earthworms
(d) Round worms

79. Once the pathogen has established in an area, steps taken to remove or destroy is known as:
(a) Exclusion (b) Eradication
(c) Protection (d) Management

80. Phycomycetes belong to:
(a) Fungi (b) Bacteria
(c) Yeast (d) None of above

81. Phytophthora belons to the family:
(a) Peronophythoraceae
(b) Pythiaceae
(c) Parenosporaceae
(d) Albuginaceae

82. Plant parasitic nematodes posses:
(a) Two-part oesophagus
(b) Three-part oesophagus
(c) Four-part oesophagus
(d) None of these

83. Plant pathogenic viroids contain:
(a) Externally seed-borne disease
(b) Externally seed-borne and soil-borne disease
(c) Air-borne disease
(d) Internally seed-borne disease

84. Powdery mildew disease is caused by:
(a) Bacterium (b) Fungus
(c) Virus (d) Nematode

85. Powdery mildew fungi belong to the family:
(a) Erysiphaceae
(b) Hypocereaceae
(c) Hypomycetaceae
(d) Nectriaceae

86. Powdery mildews are controlled by spray of:
(a) Copper fungicides
(b) Mercury fungicides
(c) Sulphur fungicides
(d) Miscellaneous fungicides

87. Rhizopus belongs to class:
(a) Oomycetes (b) Zygomycetes
(c) Trichomycetes (d) Ascomycetes

88. Root rot of papaya could be controlled by application of:
(a) Bordeaux mixture
(b) Dithane M-45
(c) Benelate
(d) Calixin

89. The founder of mycology was:
(a) Linnaeus (b) Anton de bary
(c) P.A. Micheli (d) Saccardo

90. The organisms lacking a well defined nucleus are termed:
(a) Fungi (b) Prokaryote
(c) Eukaryote (d) Protozoa

91. The plant mycoplasmal diseases are transmitted by:
(a) Aphids (b) Leaf hoppers
(c) Mites (d) Above all

92. Virus is transmitted by nematode:
(a) Globodera (b) Tylenchorhynchus
(c) Anguina (d) Xiphinema

93. Which fungus produces aflatoxin?
(a) *Penicillium notatum*
(b) *Aspergillus flavus*
(c) *Albugo candida*
(d) *Trichoderma viride*

94. Which is citrus nematode?
(a) Hoplolaimus
(b) Pratylenchus
(c) Rotylenchulus
(d) Tylenchus

95. Which is sulphur bacteria?
(a) Nitrosomonas (b) Cladethrix
(c) Thiothrix (d) Bacillus

96. Which is the root knot nematode?
(a) Angunia (b) Heterodera
(c) Meloidogyne (d) Tylenchorhynchus

97. Which of the following nematicide is a carbamate?
(a) DBCP (b) DD
(c) Phorate (d) Aldicarb

98. Which of the nutrient is supplied by mycorrhiza to the plant?
(a) Nitrogen (b) Phosphorus
(c) Sulphur (d) Iron

99. Which of them is the gram positive bacterium?
(a) Xanthomonas
(b) Pseudomonas
(c) Erwinia
(d) Corynebacterium

100. Which of them is a poisonous mushroom?
(a) Agaricus (b) Volvariella
(c) Pleaurotus (d) Amanita

101. Which one of the following is known as downy mildew fungus?
(a) Erysiphe (b) Sclerospora
(c) Taphrina (d) Urocystis

102. Which one the following tissues is most suitable for tissue culture to produce virus-free plants of potato?
(a) Leaf tissue
(b) Pollen
(c) Embryo tissue
(d) Meristem tip tissue

103. Which shows the symbiotic relationship?
(a) Rhizobium (b) Azotobacter
(c) Clostridium (d) Nitrosomonas

104. Who for the first time discovered antibiotics?
(a) Alexander flemming
(b) Stanely
(c) Anton de bary
(d) Louis Pasteur

105. Who gave the term pasteurization?
(a) Anton de Bary
(b) Louis Pasteur
(c) K.C. Mehta
(d) Leeuwenhoeck

106. Who is known as the father of nematology?

(a) N.A. Cobb

(b) G. Thorne

(c) F.G.W. Jones

(d) De-man

107. With regard to nutrient Erysiphe is:

(a) Ectoparasite

(b) Endoparasite

(c) None of the above

(d) Predator

108. ________ proposed a classification of fungi which is now widely accepted.

(a) Walker (1957)

(b) Stakman and harrar (1957)

(c) Whitaker (1969)

(d) Ainworth (1973)

ANSWERSHEET
18. Forest Pathology

---

1. **(c)**	2. **(a)**	3. **(b)**	4. **(b)**	5. **(c)**	6. **(a)**	7. **(c)**	8. **(a)**	9. **(c)**	10. **(c)**
11. **(c)**	12. **(d)**	13. **(a)**	14. **(d)**	15. **(a)**	16. **(a)**	17. **(a)**	18. **(b)**	19. **(c)**	20. **(a)**
21. **(b)**	22. **(c)**	23. **(b)**	24. **(a)**	25. **(a)**	26. **(b)**	27. **(d)**	28. **(c)**	29. **(b)**	30. **(b)**
31. **(d)**	32. **(b)**	33. **(a)**	34. **(c)**	35. **(c)**	36. **(b)**	37. **(b)**	38. **(a)**	39. **(b)**	40. **(c)**
41. **(a)**	42. **(b)**	43. **(b)**	44. **(b)**	45. **(a)**	46. **(b)**	47. **(c)**	48. **(c)**	49. **(a)**	50. **(b)**
51. **(b)**	52. **(b)**	53. **(c)**	54. **(d)**	55. **(c)**	56. **(a)**	57. **(a)**	58. **(a)**	59. **(b)**	60. **(a)**
61. **(b)**	62. **(c)**	63. **(a)**	64. **(c)**	65. **(a)**	66. **(b)**	67. **(b)**	68. **(c)**	69. **(b)**	70. **(c)**
71. **(c)**	72. **(c)**	73. **(a)**	74. **(c)**	75. **(a)**	76. **(d)**	77. **(d)**	78. **(a)**	79. **(b)**	80. **(a)**
81. **(b)**	82. **(a)**	83. **(a)**	84. **(b)**	85. **(a)**	86. **(c)**	87. **(b)**	88. **(a)**	89. **(b)**	90. **(b)**
91. **(b)**	92. **(d)**	93. **(b)**	94. **(d)**	95. **(d)**	96. **(c)**	97. **(d)**	98. **(b)**	99. **(d)**	100.**(d)**
101.**(b)**	102.**(d)**	103. **(a)**	104. **(a)**	105. **(b)**	106. **(a)**	107. **(a)**	108. **(d)**		

19

FOREST SURVEY AND ENGINEERING

KEY NOTES

- In remote sensing, reflected energy from the object is recorded by a sensor.
- Forest surveying is the art of making observations and measurements either linear or angular by applying the principles of trigonometry and geometry.
- Plane table survey is survey in which curvature of earth is not taken into account, as the survey is extend over small areas.
- Geodetic surveying is the curvature of the earth is taken into consideration. It is carried over a large area greater than 250km^2.
- Topographical survey is used to determine natural features of a country such as valleys, rivers and artificial features such as road, railways, etc.
- Cadastral survey is used to determine boundaries of field / estate
- City survey is used to locate premises, streets, water supply and drainage systems of a city
- Engineering survey is used to collect detailed data to design a project involving roads and railways etc.
- Axis of level tube is a straight line tangential to the longitudinal curve of the level tube at its center.
- Axis of level tube is also known as the bubble line
- Axis of telescope is the line joining the optical center of the object glass to the center of the eyepiece.
- Centering is the operation carried out to ascertain the fact that the vertical axis of the instrument passes through the center of the peg fixed at the required station point.
- Changing face is the operation of changing the position of vertical circle either from left to right or from right to left of the observer.
- Line of sight or line of collimation is used to indicate an imaginary line joining the optical center of the object glass and intersection of the crosshairs of the diaphragm and its continuation.
- Transitting is the operation of revolving the telescope in a vertical plane by 180° about the horizontal or transverse axis. It is also referred to as plunging or reversing.
- Levelling is defined as a method of determining the relative elevations of points on the surface of earth or an operation for finding out the difference in heights.

- A contour is an imaginary line, which joins the points of equal elevation on the ground. The plan showing the elevations and depressions of the surface of the ground is known as the contour plan or contour map.
- The vertical distance between any two consecutive contours in known as contour interval.
- The horizontal distance between two points on two consecutive contours is known as the horizontal equivalent.
- The process of placing or spacing the contour lines proportionally between the plotted ground points is known as the interpolation of contours.
- Geographic maps provide an overall idea of the mapped area in relation to climate, population, relief, vegetation, and hydrography.
- A pier is a raised walkway over water, supported by widely spread piles or pillars.

MULTIPLE CHOICE QUESTION'S

1. A broken stone as a material for road making is known as____
(a) Road Metal (b) Flags
(c) Slates (d) Limestone

2. ________ are primarily formed due to the consolidation of magma.
(a) Igneous rocks
(b) Metamorphic rocks
(c) Sedimentary rocks
(d) Secondary rocks

3. The magma solidifies on the earths' crust of the rocks is known as
(a) Extrusive (b) Intrusive
(c) irrtrusive (d) imtrusive

4. The magma solidifies beneath the surface is known as
(a) Extrusive (b) Intrusive
(c) irrtrusive (d) imtrusive

5. Identify the prominent igneous rocks
(a) Basalt (b) Sandstone
(c) Limestone (d) Quartzite

6. Bricks contains ________per cent of sand (Silica)
(a) 20-30% (b) 30-40%
(c) 40-50% (d) 50-60%

7. Bricks consists of ________per cent of Clay (Alumina)
(a) 20-30% (b) 30-40%
(c) 40-50% (d) 50-60%

8. The Standard size of the brick is
(a) 10 X 5 X 2.5 cm
(b) 10 X 7.5 X 4 cm
(c) 20 X 10 X 5 cm
(d) 20 X 12 X 6 cm

9. The crushed bricks or specially prepared round ball of clay burnt and pounded to impalpable powder is____
(a) Cement (b) Surkhi
(c) Mortar (d) Brick

10. _____ is a mixture of lime or cement, Surkhi and water.
(a) Cement (b) Surkhi
(c) Mortar (d) Brick

11. Expand RCC
(a) Resembled Concrete Cement
(b) Reinforced Concrete Cement
(c) Resembled Cement Concrete
(d) Reinforced Cement Concrete

12. ________ is the one complete layer of bricks or stone construction
(a) Course (b) Bed Joint
(c) Cross Joint (d) Lap

13. The layer of mortar between two vertical bricks or stones is ________
(a) Course (b) Bed Joint
(c) Cross Joint (d) Lap

14. ______ refers to a longitudinally cut half brick
(a) King closure (b) Queen closure
(c) Quoin (d) Bat

15. The brick cut in a manner that one end is half the width of a full brick.
(a) King closure (b) Queen closure
(c) Quoin (d) Bat

16. ______ is an external angle of a wall on construction.
(a) King closure (b) Queen closure
(c) Quoin (d) Bat

17. The horizontal distance between the vertical joints in two successive courses
(a) King closure (b) Queen closure
(c) Lap (d) Bat

18. ________ is the first step in alignment of forest roads
(a) Camber (b) Drains
(c) Reconnaissance
(d) Metalling

19. The part of the road between the edge of the actual road and the side drain is known as ________
(a) Shoulders (b) Camber
(c) Reconnaissance
(d) Drains

20. _____ is the transverse raising of the road in the centre above its edges.
(a) Shoulders (b) Camber
(c) Reconnaissance
(d) Drains

21. The mixture of clay, sand and chopped straw is ________ bricks
(a) Sundried bricks
(b) Concrete bricks
(c) Burnt bricks (d) Mixture bricks

22. The mixture of cement and sand is ________ bricks
(a) Sundried bricks
(b) Concrete bricks
(c) Burnt bricks (d) Mixture bricks

23. A Jhama brick is also known as ______
(a) Just burnt (b) Under burnt
(c) Over burnt (d) Not Burnt

24. Arrangement of wedge-shaped blocks mutually supporting each other in building construction is______
(a) Arches (b) Lintels
(c) Bricks (d) Stones

25. The main principle of surveying is to work from
(a) Higher level to lower level
(b) Part to whole
(c) Whole to part
(d) Lower level to higher level

26. Systematic errors are those
(a) Whose effect are cumulative and can be eliminated
(b) Which change rapidly
(c) Which cannot be recognized
(d) Whose character is not understood

27. The correction for sag is
(a) Always additive
(b) Always zero
(c) Always subtractive
(d) None of these

28. In chain surveying field work is limited to
(a) Linear measurements only
(b) Both linear and angular measurements
(c) Angular measurements only
(d) None of these

29. The smallest length that can be drawn in map is
(a) 0.5 mm (b) 10 mm
(c) 0.2 mm (d) 15 mm

30. Surveys which are carried to provide a national grid of control for preparation of accurate maps of large areas are known as
(a) Topographical survey
(b) Geographical survey
(c) Cadastral survey
(d) Hydrographic survey

31. Oblique offsets are used to
(a) Locate broken boundaries
(b) Check the accuracy of chain survey
(c) Locate boundary lines of property
(d) Plot the chainage

32. Cross staff is used for
(a) Setting out right angles
(b) Measuring vertical angle
(c) Measuring horizontal angle
(d) Both b & c

33. The range of dip of magnetic needle is
(a) 0-30° (b) -90° to 90°
(c) -45° to 45° (d) -30° to 30°

34. The size of thedoloite is defined by
(a) Diameter of graduated circle of lower plate
(b) Height of thedoloite
(c) Length of telescope
(d) Diameter of graduated circle of upper plate

35. Structure designed and constructed for passage of water is
(a) Dam (b) Retaining wall
(c) Road (d) Culvert

36. Staff reading over a station whose elevation is known is called
(a) Foresight (b) Intermediate sight
(c) Back sight (d) All the above

37. The relief of earth's surface can be represented by
(a) Hachures (b) Contours
(c) Shading (d) All the above

38. The line joining points of equal elevation on the surface of the earth are known as
(a) Isohytes (b) Isogonics
(c) Contours (d) Agonics

39. The method suitable for computing the area when the boundary line departs considerable from a straight line is
(a) Mid ordinate rule
(b) Simpsons rule
(c) Average ordinate rule
(d) Trapezoidal rule

40. Temporary increase in volume of sand is
(a) Drying (b) Bulking
(c) Sizing (d) Slaking

41. The time required for setting of quick setting cement is .
(a) 60 min (b) 5 min
(c) 10 min (d) 1 min

42. Mortar is applied with
(a) Trowel (b) Plum bob
(c) Hammer (d) Peg

43. In-----exposed joints of brick work are treated with lime or cement
(a) Pointing (b) Damp proofing
(c) Plastering (d) Centering

44. ------------ is extension of base structures such as walls and columns
(a) Beam (b) Foundation
(c) Lintel (d) Terrace

45. The ratio of rise to span in roofing is
(a) Spacing (b) Camber
(c) Pitch (d) Eaves

46. Vertical face of the stair is known as
(a) Riser (b) Tread
(c) Flier (d) Winder

47. Main roads connecting head quarters of ranges, divisions and important blocks are
(a) Main motorable roads
(b) Bridle paths
(c) Branch jeepable road
(d) Inspection paths

48. Process of forming surface of the roads is
(a) Laying (b) Rolling
(c) Metalling (d) Watering

49. For alignment of forest roads-------- is used
(a) Contour map
(b) Datum surface
(c) Mean sea level
(d) Line of collimation

50. Cross drainage work used for crossing natural obstacles
(a) Roads (b) Bridges
(c) Culverts (d) Retaining walls

51. Culverts are small bridges with a span less than
(a) 4 m (b) 5 m
(c) 10 ft (d) 100 ft

52. End support for super structure of bridges is termed as
(a) Abutment (b) Supporting beam
(c) Pillar (d) Column

53. Small checkdam constructed in gullies having a depth of 2.5 mts
(a) Drystone check dam
(b) Woven wire check dam
(c) Single row brush wood dam
(d) Log check dam

54. Type of spillway which is an open channel with uniform but steep slope
(a) Drop spillway
(b) Chute spillway
(c) Drop inlet spillway
(d) None of these

55. The length of Engineers chain is
(a) 20 m (b) 100 ft
(c) 66 ft (d) 30 m

56. Earth is assumed as plane upto 250 sq.km in
(a) Plane surveying
(b) Traverse Surveying
(c) Geodetic surveying
(d) Topographical surveying

57. Determination of relative heights of points on earth surface is called as
(a) Contouring (b) Surveying
(c) Levelling (d) Traversing

58. Scales in different units are called as
(a) Plain scale (b) Diagonal scale
(c) Vernier scale (d) Comparative scale

59. Horizontal distance between two consecutive contour lines
(a) Contour Interval
(b) Vertical interval
(c) Horizontal equivalent
(d) relief

60. Details like forest boundaries and encroachments are identified using
(a) Hachures (b) Form lines
(c) Map (d) Contours

61. Representative factor/ Representative fraction is
(a) Fraction expressing linear ratio or scale of map
(b) Ratio of map distance to corresponding distance on ground
(c) Fixed ratio to which ground distances have been reduced
(d) All the above

62. ------------ is attached to the wheel of cycle to record the number of revolutions
(a) Pedometer
(b) Hypsometer
(c) Passometer
(d) Sensor

63. When Sequence of connected lines does not return to the starting point it is called as

(a) Open traversing
(b) Topographic Surveying
(c) Closed traversing
(d) Triangulation

64. Line joining apex of triangle to fixed point on opposite side

(a) Tie line (b) Base line
(c) Offset (d) Check line

65. The angle between true meridian and line taken in clockwise direction is

(a) Geographical meridian
(b) Arbitrary meridian
(c) Magnetic meridian
(d) Azimuth

66. Dip of the needle is zero at

(a) Top of hill
(b) Tropic of Capricorn
(c) Tropic of cancer
(d) Equator

67. Process of reversing the telescope of theodolite about horizontal axis is known as

(a) Transiting (b) Levelling
(c) Orientation (d) Focussing

68. Mean Sea Level (MSL) adopted by survey of India for reference, is located at

(a) Mumbai (b) New Delhi
(c) Calcutta (d) Kanyakumari

69. ------------is a tunnel carrying a stream under a road or railway

(a) Bridge
(b) Culvert
(c) Retaining wall
(d) Conduit

70. Cross slope provided to raise middle of the road surface in the transverse direction to drain off rain water from road surface

(a) Crown (b) Camber
(c) Kerb (d) Shoulder

71. Type of foundation which are spread across the entire area of the building to support heavy structural loads

(a) Raft foundation
(b) Pile foundation
(c) Strip foundation
(d) Caisson foundation

72. The last reading taken from the instrument is called

(a) Back sight (b) Fore sight
(c) Intermediate sight
(d) End sight

73. Indentation or depression on the top face of a brick made with the object of forming a key for the mortar

(a) Frog (b) Plinth
(c) Closer (d) Stretcher

74. Rocks formed by temperature and pressure changes inside the Earth

(a) Igneous rock
(b) Sedimentary rock
(c) Metamorphic rock
(d) None of these

75. Hydrology helps in

(a) Predicting maximum flows
(b) Deciding the minimum reservoir capacity
(c) Forecasting the availability of quantity of water at reservoir site
(d) All the above

76. Pick up the correct equation from the following

(a) Run off = Surface run off + Ground water flow
(b) Run off = Surface run off - Ground water flow
(c) Run off = Surface run off / Ground water flow
(d) Run off = Surface run off x Ground water flow

77. Infiltration capacity of soil depends upon
(a) Number of voids present in the soil
(b) Shape and size of soil particles
(c) Compaction of the soil particles
(d) All the above

78. Absolute humidity in air
(a) Decreases at higher altitudes
(b) Increases at higher altitudes
(c) Remains constant at all altitudes
(d) None of these

79. The main factor which affects the infiltration capacity, is
(a) Thickness of saturated layer
(b) Depth of surface detention
(c) Soil moisture
(d) all the above

80. Precipitation caused by lifting of an air mass due to the pressure difference, is called
(a) Cyclonic precipitation
(b) Convective precipitation
(c) Orographic precipitation
(d) None of these

81. In India the recording type rain gauge generally used, is
(a) Weighing type
(b) Tipping type
(c) Float recording type
(d) None of these

82. In India, rain fall is generally recorded at
(a) 08.30 A.M. (b) 12.30 Noon
(c) 04.30 P.M. (d) 08.30 P.M.

83. The standard height of a standard rain gauge, is
(a) 10 cm (b) 20 cm
(c) 30 cm (d) 50 cm

84. For determination of average annual precipitation in a catchment basin, the best method is
(a) Arithmetical method
(b) Thiessen's mean method
(c) Isohyetal method
(d) None of these.

85. The initial basin recharge is equal to
(a) Interception
(b) depression storage
(c) rain absorbed by the moisture deficiency
(d) all the above

86. The run off is affected by
(a) type of precipitation
(b) rain intensity and duration of rainfall
(c) rain distribution and soil moisture deficiency
(d) all the above

87. The critical depth in a channel can be produced
(a) by raising the bottom of the channel
(b) by lowering the bottom of the channel
(c) by decreasing the width of the channel
(d) both (a) and (c) of above

88. The quantity of water retained by the sub-soil against gravity, is known
(a) Yield (b) porosity
(c) specific yield (d) specific retention

89. The rainfall cycle period in India is taken as
(a) 20 years (b) 25 years
(c) 30 years (d) 35 years

90. Non-recording rain gauge
(a) Collect the rain whose volume is measured by means of graduated cylinders
(b) Are cylindrical in shape
(c) Also known as Self recording rain guage
(d) Are generally used in hilly terrain

91. Hydrograph is a graphical representation of
(a) Surface run off
(b) Discharge flowing in the river
(c) Ground water flow
(d) Rain fall

92. Consumptive use of a crop during growth, is the amount of
(a) Interception (b) Transpiration
(c) Evaporation (d) All the above

93. Isohytes are the imaginary lines joining the points of equal
(a) Rainfall (b) Pressure
(c) Height (d) Humidity

94. The best instrument for measuring the velocity of a stream flow is
(a) Pitot tube
(b) Price's current meter
(c) Surface float
(d) Sub-surface float

95. The time required by rain water to reach the outlet of drainage basin, is generally called
(a) Time of overland flow
(b) Time of concentration
(c) Duration of the rainfall
(d) Concentration time of overland flow

96. Time of overland flow, is affected by
(a) Slope of the basin
(b) Type of the ground surface
(c) Length of the flow path
(d) All the above.

97. If the grain size of soil increases
(a) Surface area decreases
(b) Specific retention decreases
(c) Water supply in well increases
(d) All the above

98. Albedo means
(a) The ratio of the amount of solar radiation (or visible range of radiation) reflected by a surface
(b) The amount incident on it, also expressed as a percentage
(c) Both (a) & (b)
(d) None of the above

99. The average annual precipitation in India is
(a) 1100 mm (b) 1200 mm
(c) 1300 mm (d) 1400 mm

100. Equipment used to measure evapotranspiration under field condition is
(a) USWB Open pan evaporimeter
(b) Sunken screen evaporimeter
(c) Can evaporimeter
(d) Both b and c

101. The grand period of rainfall in India is
(a) South – West Monsoon
(b) South – East Monsoon
(c) North- East Monsoon
(d) North – West Monsoon

102. One of the forms of precipitation is
(a) Dew (b) Frost
(c) Sleet (d) Sublimation

103. The type of rainfall occurring in the equatorial region is
(a) Conventional
(b) Frontal
(c) Cyclonic
(d) Advectional

104. Humidity refers to
(a) Rainfall (b) Fog
(c) Clouds
(d) Atmospheric water vapour

105. Atmospheric humidity is measured by
(a) Radiometer
(b) Hydrometer
(c) Hygrometer
(d) Micrometer

ANSWERSHEET
19. Forest Survey and Engineering

1. (a)	2. (a)	3. (a)	4. (b)	5. (a)	6. (d)	7. (a)	8. (c)	9. (b)	10. (c)
11. (d)	12. (a)	13. (c)	14. (b)	15. (a)	16. (c)	17. (c)	18. (c)	19. (a)	20. (b)
21. (a)	22. (b)	23. (c)	24. (a)	25. (c)	26. (a)	27. (c)	28. (a)	29. (c)	30. (a)
31. (c)	32. (a)	33. (b)	34. (a)	35. (d)	36. (c)	37. (d)	38. (c)	39. (b)	40. (b)
41. (a)	42. (a)	43. (a)	44. (b)	45. (c)	46. (a)	47. (a)	48. (c)	49. (a)	50. (b)
51. (a)	52. (a)	53. (c)	54. (b)	55. (b)	56. (a)	57. (c)	58. (d)	59. (a)	60. (c)
61. (d)	62. (a)	63. (a)	64. (d)	65. (d)	66. (d)	67. (a)	68. (a)	69. (b)	70. (b)
71. (a)	72. (b)	73. (a)	74. (c)	75. (d)	76. (b)	77. (d)	78. (a)	79. (d)	80. (a)
81. (d)	82. (a)	83. (c)	84. (c)	85. (d)	86. (d)	87. (d)	88. (d)	89. (c)	90. (a)
91. (b)	92. (d)	93. (a)	94. (d)	95. (b)	96. (d)	97. (d)	98. (c)	99. (a)	100. (d)
101. (a)	102. (c)	103. (a)	104. (d)	105. (c)					

20

Forest Law and Policy

KEY POINTS

- CITES - Convention on International Trade in Endangered Species of Wild Fauna and Flora
- ITTA - International Tropical Timber Agreement
- WTO - World Trade Organization
- CBD - Convention on Biological Diversity
- UNFCCC - United Nations Framework Convention on Climate Change
- IPCC - The Intergovernmental Panel on Climate Change
- FORSPA – Forestry Support programme for Asia and Pacific
- 'Village forests' popularly termed for fuel forests
- LAW - Law is kind of rule or cannon wave by actions are framed.
- Reserved forest (I.F. act and F. acts) = Exclusively property of government villagers have no right
- Protected forest - Limited degree of protection
- Village forest - Assigned to a village community managed to supply forest produce to the village
- Unclassed forests - Owned by government no constituted in to RF; PF; VF.
- "scheduled animal" means an animal specified for the time being in Schedule I or Part II of Schedule II;
- "scheduled animal article" means an article made from any scheduled animal and includes an article or object in which the whole or any part of such animal
- The United Nations Framework Convention on Climate Change was signed by 154 governments in Rio de Janeiro during the United Nations Conference on Environment and Development (UNCED) in June 1992.
- Convention on Wetlands of International Importance especially as Water fowl Habitat Ramsar, 2 February 1971.
- The Paris Protocol to the Ramsar Convention on Wetlands, 1982
- The "Regina Amendments" to the Convention on Wetlands, 1987
- CITES was drafted as a result of a resolution adopted in 1963

- In 1988, the Intergovernmental Panel on Climate Change (IPCC) was created by the United Nations Environment Programme (UNEP) and the World Meteorological Organization (WMO) to assess the scientific knowledge on global warming.
- The Forest (Conservation) Act, 1980.

MULTIPLE CHOICE QUESTION'S

1. First Indian Forest Act was drafted in the year
(a) 1865 (b) 1848
(c) 1894 (d) 1927

2. The last forest policy was formulated during the year
(a) 1998 (b) 1972
(c) 1988 (d) 1952

3. Forest Conservation Act was enacted during
(a) 1880 (b) 1980
(c) 1988 (d) 1998

4. The first Inspector general of forest was
(a) Sir. Dietrich Brandis
(b) Howard
(c) Chaturvedi
(d) Alfred moller

5. The National Commission on Agriculture submitted its report during
(a) 1970 (b) 1976
(c) 1967 (d) 1980

6. The first Independent National Forest Policy was enunciated during
(a) 1950 (b) 1952
(c) 1954 (d) 1956

7. Forest and wildlife were brought under concurrent list as per
(a) 40th amendment
(b) 42nd amendment
(c) 44th amendment
(d) 48th amendment

8. The Head Quarters of Inspector General of Forest
(a) Dehra Dun (b) Missouri
(c) Delhi (d) Mumbai

9. The first Forest Policy was enunciated during
(a) 1864 (b) 1874
(c) 1884 (d) 1894

10. Declaration of national parks is covered in which section of wildlife protection act, 1972?
(a) Section 25 (b) Section 35
(c) Section 36 (d) Section 37

11. Conservation of ecological fragile ecosystem and preservation of biological diversity in terms of flora and fauna was emphasized in
(a) Forest policy, 1894
(b) Forest policy, 1952
(c) Forest policy, 1988
(d) None of these

12. Wildlife protection act was enacted during
(a) 1975 (b) 1980
(c) 1972 (d) 1995

13. The project tiger was launched in the year
(a) 1972 (b) 1995
(c) 1973 (d) 1980

14. India's First Forest Policy was enunciated during
(a) 1894 (b) 1927
(c) 1952 (d) 1988

15. Wildlife Protection Act of 1972 was amended during the year
(a) 1988 (b) 1990
(c) 1991 (d) 1980

16. In Wildlife Protection Act 1972, Nilgiri Tahr comes under
(a) Schedule I Part II
(b) Schedule I Part I
(c) Schedule II Part II
(d) Schedule III

17. World Forest Day is celebrated on
(a) 15th June (b) 1st July
(c) 1st October (d) 21st March

18. ICFRE was established during the year
(a) 1986 (b) 1980
(c) 1975 (d) 1990

19. Forest Ranger under which section has got the power for evicting any person un authorisedly occupying any land in reserve forest
(a) 68 A (b) 78 A
(c) 71 A (d) 61 A

20. The Environment (Protection) Act is attributed to the year
(a) 1972 (b) 1980
(c) 1986 (d) 1988

21. The Tamil Nadu Rosewood trees (Conservation Act) was enacted in the year
(a) 1974 (b) 1984
(c) 1994 (d) 1998

22. The Tamil Nadu Animals and Birds sacrifices prohibition act was enacted in the year
(a) 1950 (b) 1960
(c) 1970 (d) 1980

23. The Tamil Nadu Hills Areas (preservation of trees) 1955 includes
(a) Ootacamund municipal areas
(b) Coonoor municipal area
(c) Kodikanal municipal area
(d) All the above

24. The Indian board of wildlife was constitutes in the year
(a) 1952 (b) 1972
(c) 1966 (d) 1994

25. The national board of wildlife has
(a) The prime Minister as chairperson
(b) The Minister in charge of forests and wildlife as chairman
(c) The Director, ICFRE as a chairman
(d) None of these

26. The chief wildlife warden may permit any person to enter in a sanctuary for
(a) Study of wildlife
(b) Photography
(c) Tourism
(d) All the above

27. The measures of immunization against communicable diseases of livestock are taken within
(a) 1 km of Sanctuary
(b) 2 km of Sanctuary
(c) 5 km of Sanctuary
(d) None of these

28. The staffing of wild animals is
(a) Animal article
(b) Animal trophy
(c) Taxidermy
(d) None of these

29. Trade or commerce in wild animals, animal articles are tropics dealt in which section of wildlife protection act, 1972?
(a) Section 38 (b) Section 39
(c) Section 40 (d) Section 42

30. The animal coming under schedule I part II of the wildlife protection act, 1972 is
(a) Agar Monitor Lizard
(b) Black Buck
(c) Caracal
(d) Dugong

31. The Tamil Nadu Animals and Birds in urban Areas (control and regulation) act was enacted in the year
(a) 1984 (b) 1987
(c) 1994 (d) 1997

32. The animal covered in the Tamil Nadu animals and Birds in Urban Areas (control and regulation) act, 1997 is
(a) Elephant (b) Tiger
(c) Lion (d) None of these

33. The Indian Board of wildlife (IBWL) was set up by the government of India in the year
(a) 1950 (b) 1951
(c) 1952 (d) 1962

34. Bombay National History Society was started in the year
(a) 1880 (b) 1883
(c) 1948 (d) 1951

35. Dhebar Commission studied the problem of tribals during the year
(a) 1948-1949 (b) 1960-1961
(c) 1971-1972 (d) 1980-1982

36. National Forest Policy, 1988 envisages that ------------% of the total geographical area of India should be under forest
(a) 22 (b) 33
(c) 40 (d) 40

37. GATT (General Agreement on Trade and Tariff) was set up in the year
(a) 1972 (b) 1979
(c) 1984 (d) 1990

38. India's Patent Act, 1970 does not permit patenting of
(a) Herb
(b) Shrub
(c) Medicinal plants
(d) All the above

39. Sub chapter 6.2 of the Indian Forest Act, 1927 deals with the
(a) Protected forests
(b) Penalties and procedures
(c) Duty on timber and other forest produce
(d) Reserved forests

40. It shall be the duty of every citizen of India to protect and improve the natural environment including forests, lakes, rivers and wildlife and to have compassion for living creatures. In which article of the constitution of India does it appear?
(a) Article 48 A (b) Article 49 A
(c) Article 51 A (d) Article 52 A

41. Common langur is the animal coming under
(a) Schedule I (b) Schedule II
(c) Schedule III (d) Schedule IV

42. National parks and wildlife sanctuary come under which section of wildlife protection act, 1972?
(a) u /s 15 (b) u /s 25
(c) u /s 35 (d) u /s 45

43. The Tamil Nadu sandalwood transit rules were made in
(a) 1966 (b) 1967
(c) 1968 (d) 1969

44. Madras Forest act is attributed to the year
(a) 1878 (b) 1882
(c) 1865 (d) 1871

45. Big game is included in which of the following schedules of the wildlife protection act of 1972?
(a) II (b) III
(c) IV (d) V

46. Man and Biosphere (MAB) programme was started during
(a) 1971 (b) 1980
(c) 1982 (d) 1990

47. The "Charter of Indian Forestry" was based on the report given by
(a) McClelland (b) Voelcker
(c) Brandis (d) Jack Westoby

48. The challans of forest offences are prepared under section

(a) 12 (b) 14

(c) 26 (d) 36

49. Forest law is

(a) A general law

(b) A special law

(c) An international law

(d) A constitutional law

50. The forests are brought on concurrent list by 42nd amendment to the constitution in the year

(a) 1972 (b) 1976

(c) 1980 (d) 1988

ANSWERSHEET
20. Forest Law and Policy

1. **(a)**	2. **(c)**	3. **(b)**	4. **(a)**	5. **(b)**	6. **(b)**	7. **(b)**	8. **(c)**	9. **(d)**	10. **(b)**
11. **(c)**	12. **(c)**	13. **(c)**	14. **(a)**	15. **(c)**	16. **(b)**	17. **(d)**	18. **(a)**	19. **(a)**	20. **(c)**
21. **(c)**	22. **(a)**	23. **(d)**	24. **(a)**	25. **(a)**	26. **(d)**	27. **(c)**	28. **(c)**	29. **(b)**	30. **(a)**
31. **(d)**	32. **(d)**	33. **(c)**	34. **(b)**	35. **(b)**	36. **(b)**	37. **(c)**	38. **(d)**	39. **(d)**	40. **(c)**
41. **(b)**	42. **(c)**	43. **(b)**	44. **(b)**	45. **(c)**	46. **(a)**	47. **(a)**	48. **(c)**	49. **(d)**	50. **(b)**

21

Forest Economics and Marketing

KEY POINTS

- ISo – product or iso-quant curve : This curve shows that various combination of two resources which will produce given level of output.
- Iso-Cline : Iso-cline is the line joining points of same inclination on the various iso-products curves, on iso-cline the MRS = not zero.
- Least Cost Combination : Three methods to find solution to cost minimization problems :
  - Simple arithmetical calculations
  - Algebric method
  - Graphic method
- Iso-revenue line or price line : It is a line which defines all the possible combinations of two commodities which would yield an equal revenue or income.
- Isocline are straight line, passing through origin. Originally they are also scale lines.
- Price : When value is expressed in terms of money, it is called price.
- Utility : Wants satisfying power of a commodity is known as utility
- Interior good : An interior good is one in which the income effect is negative. Increase in real income reduce the quantity demanded and decreases in real income increase the quantity demanded of interior good.
- Giffen's paradox : This refers to a commodity whose demand varies directly with price.
- Demand curve: The demand curve show the different quantities of one good which the consumer will take at various possible prices, other things remaining the same.
- Demand : Demand for a good per unit of time that consumer will take off/buy/purchase the market at all possible alternative prices, other things remain constant. D=f (p,y,pr,t)
- Elastic demand : Demand is said to be elastic when with a small change in price, there is a great change in quantity demanded. ED > 1

$$Ed = \frac{\text{\% change in quantity demanded}}{\text{\% change in price}}$$

- Elasticity of demand / price elasticity of demand : Price elasticity of demand is the proportional change in quantity demanded divided by the proportional change in price

$$Ed = \frac{\text{\% change in quantity demanded}}{\text{\% change in price}}$$

- Income elasticity of demand : Income elasticity of demand is the proportional change in quantity demanded divided by proportional change in nominal income.

$$Ed = \frac{\%\text{ change in quantity demanded}}{\%\text{ change in income}}$$

- Point elasticity : When elasticity is computed at a particular point on the demand curve then it is known as point elasticity.
- Arc elasticity : When elasticity is computed between two points on demand curve than it is called Arc elasticity
- Price elasticity of supply : Price elasticity of supply is the proportional change in quantity supplied divided by the proportional change in price.
- Indifference curve : An indifference curve show the different combination of two commodities say x and y that yield equal satisfaction in the consumer or among which the consumer is indifferent.
- Monopoly : One seller & One product
- Duepoly : Two seller & One product
- Oligopoly : Few seller & One product
- Monopsony : One buyer & One product
- Duepsony : Two buyer & One product
- Oligopsony : Few more than two buyer & one product
- Marginal utility : Marginal utility is the utility, as the point where the consumer stop further consumption of a commodity

$$MU = \frac{\text{Change in the total utility}}{\text{Change in the number of unit of a commodity}}$$

Production function : Production function is a technological relationship between output and input or is a mathematical relationship describing the way in which the quantity of a particular product depends upon the quantities of particular input used. Y = f(x)

MULTIPLE CHOICE QUESTION'S

1. Economic laws are not determine as
   (a) Physical laws
   (b) Moral laws
   (c) Government laws
   (d) All of the above
2. Economics is an--------------
   (a) Art
   (b) Science
   (c) Art and Science
   (d) None of these
3. Economics is primarily a study of --------
   (a) Man (b) Wealth
   (c) Wants (d) None of these
4. ----------------- is the best measure to find worthiness of investment
   (a) Payback period
   (b) Benefit Cost ratio
   (c) Net Present Value
   (d) Internal Rate of Return

5. Principle of substitution is used in
(a) Product min
(b) Selection of enterprise
(c) Selection of practice
(d) None of these

6. The least cost principle is used in
(a) What to produce
(b) How much to produce
(c) How to produce
(d) When to produce

7. When total product means constant, the marginal product will be
(a) Zero (b) Negative
(c) Positive (d) Constant

8. When Average Product < Marginal Product, Average Product is
(a) Remains constant
(b) At maximum
(c) Increasing
(d) Decreasing

9. A production function will unitary elasticity indicates
(a) Increasing returns
(b) Decreasing returns
(c) Nothing
(d) Constant returns

10. In the rational region of production function, choice of indicator is
(a) Factor- factor price ratio
(b) Factor-product prices
(c) Product-product price ratio
(d) Factor-product quantity

11. The relationship between Total Product and Total Cost is
(a) Directly proportional
(b) No relationship
(c) Inversely related
(d) None of these

12. When Marginal Cost is at lowest, Marginal Product will be
(a) Decreasing (b) Maximum
(c) Increasing (d) Minimum

13. In balanced Zone
(a) Marginal Cost = Average Cost
(b) Marginal Cost > Average Cost
(c) Marginal Cost < Average Cost
(d) None of these

14. If Marginal Rate of product Substitution between two products is negative, then two goods are
(a) Substituting (b) Complementing
(c) Supplementing (d) Competitive

15. In case of diminishing substitution the curve will be
(a) Convex to origin
(b) Straight line
(c) Concave to origin
(d) Vertical or Horizontal

16. Optimum choice of enterprises or combination of enterprises is made based on
(a) Principle of least cost
(b) Principle of variable proportions
(c) Principle of opportunity cost
(d) Principle of cost

17. In LP, the availability and input-output coefficients as well as prices are known with certainty then assumption is
(a) Linearity (b) Additively
(c) Finiteness
(d) Single value expectations

18. In maximum returns interaction, the outgoing column will be located with the help of
(a) Highest positive value
(b) Lowest positive value
(c) Lowest negative value
(d) Highest negative value

19. Marketable surplus is given by
(a) Marketable surplus = Total production + Total Requirement
(b) Marketable surplus = Total production + Total Requirement /2
(c) Marketable surplus = Total production - Total Requirement
(d) Marketable surplus = Total Requirement - Total production

20. DMI headquarters is located at
(a) Hyderabad (b) Faisalabad
(c) Ghaziabad (d) Faridabad

21. World Standard Day is celebrated on
(a) 14th January (b) 14th November
(c) 14th October (d) 14th December

22. Co-operative Credit Societies Act was passed in
(a) 1902 (b) 1903
(c) 1904 (d) 1912

23. First Land Mortgage bank in 1920 established at
(a) Karnataka (b) Bihar
(c) West Bengal (d) Punjab

24. Nationalization of banks was carried out on
(a) 19th June, 1969
(b) 19th July, 1969
(c) 19th August, 1969
(d) 19th October, 1969

25. In case of amortized even repayment plan, the interest rate
(a) Increases (b) Declines
(c) Remains constant (d) None of these

26. If liabilities > assets, then net worth is placed at
(a) Liabilities side (b) Assets side
(c) Income side (d) None of these

27. Book value of durable assets refers to
(a) Market price of assets
(b) Purchase price of assets
(c) Production of assets
(d) Value at cost

28. Break even analysis is carried out by
(a) Break even analysis =Price per unit/Fixed cost –Variable cost
(b) Break even analysis = Fixed cost / Price per unit - Variable cost
(c) Break even analysis =Total cost/ Fixed cost - Variable cost
(d) Break even analysis = Price per unit - Variable cost / Fixed cost

29. Marginal value products of scarce resources is known by
(a) Linear programming
(b) Production function
(c) Planning
(d) Both a and b

30. Tangible benefit of project is
(a) Income distribution
(b) Standard of living
(c) National prosperity
(d) Higher yield level

31. Shadow prices of resources are used in
(a) Economic analysis
(b) Financial analysis
(c) Econometric analysis
(d) Technical analysis

32. Benefit Cost ratio is used to evaluate mostly
(a) Construction of dam
(b) Construction of building
(c) Construction of house
(d) Construction of cattle shed

33. Elasticity cannot be explained by
(a) Regression (b) GLS
(c) OLS (d) Correlation

34. 'Economics is what economist do' given by
(a) Alfred Marshall
(b) Lord Robbins
(c) J M Keynes
(d) Jacob veneer

35. Marginal utility of money cannot be
(a) Positive (b) Negative
(c) Zero (d) Both a and b

36. Indifference curves for perfect substitutes will be
(a) Convex to origin
(b) Concave to origin
(c) Perpendicular to origin
(d) Straight line

37. If income consumption curve is bending towards Y axis then commodity X is
(a) Inferior (b) Luxury
(c) Normal (d) None

38. For Giffen goods price consumption curve will be
(a) Backward going
(b) Horizontal
(c) Downward sloping
(d) Upward sloping

39. Quantity purchased will vary inversely with price when income effect is for normal goods
(a) Zero (b) Positive
(c) Negative (d) None

40. Consumer is neither worse off nor better off than before in
(a) Hicks Substitution effect
(b) Marshall Substitution effect
(c) Shitsky Substitution effect
(d) None

41. When with a change in price the total outlay on commodity remains constant, it is a case of
(a) Perfect elasticity
(b) Unit elasticity
(c) Perfect inelasticity
(d) Zero Elasticity

42. Change in quantity demand refers to
(a) Upward shift of demand
(b) Downshift of demand
(c) Movement on same curve
(d) None

43. When both a price of substitute and price of a component of commodity X rise, demand for X
(a) Falls (b) Rises
(c) No change (d) All of the above

44. Cross elasticity of demand between petrol and automobiles is
(a) Negative (b) Zero
(c) High (d) Infinite

45. For inelastic goods, fall in the price leads to
(a) Increase in Total Elasticity
(b) No change in Total Elasticity
(c) Decrease in Total Elasticity
(d) None

46. The Engel curve has all goods.
(a) Positive slope (b) Constant
(c) Negative slope (d) All of the above

47. The difference between value in use and value in exchange is
(a) Consumer surplus
(b) Marketing surplus
(c) Farm production
(d) None

48. Marginal Rate of product Substitution on the isoclines will be
(a) Positive (b) Zero
(c) Negative (d) Constant

49. Marginal Cost = Marginal Returns = Average Cost = Average Returns shows the long run, equilibrium positive of the
(a) Monopolist Firm
(b) Oligopolistic Firm
(c) Competitive Firm
(d) None

50. The concept of 'reserve army of labor' is due to
(a) David Richardo
(b) JS Mill
(c) Adam Smith
(d) Karl Marx

51. PQLI of Morris D. Morris does not include
(a) Infant mortality
(b) Life expectancy
(c) Literacy
(d) Standard of living

52. ICOR is in less developed countries
(a) Low (b) High
(c) Constant (d) Zero

53. Intangible transactions are recorded in
(a) Service account
(b) Goods account
(c) Unilateral transfers
(d) Tangible account

54. Snake in the tunnel system is followed in
(a) USA (b) SE Asia
(c) African nations (d) European Union

55. Keynes is related to which inflation theory
(a) Cost push (b) Demand pull
(c) Structural theory (d) None

56. Aggregate supply of labor in case of classical is
(a) Horizontal line
(b) Vertical line
(c) Both horizontal and vertical
(d) Intermediate

57. The difference between national income and domestic product is accounted by
(a) Net factor income from abroad
(b) Budgetary surplus/deficit
(c) Export surplus/deficit
(d) Inflow of capital

58. Gross National Product deflator is
(a) Nominal Gross National Product / Nominal Gross Domestic Product
(b) Nominal Gross National Product / Real Gross National Product
(c) Nominal Gross National Product / Real Gross Domestic Product
(d) Real Gross National Product / Nominal Gross Domestic Product

59. The demand for capital in essence is a
(a) Direct demand
(b) Derived demand
(c) Reciprocal demand
(d) Complementary demand

60. Capital formation in agriculture over the years
(a) Decreasing (b) Increasing
(c) Both a and b (d) Constant

61. Private investment is to public investment in agriculture
(a) Less than (b) More than
(c) Equal (d) Can't say

62. Project planning and budgeting are used to know---------------
(a) Commercial aspect
(b) Financial aspects
(c) Economic aspects
(d) Technical aspects of project

63. Sensitivity analysis is used in
(a) Preparation of project
(b) Appraisal
(c) Implementation
(d) Forecasting/evaluation

64. 'Economic a' is written by
(a) Marshall
(b) Aristotle
(c) Smith
(d) Mill

65. Nature of economic law is
(a) Normative (b) Destructive
(c) Exactness (d) Positive

66. Law of substitution is given by
(a) Marshall (b) Smith
(c) Gossen (d) Pigon

67. Concept of 'crystallized labor' is given by
(a) Marshall (b) Keynes
(c) Mill (d) Karl Marx

68. When Marginal Physical Product is zero
(a) Marginal Cost is Maximum
(b) Marginal Cost become vertical
(c) Marginal Cost is Minimum
(d) Marginal Cost becomes horizontal

69. When the demand curve is relatively flatter, then demand is
(a) Relatively inelastic
(b) Relatively elastic
(c) Unitary elastic
(d) All of the above

70. If the demand is elastic, for monopoly
(a) Higher prices, more profit
(b) Lower price, more profits
(c) Higher price, less profit
(d) Lower price, lower profit

71. The 'Drain Theory' about poverty in India is associated with
(a) Baldeo Ram MIridha
(b) VKRV Rao
(c) Dada Bhai Naoroji
(d) C Bose

72. Benefit cost Ratio is used to evaluate
(a) Social project (b) Public project
(c) Both a and b (d) None

73. 'Giffen' goods refers to
(a) Stock of goods (b) Lack of goods
(c) Inferior goods (d) Superior goods

74. Perfect elasticity is shown by
(a) Vertical straight line
(b) Horizontal straight line
(c) Slopping curve from left to right
(d) Slopping curve from right to left

75. Production is a function of
(a) Profit (b) Factors
(c) Price (d) Cost

76. Price theory is a part of
(a) Micro Economics
(b) Macro Economics
(c) Farm Management
(d) Marketing

77. The overhead cost is also known as
(a) Variable cost (b) Total cost
(c) Marginal cost (d) Fixed cost

78. The ultimate aim of farm management is to
(a) Increase gross income
(b) Reduce total cost
(c) Optimize factor utilization
(d) None

79. Government of India signed World Trade Organization agreement in
(a) 1981 (b) 1994
(c) 1995 (d) 1996

80. The cause of inflation is
(a) Increase in money supply
(b) Fall in production
(c) Increase in money supply and fall in production
(d) Decrease in money supply and fall in production

81. The Central Agmark Lab is located at
(a) Mumbai (b) Nagpur
(c) Bangalore (d) Kolkata

82. IBRD is synonymous to
(a) World Bank
(b) Asian Development Bank
(c) International Monetary Fund
(d) World Trade Organization

83. Which of the following is not a farming system?
(a) State farming
(b) Collective farming
(c) Rain fed farming
(d) Co-operative farming

84. Livestock insurance is included in
(a) Variable cost
(b) Special cost
(c) Fixed cost
(d) Marginal cost

85. Farm management is an
(a) Intra farm science
(b) Inter farm science
(c) Inter regional farm science
(d) International farm science

86. One of the important fixed cost on a farm is
(a) Seed cost
(b) Land rent
(c) Irrigation cost
(d) Feed cost

87. Principle of Equi marginal return is applied when
(a) The resources are unlimited
(b) The price of resources is high
(c) The price of resources vary
(d) The resources are limited

88. Who pro founded the population theory first?
(a) Adam Smith (b) Ricardo
(c) Malthus (d) Marshall

89. When Marginal Physical Product is Zero, Total Physical Product is at its maximum because
(a) Further addition to Marginal Physical Product is negative
(b) Further addition to Marginal Physical Product is positive
(c) Marginal Cost = Marginal Returns
(d) Elasticity of production is more than one

90. Regulated market ensures
(a) Procurement price
(b) Support price
(c) Remunerative price
(d) Fair price

91. Farming system approach ensures the best use of
(a) Land
(b) Land + Labour
(c) Land + Labour + capital
(d) Land + Labour + Capital + management

92. Internal economy refers to
(a) Transport
(b) Banks
(c) Communication
(d) Specialization

93. Input – Output relationship is called
(a) Cost Function
(b) Production function
(c) Price function
(d) None

94. Value Added Tax is a/an
(a) Direct Tax
(b) Indirect Tax
(c) Both a and b
(d) Cannot be classified

95. Inflation is measured by
(a) WPI (b) CPI
(c) Both a and b (d) None of these

96. Which one of the following refers to 'system of farming'?
(a) Economic (b) Operational
(c) Organizational (d) Cultural

97. An application of the accounting principles to the business of farming
(a) Farm planning
(b) Farm budgeting
(c) Farm accounting
(d) Marketing

98. Beef and hides are examples of
(a) Independent enterprise
(b) Supplementary enterprise
(c) Joint enterprise
(d) Competitive enterprise

99. Indian farming is published by
(a) Ministry of agriculture Government of India
(b) ICAR
(c) National Seed Corporation of India
(d) None of these

100. A consumer spends his income according to the law of
(a) Diminishing marginal utility
(b) Increasing cost
(c) Equi-marginal utility
(d) Least cost

101. A group of technical units are known as
(a) Farm firm
(b) Plant
(c) Economic unit
(d) All of the above

102. A market structure with large number of firms selling homogenous product is known as
(a) Monopoly market
(b) Monopolistic competition
(c) Perfect competition
(d) None

103. A single convenient unit in production for which technical co-efficient are calculated is known as
(a) Units of accounting
(b) Technical units
(c) Marginal units
(d) None

104. A table is a
(a) Material good
(b) Non-material good
(c) Free good
(d) None

105. About 15% farmers had long holding in between
(a) 2-4ha (b) 0-1ha
(c) 1-2ha (d) None

106. Air is
(a) Economic goods
(b) Free goods
(c) Transferable goods
(d) All of the above

107. As income increase the expenditure on light, fuel and rent generally
(a) Increases
(b) Decreases
(c) Remains the same
(d) None

108. At equilibrium, the slope of the indifference curve is
(a) Equal the slope of the budget line
(b) Greater than the slope of the budget line
(c) Smaller than the slope of the budget line.
(d) None

109. At present, the farming system of India has become
(a) Total fixed cost/100
(b) Total fixed cost/total variable cost
(c) Total fixed cost/output
(d) Total fixed cost/input

110. Average product is equal to marginal product when
(a) Average product is Maximum
(b) Average product is Minimum
(c) Marginal product is zero
(d) None

111. Who was the first economist to have evolve the technique of indifference curve Analysis
(a) David Richardo
(b) Bohm Bawerk
(c) F.Y. Edgeworth
(d) A.C. Pigou

112. Break-even points is determined by
(a) Marginal Cost=Marginal Returns
(b) Marginal Cost =Average Variable Cost
(c) Marginal Cost =Average Cost=Average Returns
(d) Total Returns=Total Cost

113. Building machinery and implements are examples
(a) Variable resources
(b) Fixed resources
(c) Flow resources
(d) None

114. By preparing chairs the woods creates
(a) Form utility (b) Time utility
(c) Place utility (d) Knowledge utility

115. Central co-operative Bank operates normally at
(a) District level
(b) Village level
(c) National level
(d) None

116. Chairman of planning commission is
(a) Prime Minister
(b) President
(c) Appointed by Prime Minister
(d) Appointed by President

117. Co-operatives for tribes are called as
(a) Lamps co-operative societies
(b) Multi-purpose societies
(c) Lead Bank
(d) Sativa

118. Creation of utility By way of marketing is known as
(a) Consumption (b) Production
(c) Distribution (d) Exchange

119. Criteria for goods market information is
(a) Comprehensive (b) Accuracy
(c) Relevance (d) All of the above

120. Crop farming and milk production is an example of
(a) Competitive enterprise
(b) Complementary enterprise
(c) Supplementary enterprise
(d) None

121. Decisions regarding irrigation, conservation and reclamation programs are comes under
(a) Demand change due to change in price
(b) Demand changes not due to the change of price but due to income
(c) Both a and b
(d) None

122. Demand for necessary goods is
(a) Elastic (b) Inelastic
(c) Perfectly elastic
(d) None

123. The marginal productivity theory is based on the assumption of
(a) Imperfect competition
(b) Monopolistic competition
(c) Perfect competition
(d) Elastic competition

124. Expenditure on fertilizers is called
(a) Fixed cost (b) Variable cost
(c) Marginal cost (d) Total cost

125. Extension in demand means
(a) Change in demand due to change in price
(b) Change in demand due to change in income
(c) Both a and b
(d) None

126. Extent of land forest in India is
(a) 21% (b) 20%
(c) 27% (d) 16%

127. Farm earn budget are used to
(a) Compare economic profitability of different production activities within among farms
(b) Indicate whether proposed changes will be profitable under a given set of factors
(c) Explore conditions under which certain farm practices profitable or unprofitable
(d) All of the above

128. Farm management concern with obtaining the Maximum yield per hectare and considers per unit cost therefore it is
(a) Practical Science
(b) Integrating Science
(c) Profitability Oriented
(d) Broader field

129. Farm management deals with
(a) Judicious decisions
(b) Use of scarce resources
(c) Profit maximization and family satisfaction
(d) All of the above

130. Who propounded the theory of optimum population first
(a) Sigwick
(b) Malthus
(c) Carr-Saunders
(d) Kuczyansky

131. Farming planning means
(a) Farm budgeting
(b) Cropping sequence
(c) Types of enterprise
(d) None

132. For purchasing of seeds, fertilizers, etc., the loan distributed is called
(a) Long term loan
(b) Short term loan
(c) Medium term loan
(d) All of the above

133. For purchasing the pump sets, tractor, etc., the loan is distributed for
(a) Long term (b) Short term
(c) Medium term (d) All of the above

134. Free goods have a great
(a) Value in use
(b) Value in exchange
(c) Transferability
(d) Both B and C

135. Oligopoly is a market situation which is characterized by the existence of
(a) One firm
(b) Two firms
(c) Several firms
(d) Few firms

136. Human wants are generally
(a) Static
(b) Recurring
(c) Non-recurring
(d) Limited

137. Monopsony imply the existence of a
(a) Single buyer
(b) Single seller
(c) Single seller vs Single buyer
(d) All of the above

138. If substitution ratio is equal to price ratio than cost will be
(a) Minimum (b) Increasing
(c) Maximum (d) Decreasing

139. If the income elasticity or demand is greater than one, the commodity is
(a) A necessity
(b) A Luxury
(c) An inferior good
(d) A essentially

140. If the quantity of a commodity demanded does not change when its price, the demand curve is
(a) Negatively sloped
(b) Positively slope
(c) Vertically sloped
(d) None

141. If the quantity of commodity demand remains unchanged as its price, the coefficient of price elasticity of demand is
(a) Greater than one
(b) Zero
(c) Smaller than one
(d) Equal to one

142. If two items are complementary than demand of one item will increase for in piece of another item
(a) Increase (b) Decrease
(c) No effect (d) Not known

143. In classical production function when average product is increasing
(a) Marginal product is less than average product
(b) Marginal product is equal to average product
(c) Marginal product is greater than average product
(d) None

144. In economics the term labour means
(a) Only manual labour
(b) Only mental work
(c) Manual and mental work
(d) None

145. Who wrote the book, The theory of Monopolistic competition
(a) Joan robinson (b) E.H. Chamberlin
(c) A.C. Pigou (d) Frederic Benham

146. In perfect competition, there are
(a) Large number of buyers and few sellers in the market
(b) Large number of sellers and little number of buyers in the market
(c) Number of buyers and sellers are large and nearly equal
(d) All of the above

147. In zamindari system the land revenue was charged
(a) Thorough Landlords
(b) Directly from the farmers
(c) Through Village panchayats
(d) Through either of these

148. Income tax is a
(a) Direct tax (b) Indirect tax
(c) Both a and b (d) None

149. Initial utility is the utility of the
(a) First unit (b) Last unit
(c) Marginal unit (d) Third unit

150. Man's wants are
(a) Limited (b) Unlimited
(c) Countable (d) Scarce

151. Marginal cost is equal to
(a) The average cost of the total output declared
(b) The difference between average cost and the cost of the last unit produced
(c) The cost of the last but one unit produced
(d) The extra cost of an additional unit produced

152. Per hectare calculated cost is known as
(a) Cost of cultivation
(b) Cost of production
(c) Cost of farming
(d) Total cost

153. Patents and copy rights are
(a) Material goods
(b) Non-material goods
(c) Transferable goods
(d) Scarce gods

154. Net return per hectare is
(a) Overall efficiency measure
(b) Partial efficiency measure
(c) Specific efficiency measure
(d) None

155. Market demand curved is faced by
(a) Monopolist
(b) Competitive firm
(c) Joint firm
(d) Oligopolistic firm

156. Marshall defines economics as
(a) The study of mankind in the ordinary business of life
(b) The study of human behavior as a relationship between the ends scarce means which have alternative uses
(c) An enquiry into the nature and causes of the wealth of nations
(d) Fundamentally a science of scarcity

157. Marshall definition of economics has been criticized by
(a) Richardo (b) Robbins
(c) Ruskin (d) Cartyle

158. Maximum profit will be at the point where
(a) Marginal Cost = Marginal Returns
(b) Marginal Cost is more than Marginal Returns
(c) Marginal Cost is less than Marginal Returns
(d) None

159. Percentage population of world lives in India is
(a) 16 (b) 18
(c) 20 (d) 19

160. Regional rural banks are financed by
(a) Reserve Bank of India
(b) Government of India
(c) Nationalized Banks
(d) All of the above

161. SFDA are formed on the recommendation of All India Rural Review Committee headed by
(a) M.L Danthawala
(b) C. Swaminathan
(c) K.N Raj
(d) Ashok Mehta

162. Perfect elasticity is shown by
(a) Vertical straight line
(b) Horizontal straight line
(c) Sloping curve right to left
(d) Sloping curve left to right

163. The biggest nationalized Bank is
(a) Union Bank of India
(b) Bank of Baroda
(c) Central Bank of India
(d) Bank of India

164. The Cannons of taxation was given by
(a) Adam Smith (b) Robbins
(c) Pigou (d) Marshall

165. The classical theory of interest is also called as
(a) The real theory of interest
(b) Monetary theory of interest
(c) Lon able funds theory
(d) Liquidity preference theory

166. To whom do you attribute the concept of consumers surplus
(a) A.C Pigou
(b) Alfred Marshall
(c) H.H. Gossen
(d) Barbura Wootton

167. The law which is very useful in determining the best use of limited resources is
(a) Law of opportunity cost
(b) Law of diminishing return
(c) Law of substitution
(d) None

168. The main cause of inflation is increasing money supply and decrease in production
(a) Decrease (b) Increase
(c) Remains same (d) None

169. The owner of land in state farming is
(a) Farmer (b) Capitalist
(c) Government (d) All of the above

170. Who was first economist to have propounded theory of rent?
(a) Joan Robinson
(b) R.G.D Allen
(c) Edward Chamberlin
(d) Alfred Marshall

171. The principle of taxation was given by
(a) Marshall (b) Adam Smith
(c) Hicks (d) None

172. The process of determination of wages, rent, interest and profit is called
(a) Consumption (b) Production
(c) Distribution (d) None

173. The statement "in the process of production and distribution of goods, Income is generated and distributed amongst the factors of production" was given by
(a) J.B Say (b) Classical economist
(c) Keynes (d) Marshall

174. Which of the following definitions of economics include the economic concept of Scales of preferences
(a) Wealth definition
(b) Welfare definition
(c) Scarcity definition
(d) Isolation definition

175. The total fixed curve is
(a) Increase with the level of output
(b) Remains constant at all the output levels
(c) Decreases as output increases
(d) Both b and c

176. The want satisfying power of commodity is called
(a) Demand (b) Utility
(c) Consumption (d) Production

177. Total cost is equal to
(a) Fixed cost +Variable cost
(b) Fixed cost+ variable cost/100
(c) Fixed cost-variable cost
(d) None

178. Utility in economic means
(a) Usefulness
(b) Pleasure
(c) Want-satisfying power of a commodity
(d) Satisfaction

179. Utility of the last unit is called
(a) Marginal utility
(b) Initial utility
(c) Total utility
(d) Zero utility

180. When marginal product is zero, the total product is
(a) Maximum (b) Minimum
(c) Zero (d) None

181. When total product is highest, marginal product will be
(a) Maximum (b) Minimum
(c) Zero (d) Increasing

182. When variable cost is zero, the total cost will be?
(a) Equal to variable cost
(b) Equal to fixed cost
(c) Equal to average variable cost
(d) None

183. Zamindari system was introduced by
(a) British (b) Moguls
(c) Marathas (d) All of the above

184. World Trade Organization (WTO) was established in
(a) 1992 (b) 1998
(c) 1995 (d) 1997

185. The Government established Tribal Co-operative Marketing Development Federation of India Limited (TRIFED) in
(a) 2001 (b) 1997
(c) 1987 (d) 2004

186. Reserve Bank of India was set up in
(a) 1934 (b) 1935
(c) 1904 (d) 1949

187. National Bank for Agriculture and Rural Development (NABARD) act was passed in
(a) 1980 (b) 1981
(c) 1982 (d) 1983

188. Export-import Bank of India was set up on 1st January
(a) 1980 (b) 1982
(c) 1981 (d) 1983

189. National Bank for Agriculture and Rural Development came into presence on
(a) July 12, 1982
(b) July 22, 1984
(c) October 15, 1986
(d) August 10, 1983

ANSWERSHEET
21. Forest Economics And Marketing

---

1. **(a)**	2. **(c)**	3. (a)	4. **(d)**	5. **(a)**	6. **(c)**	7. **(a)**	8. **(c)**	9. **(d)**	10. **(b)**
11. **(c)**	12. **(b)**	13. **(b)**	14. **(d)**	15. **(a)**	16. **(c)**	17. **(d)**	18. **(d)**	19. **(c)**	20. **(d)**
21. **(c)**	22. **(c)**	23. **(d)**	24. **(b)**	25. **(a)**	26. **(b)**	27. **(d)**	28. **(b)**	29. **(d)**	30. **(d)**

31. **(a)**	32. **(a)**	33. **(d)**	34. **(d)**	35. **(d)**	36. **(d)**	37. **(a)**	38. **(a)**	39. **(b)**	40. **(a)**
41. **(b)**	42. **(c)**	43. **(a)**	44. **(a)**	45. **(c)**	46.**(a)**	47. **(a)**	48. **(d)**	49. **(c)**	50. **(d)**
51. **(d)**	52. **(a)**	53. **(a)**	54. **(d)**	55. **(b)**	56. **(b)**	57. **(a)**	58. **(b)**	59. **(b)**	60. **(a)**
61. **(b)**	62. **(b)**	63. **(d)**	64. **(b)**	65. **(d)**	66. **(a)**	67. **(d)**	68. **(d)**	69. **(b)**	70. **(b)**
71. **(c)**	72. **(c)**	73. **(c)**	74. **(b)**	75. **(b)**	76. **(a)**	77. **(d)**	78. **(a)**	79. **(b)**	80. **(c)**
81. **(b)**	82. **(a)**	83. **(c)**	84. **(c)**	85. **(a)**	86. **(b)**	87. **(d)**	88. **(c)**	89. **(a)**	90. **(d)**
91. **(d)**	92. **(d)**	93. **(b)**	94. **(b)**	95. **(c)**	96. **(b)**	97. **(c)**	98. **(b)**	99. **(c)**	100. **(c)**
101. **(b)**	102. **(c)**	103. **(b)**	104. **(a)**	105. **(a)**	106. **(b)**	107. **(c)**	108. **(a)**	109. **(c)**	110. **(a)**
111. **(c)**	112. **(d)**	113. **(b)**	114. **(a)**	115. **(a)**	116. **(a)**	117. **(a)**	118. **(d)**	119. **(d)**	120. **(b)**
121. **(d)**	122. **(b)**	123. **(d)**	124. **(b)**	125. **(a)**	126. **(a)**	127. **(a)**	128. **(b)**	129. **(d)**	130. **(a)**
131. **(a)**	132. **(b)**	133. **(c)**	134. **(b)**	135. **(d)**	136. **(b)**	137. **(a)**	138. **(a)**	139. **(b)**	140. **(c)**
141. **(d)**	142. **(a)**	143. **(c)**	144. **(c)**	145. **(b)**	146. **(c)**	147. **(a)**	148. **(a)**	149. **(a)**	150. **(b)**
151. **(d)**	152. **(a)**	153. **(b)**	154. **(a)**	155. **(b)**	156. **(a)**	157. **(b)**	158. **(a)**	159. **(a)**	160. **(d)**
161. **(c)**	162. **(b)**	163. **(c)**	164. **(a)**	165. **(a)**	166. **(b)**	167. **(a)**	168. **(b)**	169. **(c)**	170. **(d)**
171. **(b)**	172. **(c)**	173. **(a)**	174. **(c)**	175. **(b)**	176. **(b)**	177. **(a)**	178. **(c)**	179. **(a)**	180. **(c)**
181. **(c)**	182. **(b)**	183. **(a)**	184. **(c)**	185. **(c)**	186.**(b)**	187. **(c)**	188. **(b)**	189. **(a)**	

22

Additional Information on Forestry

IMPORTANT MILESTONES OF INDIAN FORESTRY

1790 Tippu Sultan introduced 16 species of Eucalyptus in Nandhi hills, Mysore from Australia

1800 Commission of Forestry appointed to enquire into the availability of Teak in Malabar region. Based on the Commission's recommendation felling of Teak below 21 inches is prohibited

1805 Forest Committee set up to enquire into the capacity of forests and rights of proprietary to them based on the recommendation Royalty over Teak tree was given

1806 Captain Watson (Police officer) appointed as first conservator of Forest by Government of Madras. He establishes timber monopoly in Malabar and Travancore region

1823 Conservator of Forest post is abolished

1828 Wasteland declared as Government property

1842 Conolly, Collector of Malarbar region assisted by Chathu Menon (Sub-Conservator of forest established Nilambur Teak Plantation

1843 Afforestation work started. Champell introduces *Eucalyptus pinnata* at Wellington, Madras.

1855 Charter of the Indian Forests–Memorandum was submitted by Mc Clelland to Lord Dalhousie

1856 Cleghorn was appointed as First regular Conservator of Forest. Dietrich Brandis selected and appointed as Superintendent of Pegu Forest by Lord Dalhousie

1858 Introduction of Acacias and Eucalyptus in Nilgiris, Madras Province.

1864 British Raj established the Imperial Forest Department

1866 Dr. Dietrich Brandis, a German forest officer, was appointed Inspector General of Forests

1867 The Imperial Forestry Service was organized subordinate to the Imperial Forest Department

1869 Reorganization of forest services- CF, DCF, ACF Posts created. Teak Plantation at Makum are raised

1870 Wattles introduced in Ooty and Kodaikanal

1871 Working plan work is started. Cattle Trespass Act is passed

1875 Badan Powel and Dr. Schlich issued the first copy of The Indian Forester Journal

1878 Forest School at Dehradun is started. Indian Forest Act revised. Gamble Starts research on Non- Wood Forest Products

1879 Elephant preservation act passed

1882 Madras Forest Act passed

1883 Bombay Natural History Society established. Brandis retired from the service and Schlich becomes the Second Inspector General

1884 Forest school renamed as Imperial Forest School. *Eucalyptus grandis* introduced in kerala

1885 Training of forest officers of India started at Cooper's hill, England

1890 Dr. Voelkar starts a study on Indian Agricultural Problems. His report was the base for 1894 forest Policy

1894 First Indian forest policy is passed

1895 First Working plan Nilambur working plan is completed

1898 Vedanthangal declared as a first wildlife sanctuary in India

1901 Gass Forest Museum established in Madras Province

1905 World forestry congress set up. Systematic research in pulp and paper in India started

1906 From 1927 to 1932, forest officers were trained at the Imperial Forest Research Institute (FRI) at Dehradun, which had been established

1909 Royal commission on decentralization set up

1912 Madras forest college at Coimbatore started. Wild Birds and animals preservation act passed

1914 Destructive insects & pests Act passed. Board of forestry set up

1919 Forestry research officer appointed

1927 Indian forest act passed

1928 Royal commission on agriculture India is set up

1929 Ornithological survey in India conducted

1932 Bomboo Paper Industry (protection) Act passed. Range lands classifieds. Teak Stump planting started in Madras

1933 Imperial forest college closed down

1934 New Forest Botanical Garden established

1936 United provinces National park bill approved. First National park of India named after Hailey and established in united province.

1938 Madras Forest College Closed ; Indian Forest College (IFC) was established at Dehradun

1939 Research in Indian forest soils is started

1945 Madras Forest College reopened

1947 Import and export control act passed

1948 Forest colleges handed over to the Government of India. Central Board of Forestry is set up

1950 Vanamahotsav – The tree planting festival was started by K.M. Munshi

1954 Grassland survey in India conducted. World "Minor Forest Products" coined in world forestry conference

1955 Hill area (preservation) Act passed. Indian forest college renamed as Northern Forest Rangers college and Madras Forest College

1957 Logging branch established at FRI, Dehradun. Hailey National Park renamed as Corbett National park

1958 Wild life preservation society of India was established

1959 Tree improvement work is started at FRI, Dehradun. National Orchidarium and Botanical Garden is established in Shillong. Logging Research Institute at Batote is established

1960 Prevention of cruelty to animals act was passed

1962 Forest seed testing started in FRI, Dehradun. Large scale industrial wood plantation also gets started

1963 National Orchidarium and Botanical Garden is established at Yercaud, Tamil Nadu. G.A. Jones Conducts Aerial survey of Kulu forest

1965 Pre-investment Survey of Forest Resources (PISFR) Published the Forest Type Map of India.

1966 Modern Indian Forest Service was established, after independence, under the All India Services Act 1951. The first Inspector General of Forests, Hari Singh, was instrumental in the development of the IFS.

1968 Champion and Seth classified the Forests of India into 16 forest types. Westby coined the word "Social Forestry" in World forestry congress held in India

1970 Gujarat – first state in India to start Social Forestry Programme outside the conventional Forest area.

National Commission on Agriculture was set up.

Project Hangul was started in Jammu and Kashmir.

1972 First village Protection Committee in West Bengal was formed. National Commission on Environmental planning & Coordination set up. Project Lion started as Gir Lion Sanctuary project

1973 International Board for Plant Genetic Resources established. Project Tiger started. Sunderlal Bahughuna starts Chipko movement UNESCO started Man & Biosphere Programme. Water (Prevention & Control) Act passed. Wildlife (Stock declaration) Central Rules Passed. Wildlife (Transaction & Taxidermy) Rules also gets passed.

1974 The water prevention and control of pollution Act

1975 The Crocodile Breeding and Management projects launched with technical assistance from UNDP/FAO. Guidelines specifying that the management of protected areas should be under the remit of the wildlife wings issued by central government

1976 Forest and Wildlife brought on Concurrent list in the Seventh Schedule by 42nd Amendment to the Constitution. India becomes a member of CITES. National Commission on Agriculture Report was published.

1977 Temperate Forest Research Institute at Shimla was established.

The water (Prevention and Control of Pollution) Cess Act

1978 Central Crocodile Breeding Institute in Hyderabad was established. Joint Forest Management is started in West Bengal. Reptile skin and Elephant ivory export banned.

1979 Social Forestry Project started.

1980 Forest (Conservation) Act passed. Department of Environment constituted on 1 November 1980.

1981 Air (Prevention & Control) Act passed.

1982 Indian Institute of Forest Management in Bhopal was established. Wildlife (Protection) Act amended. Wildlife Institute of India established. Wildlife management transferred to Department of Environment in September.

1983 First National Wildlife Action Plan adopted. Memorandum imposes ban on the felling of trees at altitudes of 1,000 m and above. National Wildlife Action Plan passed. NRSA comes up with the National vegetation map. Wildlife License Rules passed.

1984 Ministry of Environment & Forest is created.

1985 B.Sc (Forestry) course started in State Agricultural Universities. Ministry of Environment and Forests, was constituted under Presidential Notification No. 74/2/185-Cab, Dated 4th January, 1985. National Wasteland Development Board established. Department of Environ-ment, Forests and Wildlife within the Ministry of Environment and Forests is created.

1986 Complete prohibition of trade in animal articles derived from Scheduled Animals. Environment (Protection) Act passed. JFM was started in Gujarat. Indira Priyadarshini Vrikshamitra Award was Instituted by the Ministry. Wildlife Action Plan amended.

1987 Forestry Survey of India published its first State of Forest Report of India. Indian Council of Forest Research & Education, established at Dehradun. Indira Gandhi national Forest Academy, Dehradun established. M.Sc (Wildlife) course started at Wildlife Institute of India. Ministry of Environment & Forest forbids the land lease to industries.

1988 Third National Forests Policy passed in the Parliament. New Seed Development Policy also passed.

1989 Biotechnological Centre for Tree Improvement established at Tirupathy. (A.P). Public sector projects of Rs. 2 billion (US$ 109 million) or more require environmental clearance from MoEF.

1991 Centre for Social Forestry and Eco rehabilitation established. FRI becomes a Deemed University. Govt. of India launches a Centrally Sponsored Scheme entitle 'Eco-development around National Parks and Sanctuaries including Tiger Reserves'. Ibex and Musk deer Projects were started. Wildlife (Protection) Act amended for the second time.

The Public Liability Insurance Act

1992 Project Elephant started.

1993 Great Indian Bustard Project started. Foundation for Revitalization and Local Health Tradition (FRLHT) established at Bangalore, Karnataka.

1995 Siberian Crane Project started at Bharatpur (Rajasthan).

The National Environment Tribunal Act

1996 Seeds Act passed.

1997 The National Environment Appellate Authority Act

1998 Export of 29 medicinal plants banned. FRLHT awarded the "Norman Borlaug Award" in the year 1998, in recognition of its pioneering efforts in conservation. Ministry of Tourism releases the Policy and Guidelines for Eco-tourism. National Zoo Policy adopted.

1999 Ministry of Environment and Forests prepares a National Policy and Macro level Action Strategy on Biodiversity through a consultative process.

2000 Biodiversity Bill, 2000 introduced in the Parliament Forest Survey of India publishes the 7th State of Forest Report (1999) of India. Kanchanjanga (Sikkim) declared as the 12th Biosphere Reserve in the Country.

2001 National Conference on Environment and Forests held at Coimbatore. Protected Area Network Cell of Wildlife Institute of India releases the list of National Parks (89 Nos.) & Wildlife Sanctuaries (486 Nos.) of India. Shri S.K. Pandey becomes the Director General of Forest.

2002 Complete ban of felling of trees in Andamans & Nicobar Islands by the Supreme Court of India. Shri Sanjay Kumar Singh (IFS Officer of 1991 Batch-Bihar Cadre) killed by the Mining Mafia in Bihar. The Prime Minister of India releases the National Wildlife Action Plan (2002-2016) at the XXI meeting of the Indian the Indian Board for Wildlife, New Delhi.

Biological Diversity Act was passed

2004 Biological Diversity Rules

2006 Forest Rights Act- "Scheduled Tribes and Other Traditional Forest Dwellers (Recognition of Forest Rights) Act" was passed

2010 National Green Tribunal Act was passed

2010 Wetlands (Conservation and Management) rules.

FIRSTS IN INDIAN FORESTRY

First Silviculturist of India	Troup	1909
First Biosphere Reserve in India	Nilgiri Biosphere Reserve	1986
First Chief Conservator of Forest	Ruther	1906
First Conservator of Forest	Watson (Police Officer)	1806
First Conservator of Forest to become a Member of Legislative Council	A.T. Shuttle Worth	
First Director of the Forest School, Dehradun	F. Bailey	1878
First National Forest Act	Indian Forest Act	1865
First Forest Entomologist	Stebbing	1901

First Forest Movement	Chipko Movement by Sunder Lal Bahuguuna	1973
First Forest Policy	Indian Forest Policy	1894
First Indian Inspector General of Forest	M.D. Chaturvedi	1949
First Inspector General of Forest	Dietrich Brandis	1864
First issue of Indian Forester issue by	Powel & Schlich	1875
First National Park	Hailey (Corbett) National Park, U.P	1936
First President of Forest Research Institute and College, Dehradun	S Eardley Wilmot	1906
First Regular Conservator of Forest	Cleghorn	1856
First Special Assistant Conservator of Forest	Ribbon trop and W. Schlich	1867
First Wildlife Sanctuary in India	Vedamthangal Wildlife Sanctuary, T.N.	1898
First Working plan in India	Nilambur Forest Working Plan	1895
First Zoo in India	Calcutta Zoo, W.B	1854

OTHER WILD LIFE PROJECT DETAILS

1970	:	Project Hangul
1972	:	Project Lion
1973	:	Project Tiger
1975	:	Project Crocodile
1991	:	Project Elephant
1991	:	Musk Deer
1991	:	Ibex
1993	:	Great Indian bustard
1995	:	Siberian crane
2009	:	Snow leapoard

SUPERLATIVES IN FORESTRY

Smallest Flowering Plant (W)	*Wolffia spp*
Smallest Flower (W)	*Wolffia spp*
Largest Flower (W)	*Rafflesia arnoldii*
Smallest Fruit (W)	*Wolffia angusta*
Largest Inflorescence (W)	*Corphya umbraculifera*
Largest Tree-Bearing Fruit (W)	Jack Fruit
Largest Leaf (W)	*Raphia regalis*

Fastest Growing Plants (W)	*Macrocystis pyrifera* (18 inches/day)
Tallest Grass	*Bambusa arundinacea*
Shortest Grass	*Mybra minima*
Largest Seed (W)	Double Coconut
Largest Palm (W)	Raffia Palm
Tree without leaf for longest period (6 months)	*Hymenodictyon excelsum*
Biggest leaf	Teak, *Dillenia*
Most Valuable root parasite	*Santalum album*
Oldest living tree (W)	*Pinus longaeva* (Bristle Cone Pine)
Oldest living shrub (W)	*Larrea tridentate* (Crosote bush)
Oldest living fossil tree (W)	*Gingko biloba* (Maiden Hair tree)
The World's most massive thing	*Sequoiadendron giganteum* (W)
Largest tree crown (W)	*Ficus bengalensis* (Banyan Tree)
Biggest buttressed tree (W)	*Ceiba pentandra* (Kapok)
Heaviest organism (W) Longest Life Span	*Populus tremuloides* (Aspan tree)
	Cedrus deodara (704 years)
	Dalbergia latifolia (600 years)
	Tectona grandis (500 years)
	Sequoia sempervirens (4000 years) – California
Biggest Tree (Girth)	*Cedrus deodara* (1646 cm girth)
	Shorea robusta (782 cm girth)
	Tectona grandis (625 cm girth)
	Sequoia sempervirens (3874 cm girth) (W)
Tallest Tree (W)	*Sequoia sempervirens* (Redwood Tree)
Biggest Family (W)	Asteraceae (Sun flower family) No. of species : 24,000
Largest Herbaceous Genera	*Astragalus* (No. of species : 2,000)
Largest Woody Plant Genera	*Euphorbia* (No. of species : 1,600)
Wood	
Lightest Wood	*Ochrama pyramidalis, Cryptomeria japonica*
Heaviest Wood	*Soymida febrifuga , Mimusops littoralis*
Weakest Wood	*Ochrama pyramidalis*
	Cryptomeria japonica
Strongest Wood	*Acacia chundra,, Mesua ferrea*
Toughest Wood	*Anogeisssus pendula*
Hardest Wood	*A.chundra, Pterocarpus santalinus*

	Olea laurifolia (African Black Wood)
Bamboo	
Largest Stem	*Dendrocalamus giganteus*
Smallest Stem	*Arundinaria microphylla*
Thickest Bamboo	*Dendrocalamus brandisii*
Strongest Bamboo	*Dendrocalamus strictus*
Weakest Bamboo	*Dendrocalamus mambraneus*
Forest Cover	
State Having Largest Forest Covers	Madhya Pradesh (77, 700 Km^2)
U.T. Having Largest Forest Cover	Andaman & Nicobar Islands (6662 Km^2)
Protected Areas	
State Having Maximum No. of National Parks	Madhya Pradesh (9)
U.T. Having Maximum No. of National Parks	Andaman & Nicobar Islands (9)
State Having Maximum No. of Wildlife Sanctuaries	Maharashtra (33)
U.T.Having Maximum No. of Wildlife Sanctuaries	Andaman & Nicobar Islands (96)
National Parks – Best covered States	Sikkim, Arunachal Pradesh, Goa, Himachal Pradesh, Jammu & Kashmir
States Without National Parks	Punjab, Tripura
Biggest National Park in India	Hemis National Park, J&K (4,100 Km^2) Desert National Park, Rajasthan (3, 162 Km^2)
Smallest National Park in India	South Button Island National Park Andaman & Nicobar (0.03km^2) North Button NP, A & N (0.44 km^2)
Biggest Wildlife Sanctuary in India	Great Indian Bustard Wildlife Sanctuary, Maharastra (8,496 km^2), Kachchh Desert Wildlife Sanctuary, Gujarat (7,506.22 km^2)
Smallest Wildlife Sanctuary in India	Goose Island Wildlife Sanctuary, Andaman & Nicobar (0.01 km^2)
Tiger Reserves	
Biggest Tiger Reserve	Nagarjuna Sagar (3568 Sq.Km), A.P
Smallest Tiger Reserve	Pench (257 Sq.Km), Maharashtra
Biosphere Reserves	
Biggest Biosphere Reserve	Gulf of Mannar, Tamil Nadu
Smallest Biosphere Reserve	Nokrek, Meghalaya
Animals	
Fastest Animal (W)	Cheetah
Slowest Animal (W)	Slow Loris
Fastest Breeder (W)	Rat

Slowest Breeder (W)	Elephant
Shortest Gestation Period (W)	Mouse (17 Days)
Tailless Primate (W)	Ape
Largest Crocodile	Estuarine Crocodile
Shortest Crocodile	Mugger Crocodile
Rarest Crocodile	Gharial Crocodile
Largest Asiatic Antelope	Nilgai
Largest Eyes (W)	African Elephant
Largest Carnivore in the World	Giant Brown Bear
Biggest Panda in the World	Giant Panda
Smallest Panda in the World	Smaller Panda (Cat-Bear)
Smallest Otter in India	Clawless Otter
Largest Indian Bat	Flying Fox
Largest Indian Wild Sheep	Great Tibetan Sheep
Largest Flying Squirrel (India)	Large Flying Squirrel
Smallest Flying Squirrel (India)	Small Travancore Flying Squirrel
Largest Indian Rhinoceros	Great Indian One horned Rhinoceros
Smallest Indian Rhinoceros	Smaller One horned Rhinoceros
Largest Indian Bovine	Gaur, Buffalo etc.
Largest Indian Deer	Sambar Deer
Most Beautiful Indian Deer	Chital
Smallest Bear (W)	Sun Bear (Honey Bear)
Smallest Dolphin (W)	Hector's Dolphin
Smallest Monkey (W)	Tiny Monkey
Smallest Deer (W)	Leaf Deer (Leaf Muntjac)
Largest Deer (W)	North American Moose Deer
Largest Whale (W)	Blue Whale
Largest Fish (W)	Whale Shark
Fastest Fish	Sailfish
Insects	
Biggest Order in Animal Kingdom	Arthropoda
Biggest group in animal kingdom	Insecta
Songster Insect	Cicada
Birds	
Smallest Bird	Tickell's Flowerpecker (India) Humming Bird (World)
Biggest Bird	Ostrich (W)
Largest Bird	Sarus Crane, Himalayan Bearded Vulture
Commonest Birds	House Crow, Sparrow, Myna, Bulbul
Rare Birds	Mountain Quail, Jerdon's Courser, Pinheaded Duck
Songster Bird	Grey winged Black Bird

Talker Bird	Hill Myna
Biggest Indian Stork	Adjutant Stork
Flight Less Bird	Penguins (World)
Longest Migratory Bird	Arctic Tern (W)
Largest Egg	Ostrich
Fastest Flying Bird (World)	Alpine Spine Tailed Swift
Biggest Order	Passeriformes

INDIAN STATE ANIMALS, BIRDS, TREES AND FLOWERS

S. No.	State/UT	Animal	Bird	Tree	Flower
1	Andhra Pradesh	Blackbuck	Indian roller	Neem	*
2	Arunachal Pradesh	Mithun	Great hornbill	Hollong	Lady slipper or chid
3	Assam	One horned rhino	White winged wood	duck	Hollong Fox-tailed orchid
4	Bihar	Gaur	India roller	Peepal	Kachnar
5	Chhattisgarh	Wild buffalo	Hill myna	*	*
6	Goa	Gaur	Black crested bulbul	Asna	*
7	Gujarat	Asiatic lion	Greater flamingo	*	*
8	Haryana	Blackbuck	Black francolin	Peepal	Lotus
9	Himachal Pradesh	Himlalayan Musk deer	Himalayan Monal	Deodar	Rhododendron
10	Jammu & Kashmir	Hangul	Black-necked crane	Chinar	Lotus
11	Jharkhand	Elephant	Asian Koel	Sal	Palash
12	Karnataka	Elephant	Indian roller	Sandal	Lotus
13	Kerala	Elephant	Great hornbill	Coconut	Kanikonna
14.	Madhya Pradesh	Swamp deer	Paradise fly catcher	*	*
15	Maharashtra	Giant squirrel	Green imperial pheasant	Mango	Jarul
16	Manipur	Sangai	Mrs. Hume's pheasant	Toon	Shirhoy lilly
17	Meghalaya	Clouded leopard	Hill myna	Gamar	Lady slipper or chid
18	Mizoram	Hoolock gibbon	Mrs. Hume's pheasant	Iron wood	Red vanda
19	Nagaland	Mithun	Blyth's tragopan	Alder	Rhododendron
20	Orissa	Elephant	Peacock	Banyan	Lotus

S. No.	State/UT	Animal	Bird	Tree	Flower
21	Punjab	Blackbuck	Northern goshawk	Sheesham	*
22	Rajasthan	Chinkara	Great Indian bustard	Khejri	Rohira
23	Sikkim	Red panda	Blood pheasant Rhododendron	Nobile orchid	
24	Tamil Nadu	Nilgiri tahr	Emerald dove Palmera palm	Kandhal	
25	Tripura	Phayre's langur	Green imperial pigeon	Agar	Nageshwar
26	Uttar Pradesh	Swamp deer	Sarus crane	Ashok	Brahm kamal
27	Uttaranchal	Himalayan Musk deer	Himalayan monal	Burans	Brahm kamal
28	West Bengal	Fishing cat	White-throated king fisher	Chatian	Shephali
Union territories					
1	Andaman & Nicobar	*	*	*	*
2	Chandigarh	*	*	*	*
3	Daman & Diu	*	*	*	*
4	Dadra & Nagar Haveli	*	*	*	*
5	Delhi	*	*	*	*
6	Lakshadweep	Butterfly fish	Sooty tern	Bread fruit	*
7	Pondichery	*	*	*	*

* To be declared

Source : *WII, 2011*

THREATENED SPECIES OF INDIA

Taxonomic group	Number of Species
Mammals	96
Birds	76
Reptiles	25
Amphibians	65
Fishes	40
Molluscs	2
Other Invertebrates	109
Plants	246
Total	659

Source : WII, 2011

LIST OF BOTANICAL GARDEN IN INDIA

S. N.	Name	Location
1	Indian Botanic Garden	Howrah
2	Botanic Garden of Indian Republic	Noida
3	Experimental Botanic Garden Andaman & Nicobar Circle	Dhanikheri
4	Experimental Botanic Garden Arunachal Field Station	Sankie View
5	Experimental Botanic Garden Arid Zone Circle	Jodhpur
6	Experimental Botanic Garden Central Circle	Allahabad
7	Experimental Botanic Garden Eastern Circle (Shillong)	Barapani
8	Experimental Botanic Garden Northern Circle	Pauri
9	Experimental Botanic Garden Northern Circle	Khirsu
10	Experimental Botanic Garden Northern Circle	Dehradun
11	Experimental Botanic Garden Sikkim Himalayan Circle	Gangtok
12	Experimental Botanic Garden Southern Circle	Yercaud
13	Experimental Botanic Garden Western Circle	Mundhwa (Pune)

INTERNATIONAL DECADES

Decade	Details of activity
1993-2002	Second Industrial Development Decade for Africa
1993-2002	Asian and Pacific Decade of Disabled Persons
1993-2003	Third International Decade to Combat Racism and Racial Discrimination
1994-2003	International Decade for the World's Indigenous People *(following 1993 International Year for World's Indigenous People)*
1995-2004	United Nations Decade for Human Rights Education
1997-2006	United Nations Decade for the Eradication of Poverty *(following 1996 International Year for the Eradication of Poverty)*
2001-2010	International Decade for a Culture of Peace and Non-violence for the Children of the World *(following 2000 International Year for the Culture of Peace)*
2001-2010	Second International Decade for the Eradication of Colonialism
2001-2010	Decade to Roll Back Malaria in Developing Countries, Particularly in Africa
2003-2012	United Nations Literacy Decade: Literacy for All *(from 8 September 2003 - International Literacy Day)*
2005-2014	Second International Decade of the World's Indigenous People
2005-2014	United Nations Decade of Education for Sustainable Development
2005-2015	International Decade for Action, "Water for Life" *(from 22 March 2005 - World Water Day)*

Decade	Details of activity
2006-2015	Decade of Recovery and Sustainable Development of the Affected Regions (relating to the Chernobyl disaster)
2008-2017	Second United Nations Decade for the Eradication of Poverty
2010-2019	United Nations Decade for Deserts and the Fight against Desertification
2011-2020	United Nations Decade of Action for Road Safety

Source: www.unac.org

INTERNATIONAL YEARS

Since 1959 the UN has designated International years in order to draw attention to major issues and to encourage international action to address concerns which have global importance and ramifications. There is a concerted effort not to designate every year - for fiscal reasons and to avoid any process of trivialization. The following is the chronological list of years designated to date

INTERNATIONAL YEAR CELEBRATION

Year	Details
1959/60	World Refugee Year
1965	International Cooperation Year
1967	International Tourism Year
1968	International Year for Human Rights
1970	International Education Year
1971	International Year for Action to Combat Racism & Racial Discrimination
1974	World Population Year
1975	International Women's Year
1978	International Anti-Apartheid Year
1979	International Year of the Child
1981	International Year of Disabled Persons
1982	International Year of Mobilisation for Sanctions Against South Africa
1983	World Communications Year
1985	International Youth Year
1986	International Year of Peace
1987	International Year of Shelter for the Homeless
1990	International Literacy Year
1992	International Space Year (endorsed, not declared)
1993	International Year for World's Indigenous People

Year	Details
1994	International Year of the Family;
	International Year of Sport & Olympic Ideal
1995	United Nations Year for Tolerance
1996	International Year for the Eradication of Poverty
1998	International Year of the Ocean
1999	International Year of Older Persons
2000	International Year for the Culture of Peace;
	International Year of Thanksgiving
2001	International Year of Volunteers
	International Year of Mobilization against Racism, Racial Discrimination, Xeno phobia and Related Intolerance
2002	International Year of Mountains;
	International Year of Culture Heritage;
	International Year of Ecotourism
2003	International Year of Freshwater
2004	International Year to Commemorate the Struggle against Slavery and Its Abolition; International Year of Rice
2005	International Year of Microcredit;
	International Year for Sport and Physical Education
2006	International Year of Deserts and Desertification
2008	International Year of the Potato; International Year of Planet Earth;
	International Year of Sanitation;
	International Year of Languages International Year of Coral reef
2009	International Year of Human Rights Learning - *from 10 December 2008* (Human Rights Day) *to 10 December 2009* International Year of Reconciliation;
	International Year of Natural Fibres;
	International Year of Astronomy
2010	International Year of Biodiversity;
	International Year for the Rapprochement of Cultures
	International Year of Youth: Dialogue and Mutual Understanding - *from 12 August 2010* (International Youth Day) *to 11 August 2011*
2011	International Year of Forests;
	International Year of Chemistry;
	International Year for People of African Descent
2012	International Year of Cooperatives

Source: www.unac.org

LIST OF IMPORTANT DAYS RELATED TO FORESTRY

February 2	:	World Wetland Day
Febuary 28	:	National Science Day
March 21	:	World Forestry Day
March 22	:	World Water Day
March 23	:	World Meteorological Day
April 7	:	World Health Day
April 18	:	World Heritage Day
April 22	:	Earth Day
April 24	:	World Lab Animals Day
May 3	:	International Sun Day
May 8	:	World Red Cross Day
May 20	:	Endangered Species Day
May 22	:	World international day for Biological diversity
May 31	:	Anti Tobacco Day
June 5	:	World Environment Day
July 11	:	World Population Day
September 16	:	World Ozone Day
September 18	:	World Bamboo Day
September 21	:	Biosphere Day
September 28	:	Green Consumer Day
October 3	:	World Habitat Day/ World Animal Day
October 1-7	:	World Wildlife Week
October 4	:	World Animal Welfare Day
October 13	:	International Day for Natural Disaster Reduction
November 10	:	Forest Martyrs Day
November 14-20	:	National Land Resources Conservation Week
November 19	:	National Integration Day/World Heritage Week (19-25)
November 21	:	World Fisheries Day
December 2	:	Bhopal Tragedy Day
December 3	:	World Conservation Day
December 14	:	National Energy Conservation Day
December 23	:	Kisan Divas (Farmer's Day) (India)

S. No.	State/ UTS	National parks	Wildlife sanctuaries	Conservation reserves	Community reserves	Protected areas	Tiger reserves
1	Andhra Pradesh	6	21			27	1
2	Arunachal Pradesh*	2	11			13	2
3	Assam*	5	18			23	3
4	Bihar	1	12			13	1
5	Chhattishgarh	3	11			14	3
6	Goa	1	6			7	-
7	Gujarat	4	23	1		28	-
8	Haryana	2	8	2		12	-
9	Himachal Pradesh	5	33			35	-
10	Jammu & Kashmir	4	15	34		53	-
11	Jharkhand	1	11			12	1
12	Karnataka	5	22	2	1	30	4
13	Kerala	6	16		1	23	2
14	Madhya Pradesh	9	25			34	6
15	Maharashtra	6	35	1		42	4
16	Manipur	1	1			2	-
17	Meghalaya	2	3			5	-
18	Mizoram	2	8			10	1
19	Nagaland	1	3			4	-
20	Orissa	2	18			20	2
21	Punjab	0	12	1	2	15	-
22	Rajasthan	5	25			30	2
23	Sikkim	1	7			8	-
24	Tamil Nadu	5	21	1		27	3
25	Tripura	2	4			6	-
26	Uttar Pradesh	1	23			24	1
27	Uttarakhand	6	6	2		14	1
28	West Bengal	5	15			20	2
Union territories							
1	Andaman & Nicobar	9	96			105	-
2	Chandigarh	0	2			2	-
3	Dadra & Nagar Haveli	0	1			1	-
4	Daman & Diu	0	1			1	-

S. No.	State/ UTS	National parks	Wildlife sanctuaries	Conservation reserves	Community reserves	Protected areas	Tiger reserves
5	Delhi	0	1			1	-
6	Pondichery	0	1			1	-
7	Lakshadweep	0	1			1	-
	Total	102	516	44	4	663	39
	% of the area	1.21 %	3.65	0.04%		4.90%	

SILVICULTURAL SYSTEM

Choice of silvicultural system

Conditions / Objectives		Silvicultural systems
Light demander species	:	Clear felling/Clear strip system
Shade Bearer species	:	Shelter wood / Selection system
Rocky & Precipitous slope	:	Selection system
Catchments of rivers	:	Selection / Indian Irregular shelter wood system
Snow prone areas	:	Selection/Indian Irregular shelter wood system
Fuel wood/small timber	:	Coppice systems
Large sized timber	:	High Forests system
Canal plantation	:	Clear felling system
Roadside plantations	:	Clear felling with reserves/simple coppice system

IMPORTANT PEST AND DISEASES

Orders & Insects

Coleoptera : Beetles

Lepidoptera : Moths, Butterflies

Isoptera : Termites

Hymenoptera : Bees, Ants, Sawflies, Wasps

Orthoptera : Grasshoppers, Crickets

Heteroptera : Bugs

Homoptera : Cicadas, Aphids, Mealy bugs, scale insects

Borers

Sal borer : *Hoplocerambyx spinicornis*

Root borer(sal) : *Lophosternus hugelli*

Bamboo borer	:	*Dinoderus brevis*
Teak borer	:	*Dichomeris eridantis*
Shisam borer	:	*Dichomeris eridantis*
Babul borer	:	*Celosterna scrabator*
Toon borer	:	*Hypsipyla robusta*
Tendu stem borer	:	*Plocaederrus ferrugineous*
Bark eating caterpillar	:	*Indarbela quadrinotata*
Defoliators		
Teak defoliator	:	*Hyblaea puera*
Sal defoliator	:	*Lymantria Mathura*
Tendu defoliator	:	*Hypocala rostrata*
Shisam defliator	:	*Plecoptera reflexa*
Deodar defoliator	:	*Ectropis deodrarae*
Gmelina defoliator	:	Calopepla leayana
Chir defoliator	:	*Lebeda nobilis*
Ailsnthus defoliator	:	*Eligma narcissus*
Others		
Sal semi-looper	:	*Ahaea janata*
Gmelina miner	:	*Pyhllocnistis amydropa*
Gmelina tingids	:	*Tingis beesoni*
Teak skeletoniser	:	*Eutectona machaeralis*
Ailanthus web worm	:	*Atteva fabriciella*
Casuarina bag worm	:	*Clania crameri*
Tendu gall former	:	*Trioza obsleta*
Zoological name of common insects & pests		
Bark eating caterpillar	:	*Indarbela quadrinotata*
Phassus borer	:	*Phassus malabarcus*
Red borer	:	*Zeuzera coffeae*
Chafer beetle	:	*Apogania coriacea*
Canker grub	:	*Dihammus cervinus*
Bihar hairy caterpillar	:	*Diacrissia oblique*
Mosquito bug	:	*Helopeltis antonii*
Some useful insects		
Honey bees	:	
Indian / Asiatic bee	:	*Apis cerana indica*
Western bee	:	*Apis mellifera*
Giant Rock bee	:	*Apis dorsata*

Little bee	:	*Apis florae*
Dammer bee	:	*Trigona iridipennis*

Silk worms (Order - Leptidoptera)

Mulberry silk worm	:	*Bombyx mori*
Indian tasar silk	:	*Antheraea mylitta*
Muga silk	:	*Antheraea assamia*
Eri silk worm	:	*Philosamia ricini*

Lac insects (Order Hemiptera : Coccidae)

Lac insect	:	*Kerria lacca (Lacciferarus lacca)*

Lac host

Palas	:	*Butea monosperma*
Ber	:	*Ziziphus jujuba*
Kusum	:	*Scheichera oleosa*

Biological control of some insects & pests

Parasites of teak defoliator

Apanteles melevolus

Cedria paradoxa

Dioctes gardeneri

Parasites & Predator of shisam defoliator

Anthia sexguttata

Microgaster plecoptera

Deiphobe incise

Disophyrys sissoo

Predator of fluted scale of Acacias

Rodolia cardinalis

Insect types

Polyphagous insects	:	Feeding on many kinds of food eg. Weevil
Oligophagous insects	:	Feeding on new kinds of food eg. Teak defoliator
Monophagous insects		Feeding on one kind of food eg. Bee hole borer of teak

IMPORTANT PLANT DISEASES AND ASSOCIATED ORGANISMS

Name of the disease	Casual organisms	Affected tree species	Effect on tree/plant
Damping-off	*Rhizoctonia, Pythium, Fusarium*	Pinus, Deodar, Eucalyptus	Death of seedlings.
Heart-rot	*Fomes caryophylli, F. badius, Hymenochaete* spp.	Sal, Khair, Teak & others	Loss in growth and damage to heartwood
Root & Butt rot	*Heterobasidium annosum*	Deodar	Death of trees
Stem Blister rust	*Cronartium ribicola*	Blue pine	Exudation o resin
Wilt disease	*Fusarium spp*	Sissoo, Casuarina	Loss of growth & death of plants
Root rot	*Ganoderma spp*	Khair, Babuil, Deodar, Siris and Chir	Loss in growth & death of plants
Pink disease	*Corticum salmonicolor*	Eucalyptus spp	Top drying & loss in growth
Spike disease	Mycopolasma like organisms	Sandal	Loss in growth and death of plants
Powdery mildew	*Unicinula tectonae*	Teak	Leaf shedding
Leaf spot	*Cercospora* spp.	Teak	White spots on the leaves

CASUAL ORGANISMS OF SOME IMPORTANT DISEASES

Wilt disease	:	*Fusarium oxysporium, Verticullum,etc.,*
Root rot	:	*Ganoderma spp*
Root & butt rot	:	*Fomes annosus, Armillaria mellea,*
Stump rot	:	*Fomes ligneous*
Heart rot	:	*Hymenomycetes (Ployporaceae)*
Punk knots	:	*Fomes caryphylli*
Stem wilt (casuarina)	:	*Trichosporium vesiculosum*
Bamboo blight	:	*Sacrocaladium orzyzae*
Important Diseases (World)		
Chestnut blight	:	*Endothia parasitica*
Dutch Elm disease	:	*Cerotocystis ulmi*
Stem blister rust (pines)	:	*Cronaritium rubicola*
Dothiostroma blight (pines)	:	*Dothiostroma pine*
Swertia felt rust (pines)	:	*Cronatium himalayense*

CLASSIFICATION OF TIMBER AND NON-TIMBER INDUSTRIES

WOOD BASED INDUSTRIES:

A) Heavy

1. Paper and pulp
2. Rayon grade pulp
3. Plywood
4. Particle board
5. Fibre board
6. Veneers
7. Railway sleepers
8. Defence products
9. Distilleries
10. Saw timber

B) Small Scale Cottage

1. Matchstick boxes
2. Toy art pieces
3. Agriculture implements
4. Charcoal manufacture
5. Furniture
6. Bamboo, Mushrooms
7. Post and telegraph poles.
8. Electricity poles
9. Sport goods.

Industries Based On By-products

A) Heavy

1. Beedi leaves
2. Lac culture
3. Sericulture
4. Apiculture, Honey and Wax
5. Gum
6. Tannin, Dye
7. Resin
8. Katha making
9. Fibre and floss
10. Grass ropes

B) Small Scale Cottage

1. Soap and perfume
2. Liquor
3. Essential oils.
4. Non essential oils.
5. Coir
6. Fertilizer / Manures

BOTANICAL NAME, COMMON NAME AND USES OF IMPORTANT FOREST PLANTS

S. No	Botanical Name	Common Name	Use or Utilities
1	*Acacia catechu*	Khair	Fodder, fuel, poles, soil conservation and Katha
2	*Acacia nilotica*	Babul	Fodder, soil conservation; agricultural implements, bark as medicine, tannin products and fuel
3	*Azadirachta indica*	Neem	Planks, tooth brush and medicine (Bark, leaves gum)

S. No	Botanical Name	Common Name	Use or Utilities
4	*Bombax ceiba, Bombax malabaricum*	Semal (Kate sawar)	Match Stiks, packing cases, play wood, soil conservation, roots used as medicines
5	*Delbergia sissoo*	Sissoo	Fodder, Timber, fuel, soil conservation
6	*Delbergia latifolia*	Rose Wood	Fodder, Timber, fuel
7	*Melia azedarach*	Bakan/ Limbara	Avenue tree, fodder, timber
8	*Tectona grandis*	Sag / Teak Wood	Structures and industrial timber, furniture, fuel, tannin, ornamental and medicinal use **KING OF TIMBEER**
9	*Terminalia tomentosa*	Asan / Ain	Industrial timber, tannin, gum, medicinal, tassar
10	*Terminalia arjuna*	Arjun	Industrial timber, small poles, fodder, timber, ornamental, tassar
11	*Acacia auriculiformis*	Australin Babhul	Denuded places, tannin, fuel
12	*Acacia chundra*	Lal khair	Timber, pole, fuel, tannin and soil conservation
13	*Aegle marmelos*	Bel	Tooth brush, fruits, medicine (Bark, gum)
14	*Albizia chinensis*	Kala sires	Timber, pole, fuel, tannin and soil conservation
15	*Albizia procera*	Safed Siris	Timber, medicine, fuel
16	*Artocarpus heterophallus*	Jack Fruit	Fruit and timber
17	*Anthocephalus chinensis*	Kadamb	Fodder, fuel, pulp wood, ornamental
18	*Emblica officinalis*	Amla	Fruit, fuel, ornamental, medicinal
19	*Madhuca latifolia*	Mahua	Liquor oil seed, timber, edible flower
20	*Mangifera Indica*	Mango	Fruit, timber, medicine.
21	*Tamarindus indica*	Imli	Fruit, fuel, medicinal, timber, leaves, flowers, fodder, poles
22	*Ziziphus jujuba*	Ber	Fruit, fodder, poles, lac
23	*Syzygium cumini*	Jamun	Fruit, fodder, poles, timber, fuel, medicinal (flowers fruits)
24	*Moringa oleifera*	Drumstick	Fruit, fodder, hand paper
25	*Sesbania grandiflora*	Hadga	Flowers, fence poles, tannin fodder, fuel, pulp
26	*Saraca indica*	Ashok	Ornamental, medicinal
27	*Santalum album*	Sandal	Industrial timber, oil, ornamental, medicinal
28	*Morus alba*	Mulberry	Fodder, fruit, sport goods, silk industries
29	*Pongamia pinnata*	Karanji	Leaves used for green manuring,branches used as tooth brush, seeds for oil

S. No	Botanical Name	Common Name	Use or Utilities
30	*Bauhinia variegata*	Kanchan	Fodder, food, timber, leaves and flowers are used as vegetables, medicine
31	*Bauhinia racemosa*	Jhangora	Medicinal plant, ornamental
32	*Bauhinia chinensis*	Kanchan	Medicinal and fuel, ornamental
33	*Michelia chamapaca*	Chaffa	Pole, industrial timber, fuel, ornamental, soil conservation, medicinal flowers are more valuable for perfume
34	*Populus deltoides*	Poplar	Pulp wood, fodder, packing cases,soil conservation, fuel match box and splints
35	*Pitheocolobiumdulce*	Vilayati Chinch	Fuel, hedge, soil conservation
36	*Prosopis chinensis*	Kabuli kikar	Fodder, fuel. soil conservation
37	*Casuarina equisetifolia*	Khadsarni or suru	Fuel, timber, good for plantation in coastal area, charco
38	*Cupressus cashmeriana*	Cypress	Industrial timber, poles, soil conservation sandy area
39	*Eucalyptus camaldulensis*	Eucalyptus	Gum, oil, tannin, soil conservation, industrial timber, paper, pulp
40	*Eucalyptus globulus*	Blue Gum	Pulp wood, fuel, timber
41	*Eucalyptus tereticornis*	Mysore Gum	Fuel and pulp wood
42	*Ficus religiosa*	Pipal	Avenue trees, fuel, fodder
43	*Ficus bengalensis*	Wad	Avenue trees, fodder, fuel, fruits, soil conservation, timber for packing cases
44	*Ficus glomerata*	Gular / Umbar	Avenue trees, fodder, fuel
45	*Delonix regia*	Gulmohar	Ornamental, avenue trees
46	*Laucaena leucocephala*	Subabul	Timber, fodder, poles, fuel, soil conservation
47	*Salix* spp.	Willow	Timber, basket making
48	*Sapindus mukorossi*	Ritha or Soapnut Tree	Fuel, soap nut for washing cloth
49	*Swietenia mahagoni*	Mahogoni	Poles, gum, tannin, industrial timber, ornamental and soil conservation.
50	*Dendrocalamus strictus*	Solid bamboo	Fodder, poles, industrial use, paper, soil conservation
51	*Bambusa arundinacea*	Hallow bamboo	Fodder, paper, poles, industrial use, soil conservation
52	*Acacia conciana*	Shikekai	Medicinal, fuel, timber, soap
53	*Prosopis juliflora*	Vilayati Babul	Medicinal, fuel, timber, fencing
54	*Prosopis cineraria*	Khejadi	Fodder, fuel, timber, fencing
55	*Terminalia chebula*	Hirda	Timber, dye, fuel, medicinal
56	*Terminalia paniculata*	Kingal	Timber, fuel

S. No	Botanical Name	Common Name	Use or Utilities
57	*Terminalia bellerica*	Behda	Medicinal, timber, fuel
58	*Polyalthia longifolia*	Drooping Ashok	Ornamental, road avenue
59	*Semecarpus anacardium*	Biba	Medicinal, fuel
60	*Cassia renigera*	Pink cassia	Ornamental
61	*Cassia fistula*	Bahaya / Amaltas	Ornamental, medicinal
62	*Diospyros melanoxylon*	Tendu	Beedi leaves
63	*Platanus orientalis*	Chinar	Timber, fuel
64	*Diosypros tomentosa*	Temburni	Ornamental, medicinal, fruits
65	*Kigelia pinnata*	Monkey Plant	Ornamental
66	*Shorea robusta*	Sal	Timber, fuel
67	*Ailanthus excelsa*	Maharukh	Fodder, match industries
68	*Quercus* spp.	Indian oak	Fodder, fuel, sericulture
69	*Lagerstroemia parviflora*	Sidha benteak	Fodder, fuel, gum, tannin, pole, timber, ornamental
70	*Hardwickia binata*	Anjan	Fodder, fuel, agricultural implements
71	*Erythrina indica*	Pangara	Fuel and soil conservation
72	*Grewia* spp.	Dhaman	Fibre
73	*Pinus roxburghii*	Chir-pine	Resin, gum, tannin, oil, pole, soil conservation
74	*Jacaranda mimosifolia*	Neel-mohar	Ornamental and soil conservation
75	*Grevillea robusta*	Silver oak	Ornamental, avenue tree
76	*Pithecolobium saman*	Rain tree	Ornamental, shade tree, fuel
77	*Spathodea campanulata*	Fountain Tree	Ornamental. shade tree, fuel
78	*Millingtonia hortensis*	Indian cork tree	Ornamental, cork industry
79	*Mimusops elengi*	Bakul	Shade tree, scented flower, rootstock for sapota, avenue
80	*Feronia elephantum*	Wood apple	Fruits, medicine, timber
81	*Annona squamosa*	Sitaphal	Fruits, fuel
82	*Cedrus deodara*	Debdar	Timber, fiuniture, flanks
83	*Glyricidia maculata*	Giripushpa	Fodder, leaf manures
84	*Butea monosperma*	Palas	Ornamental, Cottage industry, lack industry
85	*Thespesia populnea*	Ran Bhendi	Timber, fuel, rural ornamental
86	*Garcinia indica*	Kokam	Fruit, oil industry, squish, oil seed, medicine
87	*Terminalia cadappa*	Wild Almond	Ornamental, oil seed
88	*Manilkara hexandra*	Khirani	Fruits, fuel, root stock for sapota

S. No	Botanical Name	Common Name	Use or Utilities
89	*Adina cordifolia*	Haldu	Timber, fuel
90	*Gmelina arborea*	Shivan	Timber, fuel
91	*Lagerstroemia lanceolata*	Nana	Timber
92	*Callistemon lanceolatus*	Bottle Brush	Ornamental road avenue, parks
93	*Agave americana*	Agave / Ghaipath	Live fencing, rope making
94	*Pheltophorum ferrugineum*	Pelto phoorum	Ornamental road avenue tree, parks
95	*Anona reticulata*	Ramphal	Fruits, soil reclamation
96	*Buchanania latifolia*	Char oil	Fruits, soil conservation
97	*Anacardium occidentale*	Cashew nut	Cashew nut Fruits, soil conservation

Fuel Wood

Fuel wood is the most direct use of wood as a source of energy

Charcoal

Is a derivative of wood. it is converted into a more readily useable from in the forest itself before being used to private energy.

A black, porous, carbonaceous material, 85 to 98% carbon, produced by the destructive distillation of wood and used as a fuel, filter, and absorbent.

Good fuel wood tree species

Acacia nilotica
Acacia catechu
Casuarina equisetifolia
Dalbergia sissoo
Albizzia lebbeck
Adina cordifolia
Zizyphus jujuba
Chloroxyolon swietenia
Hopea parvifolia
Lagerstroemia spp.

Bad fuels and not usually used

Abies pindrow
Ailanthus excelsa
Bauhinia racemosa
Bombax ceiba
Boswellia serrate
Ficus spp.
Dalbergia paniculata
Mangifera indica
Picea smithiana
Pinus roxburghii

Fodder grasses

Cenchrus ciliaris
Panicum antidotale
panicum turgidum
Cynodon spp.
Saccharum munja
Dichanthium annulatum

Fodder legumes

Cassia tora
Indigofera cordifolia
Prosopis juliflora
Indigofera linifolia
Albizia lebbeck

Fodder Trees

Acacia catechu
Albizia lebbeck
Bombax ceiba
Dalbergia sissoo
Ficus religiosa
Gmelina arborea
Populus spp.
Acacia nilotica
Azadirachta indica
Butea monosperma
Dendrocalamus strictus
Ficus bengalensis
Leucaena leucocephala
Prosopis juliflora

- Generally occurs as sclerenchyma cells and serve to impart rigidity to th eplant
- Long cells with thick walls and small cavities
    - Soft fibres from bast or stem
    - Hard fibres (leaf or structural fibres)
    - Surface fibres borne on the surface of stems, leaves, seeds etc.
- Extraction of fibre – **Retting process**

Fibres from stems

Sterculia villosa
Grewia tiliaefolia
Bauhinia vahlii
Butea monosperma
Calotropis procera
Sterculia urens
Grewia laevigata
Hardwickia binata
Bauhinia racemosa
Trema orientalis
Sterculia foetida
Grewia oppositifolia
Acacia leucophloea
Calotropis gigantea
Ficus spp.

Fibres from Leaves

Caryota urens
Musa paradisiaca
Agave Americana
Musa spp.
Pandanus spp.
Agave cantala
Musa textilis
Agave spp.
Agave sisalana

FLOSSES

Ceiba pentandra
Cochlospermum religiosum
Populus ciliata
Bombax ceiba
Calotropis procera
Calotropis gigantea

COIR

Cocos nucifera

GRASS OILS

Palmarosa grass oil - *Cymbopogon martini*
Lemon grass oil - *Cymbopogon flexuosus*
Ginger grass oil - *Cymbopogon martini*
Citronella oil - *Cymbopogon nardus*
Khus oil - *Vetivera zizanioides*

WOOD OIL

Sandal wood - *Santalum album*
Agarwood oil - *Aquilaria agallocha*
Deodar oil - *Cedrus deodara*
Pine oil - *Pine* spp.

LEAF OILS

Eucalyptus oil - *Eucalyptus globulus*
Citriodora oil - *Eucalyptus citriodora*
Camphor oil - *Cinnamomum camphora*
Cinnamon oil - *Cinnamomum zeylanicum*
Mint oil - *Mentha arvensis*
Wintergreen Oil - *Gaultheria fragrantissima*

ROOT OIL

Costus oil - *Saussurea lappa*
Velerian oil - *Valeriana wallichii*
Sandal wood - *Santalum album*

FLOWER OIL

Keora oil - *Pandanus tectorius*

MISCELLANEOUS ESSENTIAL OIL

Boswellia serrata
Michelia champaca
Mimusops elengi
Acacia farnesiana

OIL SEEDS

Azadirachta indica
Calophyllum inophyllum
Madhuca butyracea
Garcinia indica
Madhuca indica

GUMS

Acacia gums	–	*Acacia catechu, Acacia nilotica, Acacia Senegal & Acacia modesta*
Gum kino	–	*Pterocarpus marsupium*
Bengal kino	–	*Butea monosperma*
Jhingan gum	–	*Lannea coromandelica*
Katira gum	–	*Sterculia urens, Cochlospermum religiosum, Bombax ceiba*
Moringa gum	–	*Moringa pterygosperma*
Salai gum	–	*Boswellia serrata*
Dhaura gum	–	*Anogeissus latifolia*
Semla gum	–	*Bauhinia retusa*
Miscellaneous gums	–	*Albizia chinensis, A. odoratissima, A. procera, Azadirachta indica, Mangifera indica, Terminalia aiata & Terminalia bellirica*

RESINS

Pine resins	–	*Pinus roxburghii, Pinus wallichiana, P. kesiya & P. gerardiana*
True dammar	–	*Agathis loranthifolia*
Black dammar	–	*Canarium strictum*
Rock dammar	–	*Hopea odorata*
White dammar	–	*Vateria indica*
Ral dammar	–	*Shorea robusta*

OLEO RESINS

Salai gum	-	*Boswellia serrata*
Gurjan oil	-	*Dipterocarpus turbinatus*
Piney	-	*Kingiodendron pinnatum*

GUM RESINS

Gamboge	-	*Garcinia morella*
		Commiphora mukul

Forest Tree Species Used for Different Purpose

Industrial Needs: Pulp wood, saw-timber, panel products such as, plywood and other boards, match wood, sport goods etc. constitute the industrial needs. Some of the species for meeting some of the industrial needs are given below:

1. **Construction:** *Acacia catechu, Cedrus deodara, Dalbergia sissoo, Dipterocarpus macrocarpus, Eucalyptus* spp., *Hopea parvifolra, Messua ferrea, Pterocarpus santalinus, Shorea robusta, Tectona grandis.*
2. **Tool handles:** *Acacia catechu, A. nilotica, Dalbergia latifolia, Diospyros melanoxylon, Messua ferrea, Pterocarpus marsupium, Shorea robusta, Terminalia alata.*

3. **Packing Cases:** *Abies densa, A. pindrow, Acrorcarpus* spp., *Ailanthus excelsa, Bombax ceiba, Boswellia serrata, Fiscus* spp., *Mangifera indica, Kydia calycina, Meliaazadirachta, Pinus* spp.
4. **Plywood:** *Acrocarpus fraxinifolius, Adina cordifolia, Dalbergia latifolia, D. sissoo, Dipterocarpus macrocarpus, Gmelina arborea, Mangifera indica, Michelia champaca, Schima wallichii, Shorea assamica, Tectona grandis, T. myriocarpa.*
5. **Doors and Windows:** *Tectona grandis, Cedrus deodara, Adina cordifolia, Dalbergia latifolia, D. sissoo, Dipterocarpus* spp., *Gmelina arborea, Pterocarpus dalbergioides.*
6. **Matches:** *Ailanthus excelsa, Alstonia scholaris, Anthocephalus chinensis, Bombax ceiba, Populus deltoides, Elaeocarpus* spp.
7. **Lorry bodies:** *Acrocarpus fraxinifolius, Albizia libbeck, Artocarpus chaplasha, Dalbergia sissoo, D. latifolia, Dipterocarpus macrocarpus, Pterocarpus dalbergioides, P. marsupium, Shorea robusta, Tectona grandis, Terminalia alata, T. myriocarpa.*
8. **Furniture:** *Tectona grandis, Dalbergia latifolia, D. sissoo, Pterocarpus santalinus, Adina cordifolia, Gmelina arborea, Dipterocarpus macrocarpus, Pterocarpus marsuipium, Cedrus deodara, Juglans regia.*
9. **Pulp Wood:** *Pinus* spp., *Abies pindrow, Picea smithiana, Bamboos, Eucalyptus* spp. and other fast grown hardwoods.
10. **Sport Goods**: The species for various items of sport goods.

TIMBER SPECIES RECOMMENDED FOR SPORTS GOODS

S.N.	Name of the Articles	Timber Species Recommended
1	Cricket bats	*Salix* spp.
2	Hockey sticks	*Morus albs, Celtis australis, Fraxinus* spp.
3	**Tennis and badminton rackets**	
	Bents	*Betula.* spp., *Swietenia* spp., *Fraxinus* spp., *Juglans regia, Celtis australis, Toona ciliata*
	Handle	*Melia azedarach, Michelia champaca, Gmelina arborea, Swietenia* spp., *Populus* spp., *Abies pindrow, Toona ciliata, Salix* spp.
	Wedge	*Fraxinus* spp., *Grewia tiliaefolia, Swietenia* spp., *Morus* spp., *Populus* spp., *Dalbergia sissoo, Toona ciliata, Juglans regia, Salix* spp.
4	**Squash racket frame**	
	Bents	*Betula utilis, Juglans regia*
	Handle	*Populus* spp., *Salix* spp., *Toona ciliata*
	Wedge	*Fraxinus* spp.
5	Carrom board frame and reinforcement	*Tectona grandis, Gmelina arborea, Pterocarpus marsupium, Dalbergia sissoo, Dysoxylum malabaricum*
	Carrom draughts	*Fraxinus* spp., *Adina cordifolia, D. latifolia, Juglans regia*
6	Clubs	*Anogeissus latifolia, Pterocarpus marsupium, Grewia tiliaefolia, Dalbergia sissoo, D. latifolia*
7	Balancing bench	*Fraxinus* spp, *Grewia tiliaefolia etc*

8	Cricket stumps and bails	*Quercus* spp., *Dysoxylum malalaricum, Morus* spp., *Grewia tiliaefolia, Celtis australis, Adina cordifolia*
9	Javelins	*Fraxinus* spp., *Lagerstroemia lanceolate, Morus laevigata, Grewia tillaefolia, Morus alba*
10	Hurdles	*Lagerstroemia lanceolate, Artocarpus chaplasha, Cedrus deodara, Pinus wallichiana, Dalbergia sissoo*
11	Jumping and vaulting Standards	*Anogeissus latifolia, Lagerstroemia lanceolate, Pterocarpus marsupium, Quercus* spp., *Dalbergia sissoo*
12	Wall bars	*Fraxinus* spp., *Anogeissus latifolia, Acacia nilotica, Pterocarpus marsupium, Grewia tiliaefolia, Ougeinia oojeinensis, Dalbergia sissoo, D. latifolia*

NOTES

NOTES